THE ORIGINS OF THE INFINITESIMAL CALCULUS

THE ORIGINS OF THE INFINITESIMAL CALCULUS

by

MARGARET E. BARON

DOVER PUBLICATIONS, INC.
NEW YORK

Published in Canada by General Publishing Company, Ltd.,
30 Lesmill Road, Don Mills, Toronto, Ontario.

This Dover edition, first published in 1987, is an
unabridged and unaltered republication of the work first
published by Pergamon Press Ltd., Oxford, England, in 1969.

Manufactured in the United States of America
Dover Publications, Inc., 31 East 2nd Street, Mineola, N.Y. 11501

Library of Congress Cataloging-in-Publication Data

Baron, Margaret E.
The origins of the infinitesimal calculus.

Reprint. Originally published: Oxford [Oxfordshire] :
Pergamon Press, 1969.
Bibliography: p.
Includes index.
1. Calculus—History. I. Title.
QA303.B29 1987 515′.09 87-5446
ISBN 0-486-65371-4 (pbk.)

Contents

PREFACE vii

INTRODUCTION 1

CHAPTER 1. GREEK MATHEMATICS 11

1.0 Introduction 11
1.1 The influence of Greek mathematics in the seventeenth century 14
1.2 A brief chronology of Greek mathematics 15
1.3 Early Greek mathematics and the physical world 18
1.4 The axiomatisation of Greek mathematics 25
1.5 The theory of proportion and means 27
1.6 The squaring of the circle 30
1.7 The method of exhaustion 34
1.8 The discovery method of Archimedes 46
1.9 Curves, normals, tangents and curvature 51

CHAPTER 2. THE TRANSITION TO WESTERN EUROPE 60

2.0 Introduction 60
2.1 On Hindu mathematics 61
2.2 The Arabs 66
2.3 The influence of Aristotle 69
2.4 The continuum, indivisibles and infinitesimals 71
2.5 The growth of kinematics in the West 75
2.6 The latitude of forms 81
2.7 The function concept in the fourteenth and fifteenth centuries 87
2.8 Conclusion 89

CHAPTER 3. SOME CENTRE OF GRAVITY DETERMINATIONS IN THE LATER SIXTEENTH CENTURY 90

3.0 Introduction 90
3.1 Francesco Maurolico (1494–1575) 91

3.2 Federigo Commandino (1509–1575) 94
3.3 Simon Stevin (1548–1620) 96
3.4 Luca Valerio (1552–1618) 101

CHAPTER 4. INFINITESIMALS AND INDIVISIBLES IN THE EARLY SEVENTEENTH CENTURY 108

4.1 Johann Kepler (1571–1630) 108
4.2 Indivisibles in Italy 116
4.3 Bonaventura Cavalieri (1598–1647) 122
4.4 Grégoire de Saint-Vincent (1584–1667) 135

CHAPTER 5. FURTHER ADVANCES IN FRANCE AND ITALY 149

5.0 Introduction 149
5.1 The arithmetisation of integration methods 151
5.2 First investigations of the cycloid 156
5.3 Another integration method 162
5.4 The concept of tangent 163
5.5 The composition of motions 174
5.6 The link between differential and integral processes 177
5.7 Evangelista Torricelli: tangent and quadrature 182

CHAPTER 6. CONSOLIDATION OF GAINS: FRANCE, ENGLAND AND THE LOW COUNTRIES 195

6.0 Introduction 195
6.1 Blaise Pascal (1623–1662) 196
6.2 Infinitesimal methods in England 205
6.3 Infinitesimal methods in the Low Countries 214
6.4 The rectification of arcs 223
6.5 James Gregory (1638–1675) 228
6.6 Isaac Barrow (1630–1677) 239

CHAPTER 7. EPILOGUE: NEWTON AND LEIBNIZ 253

7.0 Introduction 253
7.1 Isaac Newton (1642–1727) 255
7.2 Gottfried Wilhelm Leibniz (1646–1716) 268

BIBLIOGRAPHY 291

INDEX 299

Preface

THE impetus for this study arose some years ago when, as a result of a suggestion made by Professor McKie, I embarked on a survey of the literature relating to the Newton–Leibniz controversy. It became clear to me that many faulty judgements had arisen from failure to relate the results and methods contained in individual works to their historical background. The origins of the infinitesimal calculus lie, not only in the significant contributions of Newton and Leibniz, but also in the centuries-long struggle to investigate area, volume, tangent and arc by purely geometric methods. To understand the important changes in proof structure, method, technique and means of presentation which emerged in the later seventeenth century it is necessary to go back to Greek sources and particularly, of course, to Archimedes.

In an initial survey of material to be studied I made a great deal of use of Boyer's invaluable work, *The Concepts of the Calculus*, and the excellent bibliography provided.† Many possible lines of investigation emerged. It seemed that there was, in particular, a need for a closer study of the geometric techniques and methods developed in the seventeenth century prior to the work of Newton and Leibniz.

Many excellent individual studies now exist; for example, the work of Child on Barrow, of Scott on Wallis and of Turnbull and Scriba on Gregory. J. E. Hofmann has made exhaustive studies on the genesis of Leibniz' early mathematical work and D. T. Whiteside is in process of doing the same for Newton. The time appears to be ripe for all of these to be interpreted within the general framework of developing seventeenth-century infinitesimal methods.

Later, I found an interesting collection of material on the infinitesimal calculus contained in a posthumous work of Toeplitz published in 1949.‡ Toeplitz, however, was primarily concerned to promote a better understanding of the calculus through the introduction of historical methods. In con-

† Boyer, C. B., *The Concepts of the Calculus*, New York, 1949.

‡ Toeplitz, O., *Die Entwicklung der Infinitesimalrechnung* (ed. Köthe, G.), Berlin, 1949; *The Calculus: a Genetic Approach* (trans. Lange, L.), Chicago, 1963.

sequence the historical treatment is subordinated to the development of fundamental concepts.

In my own study the historical development is central and the methods which emerge are treated strictly within their historical context. At every stage I have explored as fully as possible both the sources on which an individual was drawing for the development of new concepts and the form and manner in which results were presented.

None the less I believe, with Toeplitz, that all of those who study or teach the calculus, and the countless physicists and engineers who make use of it in one way or another, will gain in insight and understanding through some knowledge of its historical development. I have striven, accordingly, to avoid any excesses of language or symbolism which might constitute barriers to understanding, and have consistently aimed at producing an account in a form and a manner comprehensible not only to historians of science, with no considerable knowledge of mathematics, but also to all those who are concerned with the calculus in the twentieth century and who might care to know something of its history.

Among those who have helped in providing material and ideas I am grateful to Dr. J. F. Scott who has offered me constant help and encouragement over many years: to Professor Russell of Birmingham who provided me with many insights into the work of Leibniz: to Dr. Whiteside of Cambridge who helped to introduce me to the Newton manuscripts in the Portsmouth Collection: to Dr. Tanner of Imperial College who made available to me a great deal of material not readily obtainable in England: to Professor J. E. Hofmann who offered most generous help and guidance, not only in relation to Leibniz, but also in more general matters relating to the study of the history of mathematics and who made it possible for me to attend the Seminar in the History of Mathematics at Oberwolfach.

I must also record my appreciation of the facilities afforded me by the librarians of the British Museum, the Royal Society, Cambridge University Library, University College Library, the University of London Library and the Lower Saxony State Library, Hanover.

My gratitude is due especially to Dr. A. Armitage of University College, London, for the help and encouragement he has afforded me over many years and to my colleague, Miss Joan Walmsley, for checking and commenting on the manuscript. For any errors, whether of fact or of judgement, I am, of course, solely responsible.

M. E. B.

October 1966

Introduction

> It is in Sciences as in Plants; Growth and Branching is but the Generation of the Root continued; nor is the Invention of Theoremes any thing else but the knowledge of the Construction of the Subject prosecuted. The unsoundness of the Branches are no prejudice to the Roots; nor the Faults of Theoremes to the Principles. And active Principles will correct false Theoremes if the Reasoning be good; but no Logique in the world is good enough to draw evidence out of false or unactive Principles.
>
> THOMAS HOBBES, *Six Lessons*, London, 1656

WHILST accumulating the historical material for this study it has, from time to time, appeared to me desirable that I should write an introduction defining clearly and precisely the fundamental operations of the calculus in terms of twentieth-century analysis. In an effort to order my own ideas on this matter I have consulted a number of authoritative texts and learnt a great deal about the foundations of the calculus. In the process, however, the possibility seems to have receded of saying anything simply and at reasonable length which would fulfil the dual purpose of identifying within the field of mathematics the concepts and operations associated with the infinitesimal calculus and prescribing the scope and terms of reference of any study of its origins.

It is easier to say what the calculus does than to say what it is. Practically every important development in science and mathematics from 1600 to 1900 was connected in some way or another with differential or integral methods. The infinitesimal calculus has been the principal tool in the exploitation of the earth's resources, the charting of the heavens and the building of modern technology. Applications occur wherever there exist measurable phenomena: gravitation, heat, light, sound, electricity, magnetism and radio waves.

The calculus, however, is much more than a technical tool: it is a collection of abstract mathematical ideas which have accumulated over long periods of time. The foundations and central concepts are not today what they were in the seventeenth, eighteenth, or even in the nineteenth centuries and yet, in a sense, the unifying power and richness of its field of application depends not only on what the calculus is now but on all the concepts which have contributed in one way or another to its evolution.

The history of science offers a wealth of evidence of the value of analogy as an aid to invention. Whether indeed analogy can be equally helpful in tracing the development of mathematics is another matter. Thomas Hobbes, in likening mathematics to a tree, does a valuable service in that he focuses immediate attention on the roots and origins on which all subsequent healthy growth depends. The concept of the calculus as a massive tree with roots in arithmetic, algebra and geometry and a powerful trunk supporting a vast network of interlocking branches is a pleasing enough image—the tree grows higher and the roots stronger, branches once separated from the main stem diverge further apart never to meet again. This is altogether too orderly a picture to convey anything of the stress and strain, the conflict and the tumult which have been associated with every stage in its long growth to maturity.

More apt, perhaps, is the analogy with the musical symphony† in which a number of movements of varying degrees of complexity are successively developed: at each stage we have the introduction of a theme, sometimes by a single instrument, the development and steady unfolding of a significant statement, a process of recapitulation and consolidation and finally integration of the whole with the central *underlying mood* which subsists throughout. To quote from the *Musical Companion*‡ this can be:

> a central idea such as a "motto" phrase common to some or (occasionally) all movements;
> a closely connected scheme of keys governing the four movements;
> a cousinly relationship between the main and/or secondary themes of the first movement and other themes heard subsequently;
> a scheme of instrumentation which, though containing vivid contrasts, can be said to be always in keeping with the music;
> a combination or part combination of the foregoing.

All analogies mislead and none should be taken too far. Nevertheless it seems that in this introduction it is my task to draw attention to some of the principal elements constituting the *underlying mood* of the symphony.

Although there are many themes, some complementary and some contrasting, they all contain a common element of conflict, the conflict between

† This analogy is used by Bourbaki among others: see Bourbaki, N., *Éléments d'histoire des mathématiques*, Paris, 1960.

‡ *The Musical Companion* (ed. Bacharach, A. L.), London, 1930, p. 218.

the demands of mathematical rigour imposed by deductive logic and the essential nature of the infinitely great and the infinitely small perpetually leading to paradox and anomaly. Both of these were already present with the Greeks, the first in the logic of Aristotle and the second in the paradoxes of Zeno. In Euclidean geometry the infinitely small was rejected and in the classical treatises of Archimedes we have the finest example of mathematical rigour in antiquity. Notwithstanding, in his *discovery method* we find him manipulating line and surface indivisibles skilfully, imaginatively and non-rigorously.

Discussion on the nature of the infinite, the continuum, indivisibles and infinitesimals continued with the Scholastics through the Middle Ages and Galileo was heir to the abundant legacy of mediaeval paradoxes. In the seventeenth century the conflict took on a shriller and more strident note with Kepler and Cavalieri, among others, prepared to foster methods based on intuitive concepts of line and surface elements; with Saint-Vincent, Tacquet and Gregory striving to found the new analysis on a firm basis of Archimedean proof structure, and with many of the finest mathematicians of the age so beset by problems of presentation that they were unable to find a satisfactory way of formulating their demonstrations for publication. In Newton himself, as with Archimedes, we find both the capacity to think and work in terms of the synthetic geometry and proof structure of the ancients and an intuitive appreciation of the necessity to give form, shape and substance to developing concepts of infinitesimal quantities linked with time and motion as in his fluxional calculus.

Concepts drawn from the physical world have always played a major role in stimulating imagination and providing a basis for the construction of models from which the mathematician might draw tentative conclusions. The vast edifice of Euclidean geometry, for over 2000 years an accepted model of deductive reasoning, makes notwithstanding, constant appeals to spatial imagery. Omit the diagrams, drawn or imagined, and, more often than not, the demonstration collapses. In the concepts of distance, time and motion, mathematicians have found significant ideas to grasp, to hold and to develop. Although Euclidean geometry was essentially static, kinematic solutions and moving models were found elsewhere in Greek mathematics. The spiral of Archimedes, the quadratrix of Hippias, the conchoid of Nicomedes, were all curves defined in terms of motion and in the *Collectiones* of Pappus we find many examples of a purely kinematic approach to problems. The sharp clarity of the discussions of the Merton Calculators of the fourteenth century focused attention on ideas of time, speed and instantaneous velocity and at the same time helped to disentangle and to separate some of the psychological difficulties inherent in the consideration of continuously varying quantities from those aspects of them subsequently to become amenable to mathematical manipulation.

Whilst concepts drawn from the physical world have provided many essential insights for pure mathematics, pure mathematics has in turn developed as a consequence of a response to problems posed by the physical world. Archimedes built up a theoretical basis for the foundations of statics and hydrostatics on a geometrical model; Kepler developed his infinitesimal methods in response to the practical needs of wine gaugers; the exploration of the universe, both in its terrestial and in its astronomical aspects, made enormous demands on the mathematicians of the seventeenth and eighteenth centuries. Indeed, the conceptual basis for the infinitesimal calculus has constantly been reinforced and strengthened by the interaction between pure and applied mathematics. Whilst Galileo used indivisibles to investigate motion under gravity and Huygens in studying circular motion, with Pascal and Leibniz among others we find the principle of moments playing an important part in the development of integration methods. Studies in geography, map-projection, surveying, gauging and navigation in the seventeenth century all made demands on pure mathematics in varying degrees of urgency. Practical problems stimulated the development of all manner of aids to computation and the calculation of tables in turn modified and enriched existing concepts of function and developed new ones.

The geometric model, from the static Euclidean form adopted in Greek mathematics to the sophisticated topological concepts of today, has operated not only in illuminating problems but also in determining the language and vocabulary in which the structure of mathematics has been described and the results presented. The infinitesimal calculus accordingly emerged in the form of two major classes of problem, the first concerned with the determination of arc, area, surface, volume and centre of gravity, the second with angle, chord, tangent, curvature, turning point and inflexion. Although both classes of problem are readily seen to be separately capable of inversion, thus, given the area under the curve or the tangent to the curve in terms of abscissa or ordinate, to find the curve, the relation between tangent method and quadrature is not so immediately obvious. The relation between tangent and arc ultimately became one of the most significant links between differential and integral processes and, for this and other reasons, the problem of rectification became crucial in the seventeenth century. The inverse nature of the two classes of problem was approached in terms of a geometric model by Torricelli, Gregory and Barrow but only with Newton did the relation emerge as central and general.

The development of graphical representation forged a link between the intuitive concepts of continuously varying quantities arising from physical phenomena and the geometry of the Greeks. The idea of a graph as a picture of a function or relation means that the lines on the graph can be used to interpret the relation. When the ideas of the Merton Calculators were accorded geometrical representation by Giovanni di Casali, Nicole Oresme and others the concepts of time, speed, distance and instantaneous velocity became associated with the study of curves. The shape of the curve defines the motion and the motion in turn defines the curve. The instantaneous velocity becomes associated with the tangent to the curve and the total motion or distance covered is represented by the area under the curve. Thus, a quadrature in the form of a space bounded by abscissa, two parallel ordinates and curve, becomes a general integration model for handling any continuously varying quantity; the construction of a tangent to the curve becomes a means of determining the instantaneous velocity or rate of increase of the same type of variable. Functional notions might still be limited but the basis for the geometric ideas of integration and differentiation was greatly strengthened, not only in the provision of a wider field of application but also in making available an alternative approach to the study of curves through the concepts of time and motion. In this way Galileo was able to arrive at conclusions about the path of a projectile; Torricelli in turn utilised concepts of motion to determine the areas under curves, the tangents to curves and to move from one to the other, i.e. he understood the inverse nature of differentiation and integration through the study of curves as the paths of moving points. Roberval, Barrow and subsequently Newton were among those who gained immeasurably from this concept.

The introduction of algebraic geometry by Descartes and the rapid algebraisation of the language of mathematics was not without significance for the development of the calculus in the sense that it became possible to unify methods and perceive emerging patterns expressed in terms of algebraic symbols. Nevertheless, the true distinction between algebraic geometry and the calculus is that the calculus uses infinite processes and algebraic geometry avoids them. In imposing algebraic methods in geometry, Descartes aimed to restrict the permissible processes to finite algebraic operations and to limit the study of curves to those whose equations and properties were expressible in algebraic terms.

The unification of whole classes of problems followed centuries of piecemeal investigation of special cases and was initially based on the geometric model—all problems concerned with summation, whatever the nature of the individual elements, were reduced to and expressed in terms of quadratures, all problems concerned with differential processes were reduced to tangent methods. Classification of problems was entirely absent with Archimedes. In the seventeenth century we find strong and sustained efforts to diminish the number of unresolved problems by bringing them into association either with standard problems already solved or other problems known to be awaiting solution.

Alongside of this grouping of problems we have a concomitant development of powerful general methods through which a whole range of results might be derived: Cavalieri, for example, developed the concept of the sums of the powers of the lines of a figure and from this was able to derive a series of quadratures and cubatures expressible analytically in the form $\int x^n \, dx$; Wallis, by arithmetising the work of Cavalieri and basing his integration methods on the sums of the powers of the natural numbers, was able to bring about a further generalisation in the sense that quadratures and cubatures now arose as a corollary of a single limit operation (even if that limit operation itself was still on the shaky basis of "incomplete" induction). By extending the index notation he was able to include within his general method a further set of results formerly derived by a separate and distinct approach. Newton himself derived a great deal of power from his method of infinite series and, through the expression of functions in infinite series, he hoped both to establish the existence of the functions and at the same time to develop a universal method of quadrature.

The notion of function lies at the kernel of the infinitesimal calculus. In modern mathematics a function is defined as a correspondence which associates each element of any set X with one and only one member of a set Y. For example, an expression such as x^2 has no numerical value until the value of x is assigned; x^2 is accordingly a *function* of x and we write $f(x) = x^2$. The value of $f(x)$ can be found for any integral, fractional, irrational or complex number x. The sum of a finite number of terms of an arithmetic, geometric, or power series as the case may be, is a function of the number of terms: the area of a triangle is a function of the lengths of the sides and the perimeter of a circle of the radius. The concept of function enters whenever quantities are connected by any definite physical relationship.

To Leibniz who first used the word *function* and to the mathematicians of the eighteenth century the idea of a functional relationship was limited in scope and implied the existence of a mathematical formula or prescription expressing that relationship. Indeed, the concept of function arose as a result of a long and protracted process of abstraction and generalisation in which, for the most part, special functions were investigated in isolation before the development of any general theory of functions.

From the earliest times tables of correspondences derived from observations of physical phenomena have played a major role in developing an intuitive awareness of continuously varying quantities. The astronomical tables of the Babylonians, the physical speculations of the Scholastics, the dynamics of Galileo, the astronomy of Kepler, the gravitation theory of Newton, all contributed in one way or another to suggest new ways of developing, defining and determining functional relationships.

In Greek geometry we already find raised many fundamental problems concerning the *definition* and *existence* of functional relations subsequently classified as algebraic, trigonometric and transcendental. With Euclid visual evidence was accepted as adequate proof of the existence of an area. The convexity lemmas of Archimedes, however, constituted a significant step towards recognising the necessity for the establishment of the existence of the integral (in this case, arc) on a geometric basis. The work of Valerio and later of Fermat and Gregory in the seventeenth century continued this line of development.

Throughout the whole century special functions were intensively explored and, by approaching them from a variety of different directions, understanding was gradually widened and deepened. The study of curves by kinematic methods rests, in the first instance, on an intuitive idea of a quantity continuously varying with time, and the use of Cartesian coordinates in this connection implies the idea of time as an independent parameter: thus, $x = f(t)$, $y = g(t)$. Such methods were exploited fully by Torricelli and by Barrow: later Newton's fluxions implied $\dot{x} = f'(t)$, $\dot{y} = g'(t)$ and $\dot{y}/\dot{x} = g'(t)/f'(t)$. Study of the cycloid—a mechanical curve produced by rolling—implied $x = f(\theta)$, $y = g(\theta)$ and the by-products of this work, particularly with Roberval, included the graphing and integration of trigonometric functions.

In the mid-century there is a tendency to refer in demonstrations to any curve *whatsoever* but it is not often clear what this implies—probably for the most part curves defined either by algebraic or by mechanical processes. The authority of Descartes was exercised in an attempt to ban from geometry all curves not susceptible to precise analytic definition. Fortunately, however, the impact of his views on his contemporaries was slight for they were already engaged in wholesale investigations of logarithmic, transcendental and trigonometric functions, many of which were defined in terms of processes involving imperfect limit operations and summation of series the convergence of which was taken on trust. The logarithmic function, for instance, first defined kinematically by Napier, was established on a numerical basis by the calculation of correspondences between the terms of two series, the first in geometric and the second in arithmetic progression. The area under the hyperbolic arc was investigated independently by Saint-Vincent and the connection between this function $\left(\int_0^x dt/t\right)$ and the logarithm noted by his pupil De Sarasa. The expansion of the logarithmic function in an infinite series was finally pursued by a whole host of mathematicians including Brouncker, Newton, Mercator, Gregory and Wallis.

The expression *analytic* was first used by James Gregory who defined an *analytic* quantity as one obtainable by algebraic operations together with passage to the limit. In the wild rush to develop function after function in infinite series without benefit of convergence (and in which Gregory himself joined) this sensible step was temporarily lost sight of.

From Cavalieri onwards, most of the seventeenth-century mathematicians who had a contribution to make were oppressed by the difficulties involved in setting out demonstrations in full. The burden of writing and finding time to read even a fraction of the great volume of work published bore heavily on most. The rapid adoption of algebraic notation as a means of presenting "such an abstract of the proof as is sufficient to convince a properly instructed mind",† was, under the circumstances, inevitable. Although there were many who reacted strongly against the rash of new notations and others who found the adoption of algebraic methods distasteful, the period was one in which experiments in symbolism abounded. The adoption of the familiar Cartesian notation (x, y) for the coordinates of a point in a plane and the corresponding expression of the curve equation in the form $f(x, y) = 0$, produced a recognisable pattern of results in terms of tangents and quadratures. The formulation of prescriptions, or operational rules, for deriving the tangents and quadratures directly from the initial equation was an immediate consequence. The first rules to cover differential processes were devised by Sluse and Hudde with Newton not very far behind. The invention of specialised notation to cover the results of performing the required operations according to the prescribed rules was naturally the next step.

From Cavalieri onwards the summation of an infinite number of line, or

† Whitehead, A. N. and Russell, B. A., *Principia mathematica*, Cambridge, 1910, I, Introduction.

surface, elements, indivisible or infinitesimal as the case may be, was commonly represented by the word *omnes*, abbreviated to *omn.* and to *o*. The use of single letters to represent the small increments or changes in the value of an independent variable was initiated with the 'e' of Fermat, subsequently replaced by the 'o' of Beaugrand, which was later adopted by Newton. Barrow used two separate letters to denote the small increments in x and y respectively. Although Newton was no great innovator in relation to the use of symbols he was alive to the advantages of Cartesian notation and made immediate use of the extension of the index notation developed but not actually used by Wallis. During his initial investigations on tangent methods and curvature Newton experimented with a variety of symbolic representations of the operational processes involved. Whether indeed he remained dissatisfied with the resulting forms or whether he was unconvinced of the need for new and specialised notations he never used fluxional notation consistently and appears to have regarded it as unimportant compared with the method of infinite series until challenged by Leibniz.

Leibniz, on the other hand, was throughout his whole life deeply interested in symbolic notation as such and convinced that only through its extension into every realm of human thought could comprehension and clarity be achieved. The idea of a "universal language" or a "characteristic" was his lifelong ideal and the development of an operational notation for the infinitesimal calculus was integral to this central purpose. "The true method should furnish us with an Ariadne's thread, that is to say, with a certain sensible and palpable medium, which will guide the mind as do the lines drawn in geometry and the formulas for operations, which are laid down for the learner in arithmetic."†

Perhaps no one has appreciated more clearly than Leibniz the importance of good notations in mathematics and the symbols he invented for the differential and integral calculus were so suggestive and so well chosen that it seems wellnigh impossible to conceive of mathematics without them. The meanings of the symbols have changed and so also have the very foundations of the calculus on which they are based; we now have available a vast range of other notations to represent functional relations and differential and integral processes. Nevertheless, the familiar dx, dy and $\int$ notation of Leibniz has survived nearly 300 years of trial and error and seems likely to remain with us as long as pure mathematics survives in its present form.

† *Die Philosophischen Schriften* (ed. Gerhardt, C. I.), Berlin, 1875–90 (7 vols.), VII, p. 22.

CHAPTER 1

Greek Mathematics

1.0. Introduction

Empirical methods of land measurement undoubtedly played an important part in developing the concept of area, so fundamental in the geometrical model of the integral calculus. Stone Age excavations have produced a wealth of evidence of the familiarity of early man with simple geometrical forms. Weapons, tools, utensils and jewellery have been found exhibiting symmetrical patterns made up of squares, rectangles, parallelograms, triangles (equilateral and isosceles) and regular hexagons. In many cases these are inscribed in circles. Three-dimensional solids, such as the cube, the tetrahedron, the cylinder and the sphere, seem to have become generally known at an early stage.

A common characteristic of Babylonian, Egyptian, Hindu and Chinese mathematics is in the formulation of prescription-like rules for calculating the areas of rectilinear figures. Although little is known as to how these rules were derived and although some of them are only approximately correct it is certain that the elementary idea of cutting up an area and rearranging the parts in a simpler manner must have been fundamental in all such processes. By the use of such methods the triangle and parallelogram become rectangles and the parallelepiped a cuboid (see Fig. 1.1). In all the above cases accurate results demanded the determination of the height of a right-angled triangle and the Babylonians, in particular, seem to have possessed a complete knowledge of Pythagorean numbers as well as sound methods for approximating square roots.

The determination of curvilinear areas, i.e. areas enclosed by curved lines, first appeared in attempts to approximate to the area of a circle. Knowledge of this curve developed in connection with the use of the wheel in agriculture and in pottery. The use of 3 as an approximation for π seems to have been common to most ancient civilisations and arose through taking the mean of the inscribed and circumscribed squares (see Fig. 1.2).

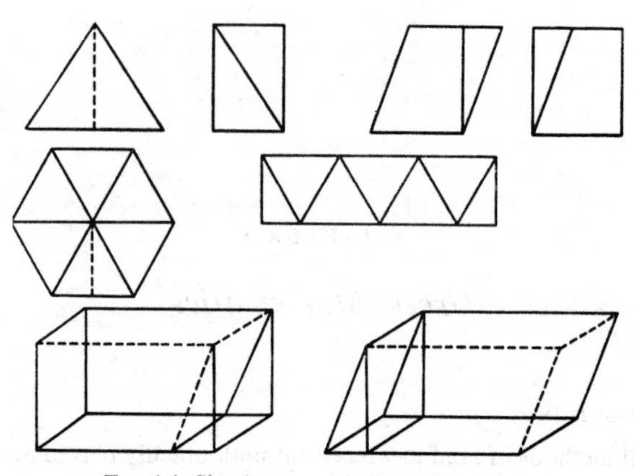

FIG. 1.1. Simple geometrical transformations.

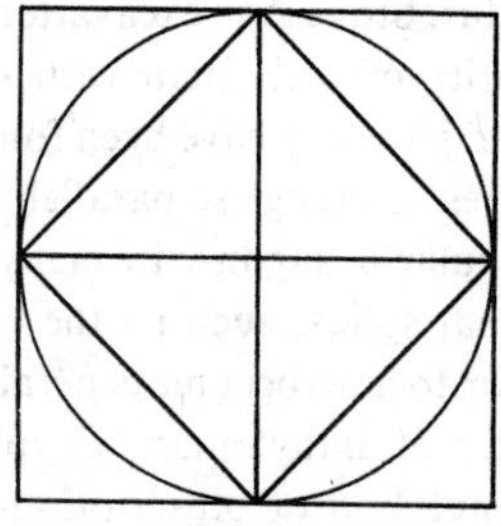

FIG. 1.2

The Egyptians used a value equivalent to $\pi/4 = (8/9)^2$ and although we have no direct knowledge as to how this was arrived at some form of geometric transformation seems likely. Use was also made of these approximations in the calculation of the volumes and surface areas of cylinders. Both the Babylonians and the Egyptians possessed correct rules enabling them to calculate the volume of the frustum of a square pyramid.

All this material constituted a background of valuable experience which, expressed in terms of practical rules for the measurement of area and volume, facilitated the development of Greek geometry on a theoretical basis. The concept of geometric transformation, already inherent in the simple rearrangements necessary to determine rectilinear areas and volumes, was essentially the foundation for a long train of development culminating in the sophisticated geometric techniques of the seventeenth century which preceded the integral calculus.

Until comparatively recently little was known of the existence of any organised body of mathematical knowledge prior to that of the Greeks and even now our information is insufficient to permit a clear account of the techniques used. Perhaps the most important gain from the painstaking researches of Neugebauer[†] and others into Babylonian mathematics has been to render more comprehensible the rapid rise to maturity of Greek mathematics during the period 600–300 B.C. Clear links have been established between much of the *geometric algebra* of the Greeks and the Babylonian rules for the solution of equations and, although in other matters the connections are more tenuous, it now seems certain that the Greek debt to the Babylonians was considerable.

The Babylonians were extremely skilful in solving problems of a finite and algebraic nature and formulated rules for solving cubic equations (for which they used tables of squares and cubes) as well as quadratic equations in the form $xy = a$, $x \pm y = b$. Although much of this work was accomplished by the tabulation of results and this in turn was facilitated by an excellent sexagesimal system of numeration, it has been natural to enquire into the possibility of the existence of an underlying proof structure. A notable effort has been made by E. M. Bruins[‡] to trace the beginnings of a deductive system in Babylonian mathematics but the existence of anything of this kind must remain a matter for conjecture.

The idea of a mathematical function has always been linked historically with developing awareness of physical correspondences and in this connection the progress made by the Babylonians in the tabulation and interpolation of astronomical data was remarkable.[§] In astronomical texts the positioning of certain numbers in columns and the use which is made of them can be interpreted as incorporating the idea of positive and negative numbers, i.e. measurements taken on opposing sides of a mean position. In its most general usage in mathematics the word function refers to any correspondence between two classes of objects and it might therefore be possible to claim for the Babylonians an important place in the history of the concept of functionality.[||] Nevertheless, because this material is too remote and because our understanding of it, even now, is so incomplete, it can scarcely be maintained that any of it has a direct bearing on the origins of the calculus. Most of what

† Neugebauer, O., *The Exact Sciences in Antiquity*, Copenhagen, 1951. Neugebauer, O. and Sachs, A., *Mathematical Cuneiform Texts*, New Haven, 1945.

‡ Bruins, E. M., On the system of Babylonian geometry, *Sumer* 11 (1955), pp. 44–9. The Babylonians appear to have had no concept of angle (other than right), proportion or similarity. Bruins suggests that they derived most of their results, including the Pythagorean theorem, from the concept of equal areas.

§ Neugebauer, O., *op. cit.*, pp. 105–9.

|| Cf. Bell, E. T., *The Development of Mathematics*, New York, 1945 (2nd ed.), p. 32.

is valuable in Babylonian and in Egyptian mathematics was assimilated by the Greeks and it is to them that we owe the first conscious and deliberate effort to place mathematical proof on a basis of deductive reasoning.

1.1. The Influence of Greek Mathematics in the Seventeenth Century

The important chain of developments which led, in the seventeenth century, to the birth of what is now termed the infinitesimal calculus, can only be understood in the light of the seventeenth-century mathematical inheritance. This consisted of a mixture of material from Greek, Roman, Hindu and Arabic sources much of which, having passed through the hands of the mediaeval *calculators*, had acquired the characteristic flavour of philosophico-mathematical scholasticism. The Greek classical tradition remained powerful throughout the century and many of the foremost mathematicians in Italy, France and England concerned themselves with the preparation of revised editions and translations of Greek works.† In the latter part of the sixteenth century Commandino (1509–75) was responsible for the publication of Latin translations of the works of Euclid, Apollonius, Archimedes, Pappus, Heron and Ptolemy. These editions and the valuable commentaries which accompanied them were carefully studied and much quoted by seventeenth-century mathematicians. New Latin editions of Euclid's *Elements* were brought out by Barrow in England, Tacquet in France and Borelli in Italy and were subsequently translated into living languages. Barrow also concerned himself with the preparation of a new Latin edition of the works of Archimedes and Wallis edited *The Sand-Reckoner* and the *Measurement of a Circle*. Leibniz, whose background of Euclidean geometry seems to have been particularly inadequate, speculated on the possibility of developing a universal language‡ in which it might be possible to exhibit the theorems of Euclid, Apollonius, Archimedes and Pappus.

† Full details of the various editions of Greek mathematical texts prepared during the seventeenth century are available in the works of Heath, T. L. and Dijksterhuis, E. J., listed below:

Heath, T. L.: *A History of Greek Mathematics*, Oxford, 1921, 2 vols. (referred to as Heath, *G.M.*).

Heath, T. L.: *The Thirteen Books of Euclid's Elements*. References given from the Dover edition, New York, 1956 (3 vols.). Referred to as Heath, *Euclid*.

Heath, T. L.: *The Works of Archimedes*, Cambridge, 1897 (Suppl. 1912). References given from the Dover edition, New York, n.d.

Dijksterhuis, E. J.: *Archimedes*, Copenhagen, 1956.

‡ *Catalogue critique des manuscrits de Leibniz*, II (ed. Rivaud, A.), Poitiers, 1914–24, no. 526 (referred to as *Cat. crit.*).

It should perhaps be emphasised that these works were not presented purely as historical research, nor even because they represented the highest level of Greek mathematical achievement but because they displayed a standard of mathematical rigour which could not, in the seventeenth century, be achieved by alternative means. Criticism of the methods of Archimedes which arose throughout the century was directed at the length and tedium of the proofs rather than at the essential logic of the proof structure.[†] At the beginning of the century any radical departure from the "methods of the ancients" was regarded not only as a regrettable lapse from mathematical rigour but also as a challenge to authority itself. In consequence most of those who dared to publish alternative methods did so somewhat apologetically, admitting freely that in the final analysis proof must rest on an Archimedean demonstration.

Although many of the violent, protracted and acrimonious controversies which characterised seventeenth-century mathematics covered a wide area of disagreement, methodological issues constituted an important element in most of them: Kepler's *Nova stereometria* was sharply attacked by Alexander Anderson;[‡] Cavalieri's *Geometria* was savagely denounced by Guldin;[§] Wallis's *Arithmetica infinitorum* was criticised by Fermat[||] who enquired "pourquoi il a préféré cette manière par notes algébriques à l'ancienne, . . .". Newton presented his *Principia* in entirely Euclidean form and in consequence avoided this particular type of controversy.[††]

1.2. A Brief Chronology of Greek Mathematics

For the sake of clarity I shall give here a brief résumé of the chronological development of Greek mathematics before discussing in greater detail the aspects of it which relate most closely to this particular study.

Greek mathematics is usually considered to have begun with Thales

† Cf. Whiteside, D. T., Patterns of mathematical thought in the later seventeenth century, *Archive for History of Exact Sciences*, 1 (3) (1961), pp. 179–388.

‡ Kepler, J., *Nova stereometria*, Linz, 1615. Anderson, A., *Vindiciae Archimedis*, Paris, 1616.

§ Cavalieri, B., *Geometria indivisibilibus continuorum nova quadam ratione promota*, Bologna, 1635 (2nd ed., 1653). Guldin, P., *Centrobaryca*, Vienna, 1635/41 (4 books).

|| See *Oeuvres de Fermat* (ed. Tannery, P. and Henry, C.), Paris, 1891/1912 (4 vols.); *Supplément* (ed. Waard, C. de), Paris, 1922, II, p. 343.

†† I do not for one moment suggest that Newton adopted this particular form in order to avoid controversy. Although this matter has been the subject of a great deal of speculation nothing conclusive has so far emerged. Dr. M. A. Hoskin recently suggested that Newton adopted this form of presentation because he hoped to publish with the *Principia* some new results of his own in the field of synthetic geometry. E. T. Bell calls the *Principia* a monument to synthetic geometry, See *The Development of Mathematics*, p. 82.

(*ca.* 585 B.C.) and the founding of the Pythagorean School (*ca.* 550 B.C.). No documentary material from this period has survived and any information available has been given by various commentators† several centuries later. Many of these accounts were based on hearsay and it is clear that, even at the time when they were compiled, the original proofs and methods were no longer available. Nevertheless, although it is impossible to identify and describe the work of individuals, the broad lines of mathematical development during this period have been fairly well established.

Within the early Pythagorean School the study of geometry was, for the first time, organised on a logico-deductive basis; in the field of arithmetic the properties of the integers were extensively explored largely by means of figurate numbers. In addition to purely mathematical developments we have an increase in rational speculation as to the nature of the physical world and its relation to human thought.

In the fifth century B.C. the systematisation of the theory of numbers and its application to the study of geometry led to the discovery of irrational numbers. Unfortunately little is known as to the date or circumstances in which this occurred.‡ We do know, however, that this discovery threatened the foundations of mathematics. The consequent impasse was resolved within the Academy of Plato when the entire system of mathematics was reorganised on an exclusively geometrical basis. Van der Waerden§ attributes the major part of this task to Theaetetus (d. 368 B.C.) and to Eudoxus (*ca.* 370 B.C.).

It was Zeuthen|| who first drew attention to the algebraic nature of the contents of Euclid (II and VI) and gave to it the title *geometric algebra* by which it has subsequently been known: Neugebauer†† was able to demonstrate the close relationship it bears to the Babylonian rules for the solution of what are now termed quadratic equations. The geometric form given to these rules is attributed to Theaetetus and the specialised geometric techniques resulting became known as the *application of areas*. The essential process here is the application of a given rectilinear figure to a straight line and the determina-

† Diels, H., *Die Fragmente der Vorsokratiker*, Berlin, 1951–2 (3 vols.) (6th ed.). Freeman, K., *The Pre-Socratic Philosophers*, Oxford, 1946.

‡ Cf. Fritz, K. von, The discovery of incommensurability by Hippasus of Metapontum, *Annals of Mathematics*, 46 (1945), pp. 242–64. Becker, O., *Eudoxos-Studien* I–V: *Quellen und Studien zur Geschichte der Mathematik, Astronomie und Physik, Studien* 2 (1933), pp. 311–33, 369–87; 3 (1936), pp. 236–44, 370–410.

§ Van der Waerden, B. L., *Science Awakening*, Groningen, 1954, pp. 165–90.

|| Zeuthen, H. G., *Histoire des Mathématiques*, Paris, 1902, pp. 34–43.

†† Neugebauer, *op. cit.*, pp. 143–4.

tion of its area in terms of a related square. The sophisticated procedures subsequently developed by later mathematicians were applied with great skill and facility to the study of two- and three-dimensional loci. An example of the extension of geometric algebra to three dimensions is found in the Delian problem† (or the duplication of the cube) which Hippocrates (*ca.* 440 B.C.) reduced to the finding of two mean proportionals between given magnitudes, one of which is twice the other. It was natural that attempts made to transform the circle into an equivalent rectilinear figure should become known as the squaring of the circle.‡

To Eudoxus is due the masterly theory of proportion which is, in fact, a theory of the real number system and is incorporated in Book V of Euclid's *Elements*. Also attributed to Eudoxus is the *Method of Exhaustion* (*Euclid*, XII) which was later so refined and developed by Archimedes that it became a powerful tool in the determination of curvilinear areas, surfaces and volumes.

During the Alexandrian Era Euclid (*ca.* 330 B.C.) systematised plane and solid geometry and organised it as a deductive system based on a given set of postulates, axioms and definitions; Archimedes (d. 212 B.C.) founded the sciences of statics and hydrostatics, basing them on an axiomatic structure similar in form to that of Euclid's *Elements*. In this work and in the area and volume determinations referred to above he made extensive use of the *Exhaustion Method;* the elegant applications which he gave to this method have been described by Heath§ as anticipations of the integral calculus and must, whether this be so or not, occupy a central position in any survey of its origins. The *Exhaustion Method* as developed by Archimedes inspired and at the same time exasperated seventeenth-century mathematicians who wished to replace it by something shorter and more direct but found themselves unable to provide the necessary degree of rigour by any alternative processes they were able to devise. Apollonius of Perga (*ca.* 230 B.C.) applied the whole complex apparatus and terminology of geometric algebra to the systematic study of conic sections. The study he made of them was so exhaustive and so detailed that most of the important metric properties of the conic sections were established for all time.

† To determine x/y, such that $x^3/y^3 = 2$; let $b/x = x/y = y/a$, $(b/a = 2)$ then $by = x^2$, $y^2 = ax$, $y^2/x^2 = ax/by$, $x^3/y^3 = b/a = 2$.

‡ In a note to Euclid I. 45, Proclus comments: "I conceive that it was in consequence of this problem that the ancient geometers were led to investigate the squaring of the circle as well. For, if a parallelogram can be found to equal any rectilineal figure, it is worth inquiring whether it be not also possible to prove rectilineal figures equal to circular." Heath, *Euclid*, I, p. 347.

§ Heath, *Archimedes*, Introd., Chap. VII (cxlii–cliv).

Of the later mathematicians the works of Diophantus (*ca.* A.D. 250) attracted a great deal of attention in the seventeenth century but have no immediate relevance for this study. More important certainly are the *Collectiones* of Pappus (*ca.* A.D. 320) who supplemented and extended the work of his predecessors besides clarifying and amplifying many difficult and obscure passages.

The Greek works available to us are, for the most part, entirely synthetic in character and, although they provide us with an adequate understanding of the scope, content and proof structure of Greek mathematics, they give little information as to the steps which led to the discovery rather than to the presentation of theorems. Of outstanding significance, therefore, was the finding by Heiberg in 1906 of a lost Archimedean treatise† entitled *The Method* in which Archimedes not only described a mechanical method for discovering theorems by treating an area as though it were made up of line elements but also hinted that similar methods might have been used by Democritus, the fifth-century mathematician and Atomist philosopher.‡ In consequence a great deal of speculation has arisen as to the extent of the role played by infinitesimal methods in Greek mathematics and also as to the relationship of such methods to the views of the materialist philosophers on the nature of the physical world.

1.3. Early Greek Mathematics and the Physical World

Early Egyptian mathematics consisted for the most part of empirical rules for performing operations on concrete objects by means of numerical processes, and generalisations which emerged seem to have resulted from the simple extension of a rule found to be valid in a number of special cases to all others. Babylonian texts, on the other hand, exhibit a much greater degree of generalisation but unfortunately we know little of the manner in which these generalisations were arrived at. With the Greeks, however, we find for the first time the idea of a general proof, valid in all cases. The developing concept of mathematics as a deductive system linked, at least in its initial assumptions and in its final conclusions with the physical world, led to increasing speculation as to the relation between the abstract elements of such a system and the concrete objects, shapes and forms belonging to the external world. Much thought was given as to the fundamental basis of knowledge and to the processes of abstraction, idealisation, generalisation and deduction through which, in one way or another, such knowledge appeared to be derived.

† Heath, *Archimedes*, Supplement (following p. 326), pp. 1–51.

‡ *Ibid.*, Supplement, p. 13.

The only physical science with any substantial basis of observation at this time was astronomy and, since in this case the objects observed were remote and their constitution unknown, it was natural that abstract and concrete elements should be regarded as interchangeable.†

By a process of abstraction the early Pythagoreans developed the fruitful concept of figurate numbers, i.e. numbers made up of discrete elements arranged in geometric forms.‡ This concept was then projected back upon the physical

† Hesse, M. B., *Forces and Fields*, London, 1961, Chap. II, pp. 29–50.
‡ See Fig. 1.3 below.

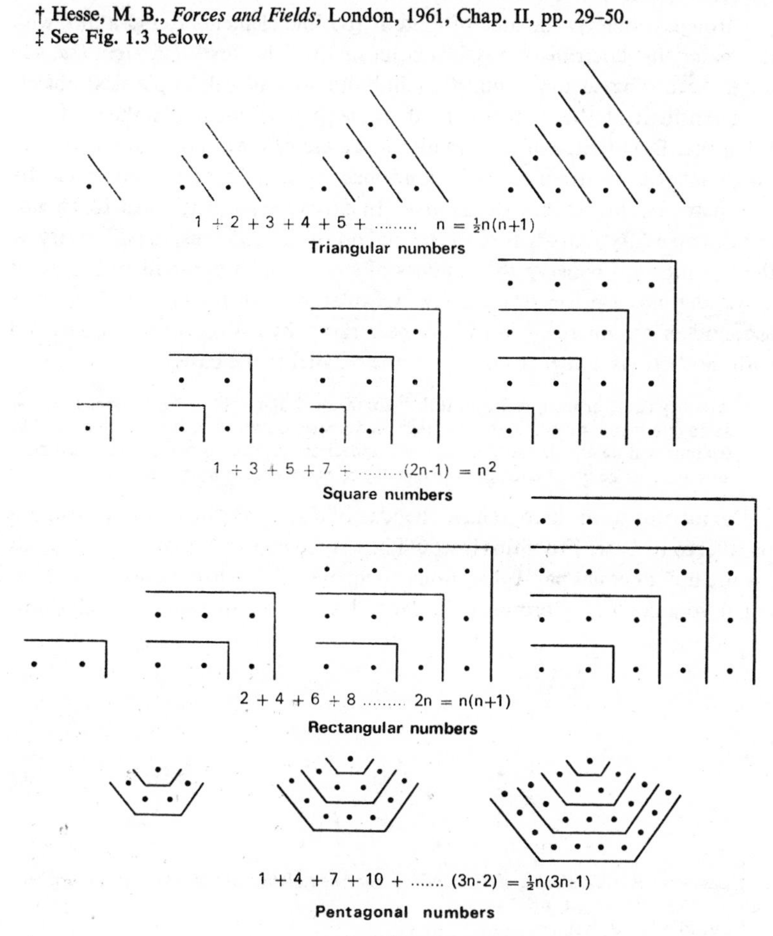

FIG. 1.3. Figurate numbers.

universe and its construction explained in terms of number. The discovery of the relation between musical harmony and the theory of proportion, attributed to Pythagoras himself, reinforced this view of the structure of the universe. The physical line segments and triangles of Pythagorean geometry could similarly be considered to be made up of discrete numerical elements until it was discovered that the diagonal of the square was incommensurable with the side. In this case, if the side be constructed of a finite number of discrete elements, how can the hypotenuse be constructed?†

Through geometry the idea of finite indivisible elements found a link with the materialist doctrine of physical atomism‡ which arose at Abdera (*ca.* 430 B.C.). According to the Atomists, mind and soul as well as physical objects were considered to be made up of indivisible particles moving in the void. The belief that Democritus of Abdera also made use of the concept of indivisibles in volumetric determinations is confirmed by certain comments made by Archimedes in his treatise *The Method.* In particular, he attributes to Democritus, some fifty years before the formal proofs of Eudoxus, the discovery of the relationships between the volumes of a cone and a pyramid and those of a cylinder and a prism respectively.§ Of interest in this context is Plutarch's account|| of the queries said to have been raised by Democritus in connection with the sections of a cone cut by a plane parallel to the base.

> Are they equal or unequal? For, if they are unequal, they will make the cone irregular as having many indentations, like steps, and unevennesses; but, if they are equal, the sections will be equal, and the cone will appear to have the property of the cylinder and to be made up of equal, not unequal, circles, which is very absurd.

Democritus seems here to have the idea of a solid as "made up" of sections parallel to its base. From this it would be easy to conclude that any two solids "made up" of equal parallel sections at equal distances from their bases have equal volumes. (This proposition, derived by means of similar conclusions,

†

‡ Lasswitz, K., *Geschichte der Atomistik vom Mittelalter bis Newton*, Hamburg and Leipzig, 1890 (2 vols.), I, pp. 14–18.

§ Euclid XII. 10, XII. 5.

|| Given from Heath, *G.M.*, I, p. 180.

was first published by Cavalieri† in 1635.) Since the volumes of certain pyramids can be readily obtained by the dissection of a prism (see Fig. 1.4) it becomes possible to extend the results obtained in this way to all pyramids

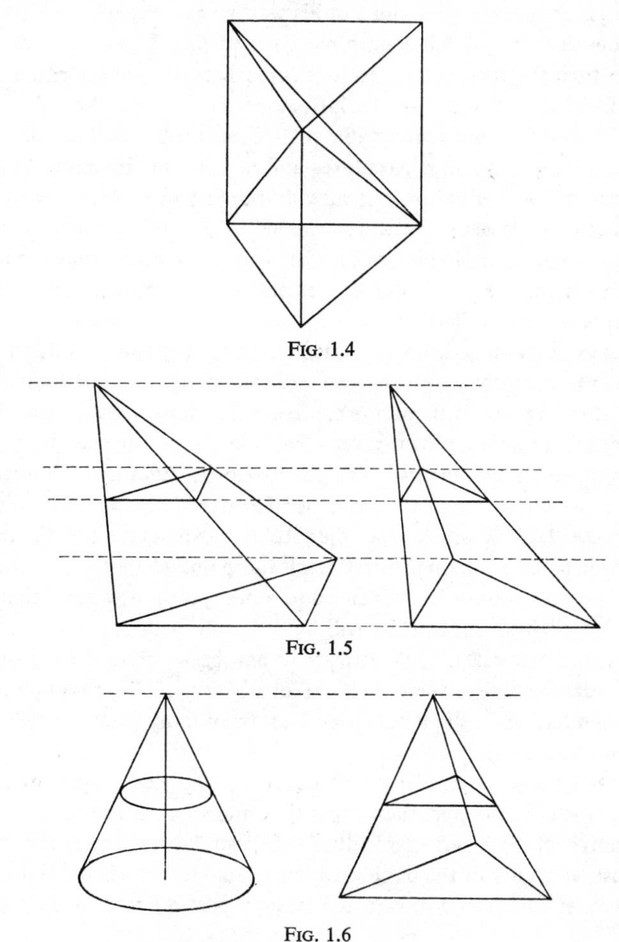

FIG. 1.4

FIG. 1.5

FIG. 1.6

with equal bases and equal heights (see Fig. 1.5) and since the theorem applies to pyramids with any type of polygonal base it can, by inference, be extended to the cone (see Fig. 1.6). Alternatively, if each circular section of the cone

† Cavalieri, B., *Geometria indivisibilibus*, *op. cit.*, Book VII, Theorem I, Prop. I, pp. 4–9.

be replaced by a triangle of equal area, the volume of the cone will be equal to that of a triangular pyramid with equal base and equal height. Unfortunately all this must remain conjectural for, although Democritus is known to have written on mathematics, none of his works has survived.

Any consideration of the conceptual basis of Greek mathematics rather than of its formal structure must include some discussion of the paradoxes of Zeno of Elea† (fl. 460 B.C.). In the history of the Pre-Socratic philosophers there have been few more intractable or difficult problems than that of the interpretation of the famous paradoxes and it is not my intention to add to the number of conflicting theories already in existence. Nevertheless, since some of the more recent conjectures have led to the most extravagent claims for Zeno, both in regard to his place in the history of mathematics and in relation to the origins of the calculus, it is essential that some account should be given of recent work in this field.

A paradox in mathematics arises when contradictory conclusions are deduced from an apparently impeccable chain of reasoning. In the *Dichotomy* Zeno argues that before a body in motion can reach a certain point it must traverse half the distance; before it can traverse half the distance it must traverse the quarter; and so on, *ad infinitum*. The paradox of Achilles and the tortoise is of a similar nature. If the tortoise and Achilles start from the same point and if the tortoise starts before Achilles then Achilles can never catch the tortoise for, in order to do so, he must first reach the point where the tortoise was; by this time the tortoise has advanced to a new position; when Achilles has reached this position the tortoise will have moved further, and so on. Hence Achilles can never reach the tortoise. It has been inferred by Tannery,‡ Heath,§ and others that these two paradoxes rest on the assumption that space is infinitely divisible; a conclusion contrary to experience implies that this assumption is falsely based.

In the *Arrow* and the *Stadium* Zeno deduces equally contradictory conclusions from the contrary hypothesis, that time and space are composed of ultimately indivisible elements. Difficulties in interpretation have arisen primarily because accounts of the paradoxes are available only from Plato, Aristotle and other commentators who, for the most part, quote Zeno only in order

† Cf. Cajori, F., History of Zeno's arguments on Motion, *Amer. Math. Monthly* 22 (1915), pp. 1–6, 39–47, 77–82, 109–15, 143–9, 179–86, 215–20, 253–8, 292–7. Schramm, M., *Die Bedeutung der Bewegungslehre des Aristoteles für seine beiden Lösungen der zenonischen Paradoxie*, Frankfort-on-Main, 1962.

‡ Tannery, P., *Pour l'histoire de la science hellène*, Paris, 1887.

§ Heath, *G.M.* I, pp. 273–83.

to refute something which he is thought to have asserted. Nevertheless, it is easy enough to recognise that the resolution of the first two paradoxes is dependent on the summation of certain geometric series and involves the notions of limit, convergence, continuity and other concepts which were scarcely placed upon a satisfactory mathematical basis until the nineteenth century. Because Zeno was the first to emphasise the difficulties inherent in such considerations Bertrand Russell,[†] among others, has claimed that Zeno anticipated Weierstrass.

By his contemporaries Zeno seems to have been regarded as a supporter of Parmenides, the Sophist, who denied the existence of movement. In this case the paradoxes can be considered as demonstrating that, whatever view be held as to the divisibility of time and space, the same conclusion is reached, i.e. that movement is impossible. In this case Zeno must be regarded simply as an Eleatic philosopher and his arguments as directed against other philosophers. It was Tannery,[‡] however, who first interpreted the paradoxes as an attack on the Pythagoreans who constructed solids, surfaces and lines from indivisible points. Although this suggestion was at first treated with some reserve it was subsequently expanded by Hasse and Scholz[§] who put forward the plausible conjecture that the early Pythagoreans, in consequence of the discovery of irrational numbers, had been forced into operations which involved infinitely small quantities. Zeno's devastating critique obliged them to reform their methods and to place them upon a sound mathematical footing: hence was developed the *Method of Exhaustion*, and infinitesimal processes—if not entirely eliminated—were at any rate driven so far below the surface that no trace of them remains in the formal development of Greek mathematics.

This view of Zeno has a certain dramatic quality which has assured for it a place in most of the more popular histories of mathematics. The formulation of Zeno's paradoxes is spoken of as one of the cardinal turning points in the history of mathematics and Zeno himself moves into a commanding position as the man who saved Greek mathematics in its hour of need.[||] Aristotle is quoted in support in that he says of mathematicians that they made no use of magnitudes increasing or diminishing *ad infinitum* but contented themselves with finite quantities that can be made as great, or as small as we please.[††]

† Russell, B. A. W., *The Principles of Mathematics*, I, Cambridge, 1903, pp. 346–53.

‡ Tannery, P., *Pour l'histoire de la science hellène*, pp. 255–70.

§ Hasse, H. and Scholz, H., *Die Grundlagenkrisis der Griechischen Mathematik*, Berlin, 1928.

|| Bell, E. T., *The Development of Mathematics*, p. 56.

†† Arist. *Phys.* III. 7. 207^b, 31. See Heath, *G M.* I, p. 272.

Van der Waerden,† however, after a careful study of the textual references combined with a survey of the chronology and development of early Greek mathematics, has concluded that the paradoxes of Zeno neither precipitated nor resolved any crisis in Greek mathematics. Unfortunately space does not permit more than a brief review of his findings:

1. There is no evidence that the Pythagoreans either used, or required to use, infinitesimal methods. Any problems which might have engendered such methods were on the periphery rather than in the centre of Pythagorean mathematics and therefore could not have precipitated a crisis.
2. Hippocrates and Archytas, more than a generation later, were satisfied to use the Pythagorean theory of proportion (for commensurable magnitudes). Hence, whenever the discovery of incommensurable magnitudes occurred, it certainly precipitated no crisis before the end of the fifth century.
3. The Atomist philosophy grew up partly as a reaction against the Eleatics and a generation after Zeno. Hence the queries raised by Democritus and the results obtained by him (possibly obtained by the use of indivisibles) followed after Zeno and cannot therefore have provoked an attack by him.
4. The text suggests that Zeno's primary purpose was rather to defend Parmenides by denying the possibility of motion than to attack any particular ideas on the divisibility of space and time.

In consequence of these conclusions Van der Waerden has made a determined effort to banish Zeno from the history of mathematics and his recent work, *Science Awakening*, contains no mention of Zeno. Such an omission can only be justified if a somewhat narrow view be taken of the history of mathematics. Although the paradoxes may well have been formulated outside any recognisable stream of mathematical thinking there can be no question that they have subsequently proved a focal point‡ for much fruitful discussion of concepts of space and time, the nature of the continuum, infinitesimals and indivisibles. Before any resolution of such problems could be attempted it was necessary to draw a clear distinction between the psychological difficulties inherent in the formation of adequate concepts of space, time, and instantaneous velocity, and their precise definition in terms of limit. Nevertheless, intuitive ideas, arising from experience of the physical world, have always played an important part in mathematical discovery and the historian who is concerned to understand and explain development as well as to analyse achievement can scarcely afford to ignore the background of philosophical

† Van der Waerden, B. L., Zenon und die Grundlagenkrise der Griechischen Mathematik, *Math. Ann.* 117 (1940), pp. 141–61.

‡ Bertrand Russell comments: "These arguments are in any case well worthy of consideration, and as they are, to me, merely a text for discussion their historical correctness is of little importance." See *The Principles of Mathematics* I, p. 348.

discussion.[†] The main difficulty, however, lies in making judgements of relevance, for imagination and intuition operate according to no simple, straightforward and immediately recognisable laws. At all events, during the third century, through the axiomatic systems established by Euclid and Archimedes, mathematics was finally separated from philosophy and its relationship to the physical world substantially clarified.

1.4. The Axiomatisation of Greek Mathematics

The final axiomatic structure of Greek mathematics was laid down in the third century B.C. and has been available for detailed scrutiny for over 2000 years in the *Elements* of Euclid, the treatises of Archimedes and the *Conics* of Apollonius.

It is never easy to establish with any degree of certitude the relationship between logic and philosophy on the one hand, and the structure of mathematical and scientific systems on the other. For this reason alone any attempt to assess the influence of the philosophy of Plato and the logic of Aristotle on the work of systematisation undertaken by Euclid and culminating in the thirteen books of the *Elements* is fraught with difficulty. Between the logic of Aristotle and the geometry of Euclid, at any rate, the interaction seems to have been mutual, for the impact of mathematics on Aristotle's thought is only paralleled by the respect for formal logic displayed by Euclid in the *Elements*. This monumental work rapidly established itself as one of the supreme models of rigorous reasoning and has influenced all views on the nature of mathematical form and structure to the present day. The axiomatic or postulational approach laid down by Euclid was extended and developed in the field of geometry by Archimedes who not only showed himself a supreme master of the Euclidean method but made substantial and original advances by establishing the study of the physical sciences of statics and hydrostatics on a similar axiomatic basis.

In the *Elements*[‡] Euclid unified a collection of isolated discoveries and theorems into a single deductive system based upon a set of initial postulates, definitions and axioms. The definitions, although in some respects obscure and unhelpful, nevertheless suffice to identify the points, lines, planes and

† Cf. Becker, O., *Das mathematische Denken der Antike*, Göttingen, 1957, *passim*. Lasswitz, *op. cit.* I, pp. 1–8, 175–84, 208–10. Brunschvicg, L., *Les Étapes de la philosophie mathématique*, Paris, 1912, pp. 153–9.

‡ It is known that others before Euclid wrote on the *Elements*. Unfortunately only fragmentary material has survived so that it is not possible to say how far Euclid advanced beyond his predecessors.

circles of physico-geometric experience and the postulates describe the technical requirements for their construction: the axioms (or common notions) represent rules of logic by means of which consequences may be deduced from the postulates. Although both Euclid and Archimedes make use of a great many logical processes not listed among Euclid's axioms, those which are given constitute a satisfactory definition of equivalence of measure, a notion of fundamental significance in the construction of any formal deductive system.

From the outset it is clear that our concern is with the real world and not with any abstract system. All concepts not immediately comprehensible in simple concrete terms such as infinitesimals, instantaneous velocity, divisibility to infinity are completely eliminated. The infinite is never invoked. To quote from Aristotle† who writes of mathematicians:

> In point of fact they do not need the infinite and do not use it. They postulate only that the finite straight line may be produced as far as they wish.

The entire content of all thirteen books is expressed in geometric terminology and involves constant recourse to spatial imagery. The geometric forms are perfect, rigid and wholly static. With Euclid the Greeks rejected, formally at any rate, the fruitful concept of lines as collections of points and surfaces as collections of lines: a surface became merely the boundary of a solid: a line became the boundary of a surface: a point became the extremity of a line. In the light of twentieth-century mathematics when it is scarcely possible to embark on any serious mathematical investigations without using the term *set* and when, for example, all consideration of plane figures, lines and volumes is made in terms of point set theory, it is not easy to view this particular restriction imposed by the Greeks on their own mathematical methods dispassionately. Nevertheless, to describe the imposition of this restriction as voluntary mutilation‡ and to ascribe to it the subsequent decline in Greek mathematics seems to be going altogether too far. There is absolutely no evidence either that such indivisible methods were, at any time, extensively used, or that they ever represented more than a naïve intuitional basis for exploring the properties of number and space. Archimedes, at any rate, was certainly not deterred by any formal restrictions from using such *mechanical methods* as a means of discovery.

† Arist. *Phys.* III. 7. 207^b, 29–31.

‡ Cf. Lacombe, D., L'Axiomatisation des mathématiques au IIIe siècle avant J.C., *Thalès* 6 (1949–50), pp. 37–58.

1.5. The Theory of Proportion and Means

An important result of the geometrisation of the foundations of mathematics arose from the fact that the line segments, in terms of which all magnitudes were now expressed, could not, because of the existence of incommensurable lines, be considered numerically measurable. Of such quantities it is only possible to say that they are greater than some other quantity, or that they stand in a given relation to some other quantity. It is impossible to make direct statements about areas and volumes. For example, a statement about the area of a circle cannot be made in the form, $A = kd^2$. It is necessary to use the form, $A_1/A_2 = d_1^2/d_2^2$, i.e. two circles are to one another as the squares on their diameters.†

In consequence, the theory of proportion, attributed to Eudoxus and expounded fully in Euclid V, is of fundamental importance, not only in that it represents essentially a formal theory of the real number system, but also in that the technical manipulations which developed in association with it dominated mathematics until the seventeenth century.

The key definition is worded as follows:

> Magnitudes are said to *be in the same ratio*, the first to the second and the third to the fourth, when, if any equimultiples whatever be taken of the first and third, and any equimultiples whatever of the second and the fourth, the former equimultiples alike exceed, are alike equal to, or alike fall short of, the latter equimultiples respectively taken in corresponding order. (Euclid V, Def. 5)

In modern notation, let $\alpha = a/b$ and $\beta = c/d$, then $\alpha = \beta$ provided that, for all integers, m, n, $na \gtreqless mb$, if, and only if, $nc \gtreqless md$.

Alternatively the Eudoxian definition can be expressed as follows: let α and β be two numbers (rational or irrational) and let them divide the rationals into two sets A and B, A' and B' such that, if $n/m \leqslant \alpha$, $n/m \in A$, and if $n/m > \alpha$, $n/m \in B$; similarly, if $n/m \leqslant \beta$, $n/m \in A'$ and if $n/m > \beta$, $n/m \in B'$. Then, if $A = A'$ and $B = B'$, $\alpha = \beta$. Hence this definition of Eudoxus divides the rational numbers into two coextensive classes and is therefore formally equivalent to Dedekind's theory of irrationals.

The main operations of the theory of proportions are fully developed in Euclid V, and they are used with great skill and dexterity by Archimedes and Apollonius. In the seventeenth century much of the transitional geometric work by means of which the related problems of tangents and quadratures ultimately became linked hinged on processes involving the construction of

† Euclid XII. 2.

arithmetic and geometric means. These constructions were derived from proportion theory.

In view of the importance of these operations the principal processes are listed below:

1. The proportion, $a/b = c/d$, can be transformed by the operations:

permutando (or alternando) $a/c = b/d$

invertendo $b/a = d/c$

componendo $\dfrac{a+b}{b} = \dfrac{c+d}{d}$

separando $\dfrac{a-b}{b} = \dfrac{c-d}{d}$

convertendo $\dfrac{a}{a-b} = \dfrac{c}{c-d}$,

provided $a > b$ and $c > d$.

A combination of these leads to, $a/b = c/d = (a+c)/(b+d)$, and consequently, for any number of terms,

$$a/b = c/d = e/f = \ldots = \frac{a+c+e+\ \ldots}{b+d+f+\ \ldots}.$$

The above-mentioned operations, with suitable changes of sign, are frequently also applied to inequalities.

2. The use of arithmetic, geometric and harmonic means dates from the early Pythagoreans. The definitions can be expressed as follows:

Arithmetic mean: b is the arithmetic mean between a and c, provided

$$a-b = b-c \quad \text{(A.M.)}.$$

Geometric mean: b is the geometric mean between a and c, provided

$$a/b = b/c \quad \text{(G.M.)}.$$

Harmonic mean: b is the harmonic mean between a and c, provided

$$1/a - 1/b = 1/b - 1/c \quad \text{(H.M.)}.$$

From these relations follows easily

$$\text{(A.M.)} \times \text{(H.M.)} = \text{(G.M.)}^2$$

The Pythagoreans understood the summation of arithmetic progressions which they handled by means of figurate numbers.

Thus:

$$1+2+3+\dots n = \tfrac{1}{2}n(n+1) \quad \text{(triangular numbers)}$$
$$1+3+5+\dots (2n-1) = n^2 \quad \text{(square numbers)}$$
$$2+4+6+\dots 2n = n(n+1) \quad \text{(rectangular numbers)}$$
$$1+4+7+\dots (3n-2) = \tfrac{1}{2}n(3n-1) \quad \text{(pentagonal numbers)}$$

From these basic results it is not too difficult to arrive at the sums of the squares and the cubes, i.e.

$$1^2+2^2+3^2+\dots n^2 = \frac{n(2n+1)(n+1)}{6}$$

$$1^3+2^3+3^3+\dots n^3 = \frac{n^2(n+1)^2}{4}$$

The sum of the squares was known to the Babylonians and is used frequently by Archimedes. Much less is known of the history of the sum of the cubes which appeared in Roman times and most probably came from Greek sources.

3. What we now term a geometric progression is, with Euclid, a series of terms in continued proportion, i.e.

$$a/b = b/c = c/d = \dots$$

a/c is termed the *duplicate* ratio of a/b, i.e. $a/c = (a/b)^2$
a/d is termed the *triplicate* ratio of a/b, i.e. $a/d = (a/b)^3$
b, c, d, are termed *geometric means* between a and e.

The sum of a geometric progression is given as follows:

> If as many numbers as we please be in continued proportion and there be subtracted from the second and the last numbers equal to the first, then, as the excess of the second is to the first, so will the excess of the last be to all those before it.
>
> (Euclid VIII. 35)

Thus, if $$a_1/a_2 = a_2/a_3 = a_3/a_4 = \dots a_n/a_{n+1},$$

we have, $$\frac{a_2-a_1}{a_1} = \frac{a_{n+1}-a_1}{S_n}, \quad \text{where} \quad S_n = \sum_1^n a_r.$$

4. The relation between the arithmetic mean and the geometric mean is established in Euclid V. 25.

If four magnitudes be proportional, the greatest and the least are greate than the remaining two.

Thus, if $a/b = c/d$, then $a+d > b+c$, provided $a > b \geqslant c > d$.
A special case of this is where $b = c$, when $a+d > 2b$

$$\frac{a+d}{2} > b,$$

i.e. the arithmetic mean is greater than the geometric mean. The extension of this result to the generalised arithmetic and geometric means, i.e.

$$\frac{a_1+a_2+a_3+\ldots a_n}{n} \geqslant (a_1 a_2 a_3 \ldots a_n)^{1/n},$$

provided a powerful base for seventeenth-century techniques.

Although the Eudoxian theory of proportion provoked a great deal of criticism in the seventeenth century much of this appears to have been due to obscurity of wording and poor translations. At all events no attempt was made to alter it in any fundamental way. Isaac Barrow, in particular, thought very highly of it.†

1.6. The Squaring of the Circle

Although the three famous problems of antiquity, the trisection of the angle, the doubling of the cube and the squaring of the circle all, in one way or another, substantially influenced the progress of mathematics, it was the last of these that retained throughout its historical development the closest and most direct relationship with the concept of integration.‡

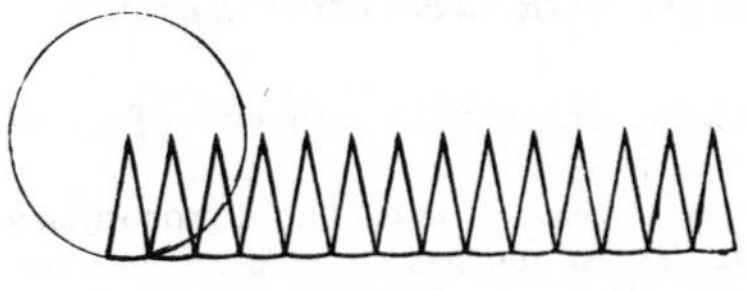

Fig. 1.7

The circular arc is one which can by a mechanical process be unrolled and stretched along a straight line. In this way the relation between area and arc becomes intuitively clear, i.e.

$$\text{area} = \tfrac{1}{2}\,\text{circumference} \times \text{radius} \quad \text{(see Fig. 1.7).}$$

† Barrow, I., *Lectiones mathematicae*, London, 1683, *passim* (given in Cambridge, 1666).

‡ An interesting early account of this problem is that of Montucla: see Montucla, J. E., *Histoire des recherches sur la quadrature du cercle*, Paris, 1754 (2nd ed. 1831).

This relation was first established on a rigorous basis by Archimedes† but was known from the fourth century B.C. From then on the attempt to evaluate surfaces enclosed by curved lines and volumes enclosed by curved surfaces was a dominating feature in the development of what are now called integration methods. Moreover, the early quadratures and cubatures were important formative influences in that each new effort helped to clarify the concepts of arc, area and volume.

According to Plutarch,‡ Anaxagoras of Clazomenae (*ca.* 420 B.C.) occupied himself whilst in prison with the squaring of the circle. Antiphon (*ca.* 430 B.C.) developed a method by which he inscribed successive polygons in a circle. Simplicius§ suggests that he expected in this way to arrive ultimately at a polygon whose sides would actually correspond with the circle. Conflicting accounts are given of Bryson (*ca.* 410 B.C.) who appears to have been the first to introduce circumscribed, as well as inscribed, polygons. So little, however, is known of these early methods that it is worth mentioning them only to emphasise the persistence with which the problem seems to have been attacked.

A different approach arose from the work of Hippocrates of Chios (*ca.* 430 B.C.) on the quadrature of lunes, i.e. curvilinear areas enclosed between circular arcs. If semicircles be described on each of the equal sides of an isosceles right-angled triangle it can be shown that the two lunulae so formed are together equal to the area of the original triangle (see Fig. 1.8).

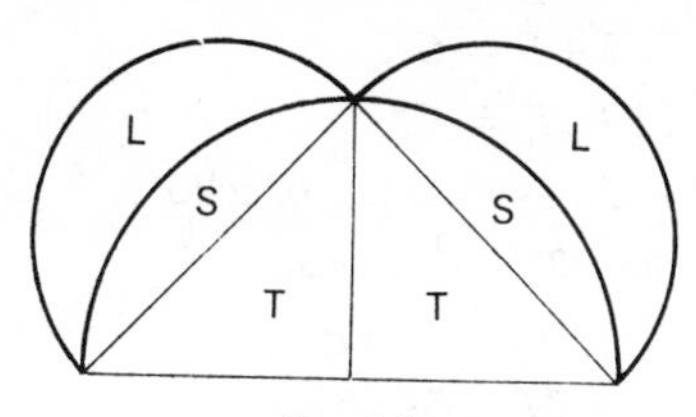

Fig. 1.8

If the enclosed areas be denoted by L, S, T, as indicated in the figure,

and if

$$L+S = C_1,$$
$$2S+2T = C_2,$$

† *Measurement of a circle*, Prop. I.

‡ Cf. Heath, *G.M.* I, p. 173.

§ See Rudio, F., *Der Bericht des Simplicius über die Quadraturen*, Leipzig, 1907. Also Rudio, F., Das Problem von der Quadratur des Zirkels, *Vierteljahrsschrift der Naturforschenden Gesellschaft in Zurich* 35 (1890), pp. 1–51.

since $$\frac{C_1}{C_2} = \frac{1}{2},\dagger$$

then
$$2S+2T = 2C_1,$$
$$S+T = C_1$$
$$= L+S,$$
and
$$T = L.$$

This result established the exact quadrature of a particular curvilinear area and encouraged Hippocrates and others to believe that further work along these lines might produce results more directly related to the circle itself.

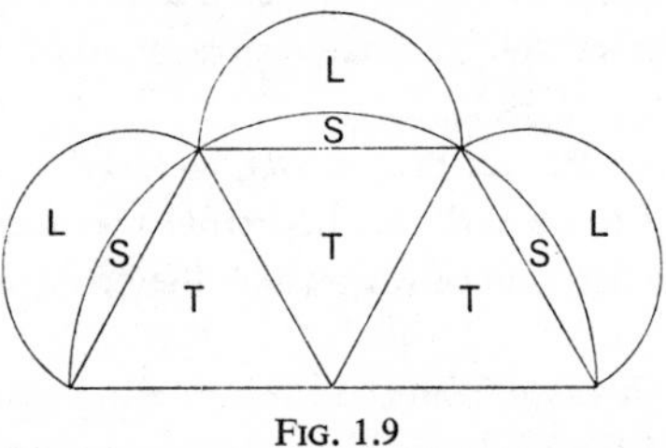

FIG. 1.9

Let a regular hexagon be inscribed in a circle, and let semicircles be described (outwards) on three consecutive sides: if the enclosed areas be denoted by L, S, T (see Fig. 1.9),

and if
$$L+S = C_1,$$
$$3S+3T = C_2,$$
since
$$C_1/C_2 = 1/4 \quad \text{(Euclid XII. 2)},$$
then
$$3S+3T = 4L+4S,$$
$$3T = 4L+S,$$
$$3T = 3L+C_1.$$

Hence, if the lune be quadrable so also is the circle. But Hippocrates seems to have recognised that this particular lune was not quadrable and, although he was able to investigate other special cases, made no further contribution towards the quadrature of the circular area itself.

Yet another line of enquiry was opened up by Dinostratus (*ca.* 350 B.C.) who made use of the *quadratrix* of Hippias of Elis (*ca.* 430 B.C.) to rectify the circular arc. These, and other methods which relate to the use of special

† Euclid XII. 2: this result was used by Hippocrates although the formal proof is attributed to Eudoxus.

curves, will be discussed in a later section. The classic treatment of this problem, however, was laid down by Archimedes in his treatise, the *Measurement of a Circle*,† and all subsequent quadratures up to the seventeenth century followed the same general lines of development.‡

The determination of the circle area was for the first time brought into formal association with the rectification of its arc by the provision of a rigorous proof that the area of a circle is equal to the area of a right-angled triangle with the circumference as base and the radius as its perpendicular height. Approximate bounds for the perimeter of the circle were then established by successively inscribing and circumscribing 6-, 12-, 24-, 48-, and 96-sided polygons within and without the circle: if p_n and P_n be the perimeters of an inscribed and circumscribed polygon respectively, we have

$$p_6 < p_{12} < p_{24} < p_{48} < p_{96} < \text{perimeter of circle} < P_{96} < P_{48} < P_{24} < P_{12} < P_6.$$

Essentially, therefore, the required magnitude is *compressed* between the terms p_n and P_n, where p_n is a monotonically ascending sequence and P_n is a monotonically descending sequence. The existence of the circular arc, however, is taken as obvious from intuition alone and, at each stage in the computation, its length is assumed to lie between p_n and P_n. As each successive result shows that $p_{2n} > p_n$ and $P_{2n} < P_n$ the bounds between which the perimeter lies can be successively narrowed without any direct consideration of the convergence of the sequences.

It is not difficult to establish that, in the sequence

$$P_n, p_n, P_{2n}, p_{2n}, P_{4n}, p_{4n}, \ldots$$

starting from the third term, every P_{2k} is the harmonic mean and every p_{2k} the geometric mean of the preceding two terms,

i.e. $$P_{2k} = \frac{2p_kP_k}{p_k+P_k}, \quad p_{2k} = \sqrt{(p_kP_{2k})}.$$

This interpretation of the process gives to it an appearance of generality which it did not, in fact, possess and it was certainly not presented in this form by Archimedes.

† See, Heath, T. L., *Archimedes*, pp. 91–8. Dijksterhuis, E. J., *Archimedes*, pp. 222–40.

‡ Cf. Hobson, E. W., *Squaring the Circle*, Cambridge, 1913 (2nd ed., New York, 1953, p. 21). See also Dörrie, H., *Triumph der Mathematik*, Breslau, 1933, pp. 183–7. The most recent discussion of the method in general is found in Dijksterhuis, E. J., *Archimedes*, pp. 222–40.

1.7. The Method of Exhaustion

The term *Exhaustion* was first introduced by Grégoire de Saint-Vincent† to describe the process devised by Eudoxus to avoid the direct passage to the limit. This method appears in its simplest form in Euclid. XII, where it is used to prove that the areas of circles are to one another as the squares of their diameters‡ and also that the volumes of triangular pyramids of equal height are proportional to the areas of their bases.§ Archimedes subsequently exploited the exhaustion techniques with consummate skill and complete mastery in a variety of ways for the determination of curvilinear areas, volumes, surfaces and arcs, and his work in this field has frequently been referred to as genuine integration. Indeed, D. T. Whiteside, who recently utilised the resources of twentieth-century analysis to expose the essential proof structure inherent in the generalised seventeenth-century exhaustion technique, refers to it as "equivalent to a Cauchy–Riemann definite integral defined on a convex point-set".||

The proofs of Archimedes have been presented in innumerable ways ranging from literal translation to the full use of modern and even special symbolism.†† Although much has been gained from these studies, both in insight and understanding, I am myself inclined to feel that the spirit and the form are best presented if the spatial imagery is retained throughout and if the modern notation, the use of which is inevitable on the grounds of length alone, is of a simple unspecialised nature. It also seems to me that an adequate impression of the power, as well as the limitations of the method, is best conveyed by consideration of selected examples; its richness lies rather in the variety of suggestive geometric approaches than in the standardised *reductio* techniques through which the conclusions were finalised.

Although neither in *Euclid* nor in the works of Archimedes are any direct statements made as to limiting values, there are in both certain important notions which, in a sense, prescribe the conditions under which operations with limit processes may be conducted. These notions can be summarised as follows:

1. Any finite quantity, however small, can be made as large as we please by

† Gregorius a S. Vincentio (1584–1667) (Grégoire de Saint-Vincent), *Opus geometricum guadraturae circuli et sectionum coni*, Antwerp, 1647, p. 739.

‡ Euclid XII. 2.

§ Euclid XII. 5.

|| Whiteside, *op. cit.*, p. 331.

†† Cf. Dijksterhuis, *op. cit., passim.*

multiplying it by a great enough number; *or*, given two unequal magnitudes α and β ($\beta < \alpha$) there exists,

(i) a number n such that $n\beta > \alpha$ (Euclid V, Def. 4),

(ii) a number n such that $n(\alpha-\beta) > \gamma$, where γ is any magnitude whatsoever of the same kind (*The Axiom of Archimedes*,† *On the Sphere and Cylinder*, Book I).

2. Any finite quantity can be made as small as we please by repeatedly subtracting from it a quantity greater than, or equal to, its half; *or*, given two unequal quantities α and β ($\beta < \alpha$), there exists a number n, such that $(1-p)^n\alpha < \beta$, where $p \geqslant \frac{1}{2}$ (Euclid X. 1).

A simple example of the use of the above notions in a *reductio* proof is found in Euclid XII. 2. It is known that the areas of similar polygons inscribed in circles are as the squares of the diameters:‡ the extension to the areas of the circles themselves by increasing the number of sides indefinitely is intuitionally obvious. The formal proof may be represented as follows: let C, C', d, d', denote the areas and diameters respectively of two circles, then, *either*, $C/C' < d^2/d'^2$, $C/C' > d^2/d'^2$, or $C/C' = d^2/d'^2$.

(i) Suppose $C/C' < d^2/d'^2$; then, for some magnitude S, let $C/S = d^2/d'^2$. Hence, $C/C' < C/S$ and $S < C'$.§

(ii) Starting with a square and proceeding by repeated bisection of the arc let a sequence of polygons, areas $p_1', p_2', p_3', \ldots, p_n'$, be inscribed in the circle, area C', so that,

$$C'-p_1' < \tfrac{1}{2}C',$$
$$C'-p_2' < \tfrac{1}{2}(C'-p_1'),$$
$$C'-p_3' < \tfrac{1}{2}(C'-p_2'),$$

and, in general,

$$C'-p_n' < \tfrac{1}{2}(C'-p_{n-1}').$$

Hence, by Euclid X. 1, $C'-p_n'$ can be made less than any given area.

(iii) By suitable choice of n, let $C'-p_n' < C'-S$, hence $S < p_n'$.

(iv) Now, if p_n denote the area of a similar polygon inscribed in the circle, area C, we have,

$$\frac{p_n}{p_n'} = \frac{d^2}{d'^2} = \frac{C}{S}, \quad \text{and} \quad \frac{p_n}{C} = \frac{p_n'}{S}.$$

† Listed as an unproved fundamental proposition (Assumption V). Discussed by Dijksterhuis (*op. cit.*, pp. 146–9) and by Heath (*op. cit.*, Introduction, pp. xlvii–xlviii).

‡ Euclid XII. 1.

§ Existence of a 4th proportional assumed.

Since, however, $p_n < C$, $p'_n < S$, and the conclusion in (iii) above is contradicted. Hence, $C/C' \nless d^2/d'^2$.

(v) A similar argument is then applied to the assertion, $C/C' > d^2/d'^2$, which also reduces to a contradiction.† Hence, $C/C' \ngtr d^2/d'^2$.

(vi) Finally, if $C/C' \nless d^2/d'^2$, and if $C/C' \ngtr d^2/d'^2$, then $C/C' = d^2/d'^2$.

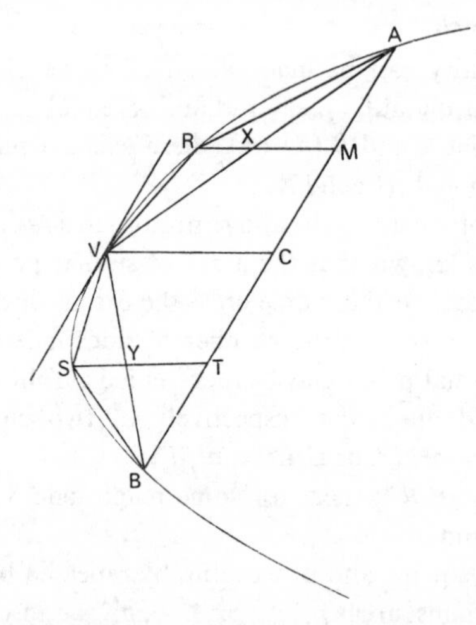

FIG. 1.10. CV is a diameter bisecting the chord AB: RM, ST are diameters bisecting AC and CB, respectively. It is easily established that $RX = SY = \frac{1}{4}VC$. Hence, $\triangle VRA = \frac{1}{4}\triangle VCA$, $\triangle VSB = \frac{1}{4}\triangle VBC$, $\triangle VRA + \triangle VSB = \frac{1}{4}\triangle VBA$.

In one of the methods he supplies for the quadrature of the parabola‡ Archimedes uses an approach similar to that in Euclid XII. 2. By repeated bisection of the arc he builds up an inscribed polygon as the sum of a sequence of triangles. Thus, if $\Delta AVB = \Delta$ (see Fig. 1.10), then,

$$\Delta + \frac{\Delta}{4} + \frac{\Delta}{4^2} + \frac{\Delta}{4^3} + \ \dots \ + \frac{\Delta}{4^{n-1}} < \text{segment } AVB.$$

Since, however,

$$\sum_1^n \frac{\Delta}{4^{r-1}} = \frac{4}{3}\Delta\left(1 - \frac{1}{4^n}\right),$$

† The argument presented in (v) above can be given in a more direct form by the use of circumscribed as well as inscribed figures and the process is frequently adopted by Archimedes.

‡ Archimedes, *Quadrature of the Parabola*, Prop. 24.

the conclusion, segment $AVB = \frac{4}{3}\Delta$, appears to be intuitively obvious. The standard *reductio* proof, however, follows once again. A similar approach is used for the area of the ellipse.†

It is, however, in the use which he makes of the *Compression* method that Archimedes demonstrates the full power and elegance of the exhaustion techniques and it is here that their relation to the Cauchy–Riemann integral is most clearly exhibited.

The method consists essentially in the establishment through some specialised geometric construction, of a monotonically ascending sequence, I_n, and a monotonically descending sequence C_n, between which lies the magnitude S whose value is to be investigated. The terms I_n and C_n consist of the perimeters, surface areas or volumes of inscribed and circumscribed figures respectively and the relation

$$I_1 < I_2 < I_3 < \ldots < I_n < S < C_n < C_{n-1}, \ldots, C_2 < C_1$$

is validated by a whole series of important convexity lemmas‡ such as:

(1) of lines which have the same extremities the straight line is the least;
(2) lines (and surfaces) are concave in the same direction provided that all straight lines formed by joining any two points on the line (or surface) lie on the same side of the line (or surface);
(3) if two lines (or surfaces) which have the same extremities be concave in the same direction then that line (or surface) which is either wholly or partly enclosed within the other is the lesser of the two (see Fig. 1.11).

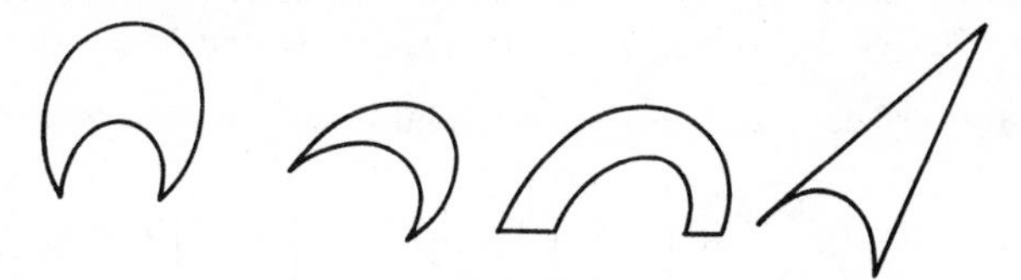

FIG. 1.11

By means of the particular construction adopted for I_n and C_n it is now shown that the difference $(C_n - I_n)$ can be made less than any assigned magnitude *or* that the ratio C_n/I_n can be made less than the ratio of the greater of any two assigned magnitudes to the lesser§ by suitable choice of n. It is now necessary to determine some quantity K, such that $I_n < K < C_n$ for all values of n.

† Archimedes, *On Conoids and Spheroids*, Prop. 4.
‡ Archimedes, *On the Sphere and Cylinder*, Book I, Definitions.
§ See Dijksterhuis, *op. cit.*, pp. 130–1.

We have thus,

$$I_n < S < C_n,$$
$$I_n < K < C_n.$$

Hence, either $S = K$, or $S > K$ or $S < K$. The proof is completed by *reductio ad absurdum*.

The interesting ratio form is adopted in the important series of propositions in which Archimedes determines the surface areas and volumes of a sphere and a segment of a sphere.† The results obtained here are formally equivalent to the summation of certain trigonometric series and the limiting values of these sums to the definite integrals of trigonometric functions.‡ The proofs are fascinating and suggestive and they attracted a great deal of attention as the basis for geometric integration techniques in the seventeenth century. For this reason I think it worth while to give a brief account of the series of propositions by which the surface area of a sphere is shown to be equal to the area of a circle the radius of which is equal to the diameter of the sphere.

In considering the surface area of a sphere Archimedes inscribes within it a regular polygon ($2n$ sides) and rotates the great circle containing it about a diameter (see Fig. 1.12). The surface area of the resulting solid is then determined by considering it as made up of successive conical frusta, each replaced by an equivalent circular area and these in turn by a single circle.

If the radii of the frustum of a cone (see Fig. 1.13) be s_1 and s_2 respectively and the slant height be l then the surface area equals $\pi(s_1+s_2)l$. The Greeks, of course, had no means of expressing such an area other than by replacing it by an equivalent circle, radius R_1, thus $(s_1+s_2)l = R_1^2$, and $R_1 = \sqrt{((s_1+s_2)l)}$.

Hence, the surface area of the inscribed solid $= \pi \sum_0^{n-1} R_r^2 = \pi R^2$,

$$\text{and} \quad R = \sqrt{\left[\sum_0^{n-1} R_r^2\right]} = \sqrt{\left[\sum_0^{n-1} l(s_r+s_{r+1})\right]} = \sqrt{\left[2l \sum_0^{n-1} s_r\right]},$$

where $s_0, s_1, s_2, s_3, \ldots, s_n$, are semi-chords of the circle drawn through the vertices of the polygon perpendicular to the axis of rotation (see Fig. 1.12, $s_0 = s_n = 0$).

But, from similar triangles (see Fig. 1.12)

$$\frac{s_1}{c_1} = \frac{s_2}{c_2} = \frac{s_3}{c_3} = \frac{s_4}{c_4} = \ldots \frac{s_{n-1}}{c_{n-1}} = \frac{BC}{l}.$$

† Archimedes, *On the Sphere and Cylinder*, Book I.
‡ See Heath, *Archimedes*, Introduction, pp. cxliv–cxlvii.

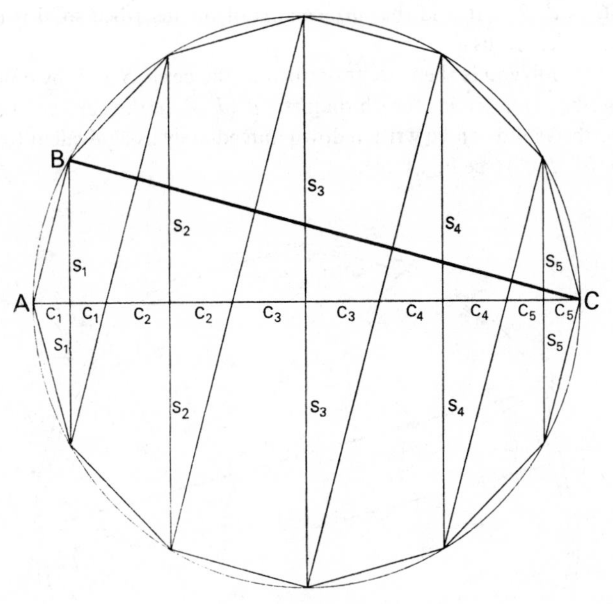

FIG. 1.12

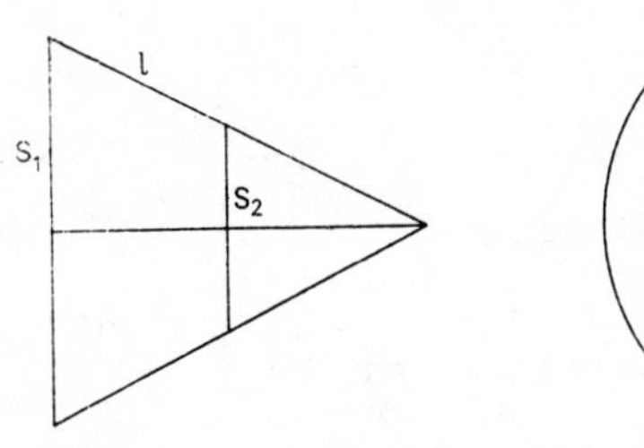

FIG. 1.13

Now, if d be the diameter of the circle, then $2\sum_{1}^{n-1} c_r = d$. Hence, adding,

$$\frac{2\sum_{1}^{n-1} s_r}{d} = \frac{BC}{l}, \quad 2l\sum_{1}^{n-1} s_r = d.BC,$$

and $$R = \sqrt{(d.BC)}.$$

Since $BC < d$, $R < d$, and the surface area of the inscribed solid is less than that of a circle, radius d.

A similar polygon is then circumscribed to the circle S and, by considering it as inscribed in a circle S' with diameter d' $(d' > d)$, the surface area of the circumscribed solid can be written down immediately as equivalent to a circle, radius $\sqrt{(d'.B'C')}$ (see Fig. 1.14).

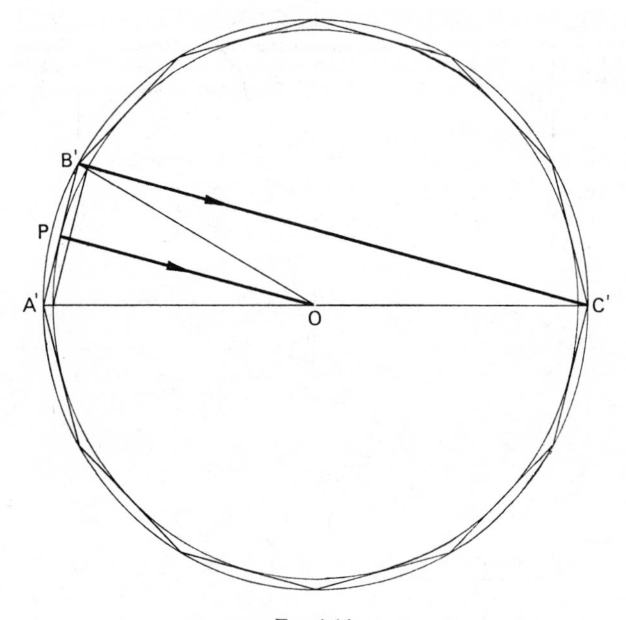

FIG. 1.14

In $\triangle A'B'C'$, $A'P = PB'$,
$A'O = OC'$.

Hence, by the mid-point theorem,

$$B'C' = 2PO$$
$$= d.$$

Since, however, $B'C' = d$, the surface area of the circumscribed solid is greater than that of a circle, radius d. The proof then follows by *reductio ad absurdum*.

Let S_n and S_n' denote the surface areas of the inscribed and circumscribed solids respectively, let S denote the surface of the sphere and K the area of a circle, radius d. Then, we have,

$$S_n < S < S'_n,$$
$$S_n < K < S'_n.$$

(i) Let $S < K$ and let β, γ, be such that $\beta/\gamma < K/S$† $(\beta > \gamma)$ and also let n be such that

$$\frac{S'_n}{S_n} = \frac{l'^2_n}{l^2_n} < \frac{\beta^2}{\delta^2}\left(= \frac{\beta}{\gamma}\right) \qquad \text{(where } \delta^2 = \beta\gamma\text{)}.$$

Hence, $\dfrac{S'_n}{S_n} < \dfrac{K}{S}$, *which is impossible*, as $S'_n > K$ and $S_n < S$.

(ii) Let $S > K$ and let β, γ, be such that $\beta/\gamma < S/K$† $(\beta > \gamma)$ also, as before, let n be such that

$$\frac{S'_n}{S_n} = \frac{l'^2_n}{l^2_n} < \frac{\beta^2}{\delta^2}\left(= \frac{\beta}{\gamma}\right) \qquad \text{(where } \delta^2 = \beta\gamma\text{)}.$$

Hence, $\dfrac{S'_n}{S_n} < \dfrac{S}{K}$, *which is impossible*, since $S'_n > S$ and $S_n < K$.

(iii) Since $S \nless K$ and $S \ngtr K$, then $S = K$.

Many of the important results obtained by Archimedes arise from the skilful use of inequalities derived from the consideration of expressions for the sums and the sums of the squares of the natural numbers.

Let $$S(n) = 1+2+3+ \dots +n = \frac{n(n+1)}{2}$$

and $$S(n-1) = \frac{n(n-1)}{2},$$

hence $$S(n) > n^2/2 \quad\text{and}\quad S(n-1) < n^2/2. \qquad \text{(i)}$$

Similarly, let $$S(n^2) = 1^2+2^2+3^2+ \dots +n^2 = \frac{n(n+1)(2n+1)}{6}$$

and $$S(n-1)^2 = \frac{(n-1)(n)(2n-1)}{6}$$

hence $$S(n^2) > n^3/3 \quad\text{and}\quad S(n-1)^2 < n^3/3. \qquad \text{(ii)}$$

† Justified from Props. 2 and 3, *On the Sphere and Cylinder*, Book I.

A simple example of the use of the inequalities (i) above is in the proof that the volume of a conoid or paraboloid of revolution is equal to one half of that of the enveloping cylinder.† Let the conoid and its enveloping cylinder be cut by n equidistant planes perpendicular to the axis and let figures be inscribed and circumscribed to the conoid so that each is made up of thin cylinders, i_r and c_r, respectively (see Fig. 1.15).

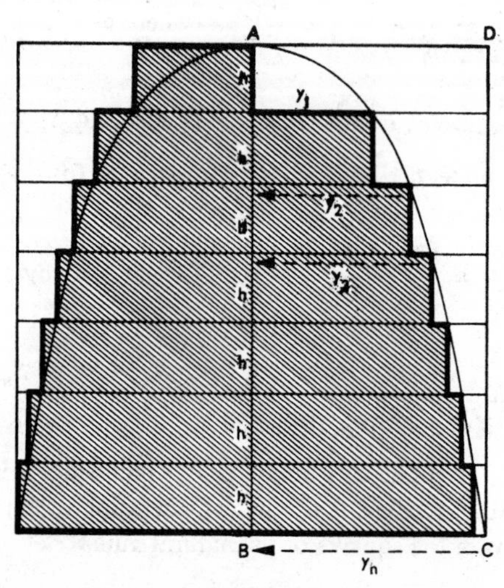

Fig. 1.15. The *conoid* is the solid formed by rotating the curve about the axis AB.
The *inscribed solid* I_n is formed by rotating the shaded figure on the right about the axis AB.
The *circumscribed solid* C_n is formed by rotating the shaded figure on the left about the axis.
The *enveloping cylinder* is formed by rotating the rectangle $ABCD$ about the axis AB.

Let
$$I_n = i_1+i_2+i_3+\dots+i_n,$$
$$C_n = c_1+c_2+c_3+\dots+c_n,$$
and
$$S_n = s_1+s_2+s_3+\dots+s_n,$$

where I_n is the volume of the inscribed solid, C_n is the volume of the circumscribed solid and S_n is the volume of the enveloping cylinder. But (see Fig. 1.15), $i_1 = 0$, $c_1 = i_2$, $c_2 = i_3$, and in general, $c_{r-1} = i_r$.

† See Archimedes, *On Conoids and Spheroids*, Props. 19–22.

Hence,

$$\begin{aligned} C_n - I_n &= (c_1+c_2+c_3+\ \dots\ +c_n)-(i_1+i_2+i_3+\ \dots\ +i_n) \\ &= (c_1-i_2)+(c_2-i_3)+\ \dots\ +(c_{n-1}-i_n)+c_n \\ &= c_n, \end{aligned}$$

which quantity can be made as small as we please by taking n great enough.

Now, if the ordinates be denoted by $y_1, y_2, y_3, \dots, y_n$, and the abscissae by $h, 2h, 3h, 4h, \dots, nh$, from the property of the parabola we have, $y_r^2 \propto rh$, hence $i_r \propto (r-1)h$, $c_r \propto rh$, $s_r \propto nh$.

Thus,

$$\frac{s_r}{i_r} = \frac{nh}{(r-1)h} = \frac{n}{r-1},$$

and, summing,

$$\frac{\sum_1^n s_r}{\sum_1^n i_r} = \frac{\sum_1^n n}{\sum_1^n (r-1)} = \frac{2n^2}{n(n-1)} > 2.$$

Similarly,

$$\frac{s_r}{c_r} = \frac{nh}{rh} = \frac{n}{r},$$

$$\frac{\sum_1^n s_r}{\sum_1^n c_r} = \frac{\sum_1^n n}{\sum_1^n r} = \frac{2n^2}{n(n+1)} < 2.$$

Hence,

$$I_n < \frac{S_n}{2} < C_n.$$

Also, from the convexity lemmas,†

$$I_n < \text{Volume of conoid} < C_n,$$

and the proof is completed by *reductio ad absurdum*.

An analogous proof is given for the area bounded by the first turn of a spiral arc‡ ($R = a\theta$) and the initial line. In this case the area is divided into n equi-angular sectors, to each of which circular sectors are inscribed and circumscribed as indicated (see Fig. 1.16).

† See p. 37 above.

‡ Archimedes, *On Spirals*, Props. 25–7.

As before let

$$I_n = \sum_1^n i_r \quad \text{and} \quad C_n = \sum_1^n c_r, \quad S_n = \sum_1^n s_r,$$

where i_r is a sector of the inscribed figure, c_r is a sector of the circumscribed figure and s_r a sector of a circle, radius a. Since $i_1 = 0$, $c_1 = i_2$, $c_2 = i_3$, $\ldots$, $c_{n-1} = i_n$, we have, $C_n - I_n = c_n$, which again can be made less than any assigned quantity by suitable choice of n.

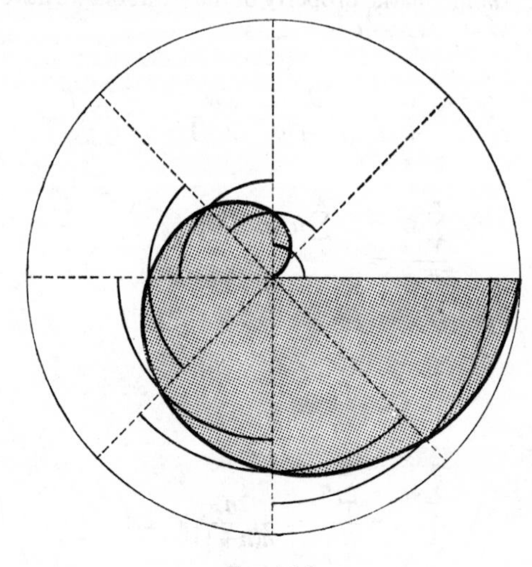

FIG. 1.16

Also, the areas of equi-angular circular sectors are proportional to the squares of their radii and the radii are proportional to the angles. Hence, if $n\theta = 2a\pi$, we have

$$\frac{\sum_1^n s_r}{\sum_1^n i_r} = \frac{\sum_1^n a^2(n\theta)^2}{\sum_1^{n-1} a^2(r\theta)^2} = \frac{n^3}{\sum_1^{n-1} r^2} = \frac{6n^3}{(n-1)\,n(2n-1)} > 3.$$

Similarly,

$$\frac{\sum_1^n s_r}{\sum_1^n c_r} = \frac{\sum_1^n a^2(n\theta)^2}{\sum_1^n a^2(r\theta)^2} = \frac{n^3}{\sum_1^n r^2} = \frac{6n^3}{n(n+1)(2n+1)} < 3,$$

and the proof is completed by *reductio ad absurdum*.

In a second formal method for the quadrature of the parabola Archimedes reduces the problem to similar considerations but in this case builds up a system of inequalities by an ingenious weighing method.† Other results for the hyperboloid of revolution‡ and the spheroid§ are arrived at by using a combination of $S(n)$ and $S(n^2)$. It is necessary, however, to remark that, although for convenience I have considered the above methods in an order which brings out the common dependence of certain of them on particular series, Archimedes himself makes no move to relate such results. The volumetric determinations for the cone and the sphere, and the area determinations for the spiral and the parabola are all expressible in the notation of the calculus in terms of $\int x^2\,dx$ but Archimedes preferred to make use of any special geometric relations which seemed convenient rather than to unify his approach. The proofs which I have quoted and many others besides are scattered around in a number of different treatises and no systematic ordering of the whole is attempted.

In the formal geometric sense the build up for each proposition or set of propositions is complete and rigorous. At no stage is any significant attempt made to reduce the tedium of constant repetition by supplying a general proof with suitable conditions of validity and thenceforth deducing special results from it. Each separate proof is given in its entirety and there is nowhere any formal treatment which permits of the use of analogous problems already solved in the derivation of new results. Although the building of Riemann-sums and the use of a *reductio* proof to establish the identity of upper and lower bounds is common procedure, special artifices predominate not only in influencing the processes for building such sums but also in exhibiting the common limit of upper and lower sums. Is this the integral calculus? If it were, then the seventeenth-century struggle to extract from these methods something straightforward and direct which would provide within its structure for the reduction of new problems to those already known, would have been unnecessary. Is it an anticipation of the integral calculus? In a limited sense, yes, for it contains the fundamental geometric concept of integration expressed in a manner which is geometrically rigorous and formally equivalent to the Riemann integral. In a wider sense, certainly not, for the essential property of any calculus is surely that it should provide a set of rules for calculation. This idea never entered into any of the formal treatises of Archimedes.

† Archimedes, *Quadrature of the Parabola*, Props. 14–16.
‡ Archimedes, *On Conoids and Spheroids*, Props. 25, 26.
§ *Ibid.*, Props. 27–30.

Something of this kind is, however, provided in the treatise *The Method* and any further discussion should perhaps be deferred until this has been briefly reviewed.

1.8. The Discovery Method of Archimedes

A dangerous situation is created in mathematics when a gap develops between the formal standards required for rigorous proof and the less refined but vigorous methods developed for the investigation of new problems. When the gap is sufficiently narrow publication of formal proofs gives clear guidance as to the methods of discovery and leads easily to further investigations. When the gap widens then a difficult decision must be made, either to sacrifice rigour and publish an account likely to promote further mathematical invention or to re-cast the proofs and present them in a formally rigorous manner offering neither help nor guidance in further development.† Save for the single Archimedean treatise known as *The Method* the Greeks chose the latter course and their flawless proofs are entirely synthetic in character. Archimedes, however, supplemented his classical treatises with a full and satisfying description of "a certain special method" which he used "to investigate some of the problems in mathematics by means of mechanics".‡ In the preface to the treatise he distinguishes clearly the investigatory methods employed "to supply some knowledge of the questions" and the formal methods subsequently required to "furnish an actual demonstration". His purpose in publishing such an account is made abundantly clear§ when he says:

> I now wish to describe the method in writing, partly, because I have already spoken about it before, that I may not impress some people as having uttered idle talk, partly because I am convinced that it will prove very useful for mathematics; in fact, I presume there will be some among the present as well as future generations who by means of the method here explained will be enabled to find other theorems which have not yet fallen to our share.

The expressed wish of Archimedes to further the progress of mathematics by the communication of this method appears to have remained unfulfilled. It is referred to in Heron's *Metrica*|| and the manuscript discovered by Heiberg in Constantinople in 1906 is in the hand of a tenth-century copyist. Beyond this, for over 2000 years the manuscript seems to have remained unread and

† Bourbaki, N., *Éléments d'histoire des mathématiques*, Paris, 1960, p. 181.
‡ Archimedes. *The Method* (Introductory note).
§ Archimedes, *The Method* (translation given from Dijksterhuis, *op. cit.*, p. 315).
|| Cf. Heath, *Archimedes*, Suppl., p. 6; also Dijksterhuis, *op. cit.*, pp. 44–5.

unnoticed. Notwithstanding, similar concepts were developed during the mediaeval period† and were widely publicised in the works of Galileo and Cavalieri in the seventeenth century although at this time only the classical treatises of Archimedes were known.

The fundamental basis of the discovery method is simple and depends upon the imaginary balancing of magnitudes against one another on the arm of a lever. If masses m_1 and m_1' balance at distances d and d' respectively from the point of suspension, then

$$m_1 d = m_1' d_1'.$$

If also,

$$m_2\, d = m_2'\, d_2' \quad \text{and} \quad m_3\, d = m_3'\, d_3', \ldots$$

then,

$$d \sum_1^n m_r = \sum_1^n m_r'\, d_r'.$$

The masses $m_1, m_2, m_3, \ldots m_n$ are thus conceived as placed at a common point, H, at a distance d from the point of suspension and the masses m_1', m_2', $m_3', \ldots, m_n'$, at distances $d_1', d_2', d_3', \ldots, d_n'$.

Hence, if

$$\sum m_r = M, \quad \sum m_r' = M' \quad \text{and} \quad \sum m_r' d_r' = M'd', \quad \text{then} \quad \frac{M}{M'} = \frac{d'}{d}.$$

If M', d' and d be known then M can be determined and if the above method extended only to finite sums it would be unexceptionable. It is, however, in the interpretation given to the masses m_r and m_r' and their sums $\sum m_r$ and $\sum m_r'$ that the character of the process, unique in Greek mathematics, becomes apparent. The initial balancing of the line and surface elements of two figures, plane or solid, leads to the development of a relation between the figures themselves.

For example, to find the area of the parabolic segment APB‡ (see Fig. 1.17) it is conceived as *made up* of line segments OP which are balanced about C (along the lever arm CB) against corresponding line elements MO of a known area FAB. FB is tangent to the curve at B; FA, MO, PO are parallel to the diameter, ST. From geometric considerations a relation of the form $OP \,.\, CB = OM \,.\, CU$ is established, where $CH = CB$ and CU is the distance of OM from C. The line elements OP are then, in imagination, transferred to

† See Chap. 2.

‡ Archimedes, *The Method*, Prop. 1.

the point H so that we have $OP \,.\, CB = OP \,.\, CH = OM \,.\, CU$ and, summing for all such elements, $CH \sum OP = \sum OM \,.\, CU$.

But, if the area FAB be considered as *made up* of the line elements OM, then $\sum OM \,.\, CU = CX \,.$ area AFB, where CX is the distance of the mass centre of AFB from C.

Hence the area of the segment APB *placed at* H balances the area AFB *placed where it is*, and consequently

$$\frac{\text{Area } APB}{\text{Area } AFB} = \frac{CX}{CH} = \frac{1}{3}.$$

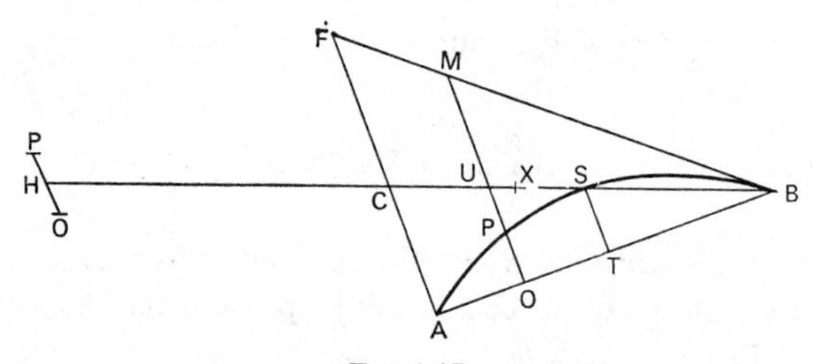

FIG. 1.17

In a number of volumetric determinations the sphere, cone and cylinder are considered as *made up* of circles which, in a variety of different ways can be arranged to balance each other. For example, the volume of the conoid is determined as follows:† let APC be a parabolic arc (see Fig. 1.18) and let PS and CD be the ordinates at P and C, respectively; let $HA = AD$.

From the property of the parabola,

$$\frac{DC^2}{SP^2} = \frac{DA}{AS} = \frac{HA}{AS}.$$

DC, however, is the radius, SN, of the enveloping cylinder so that

$$\frac{DC^2}{SP^2} = \frac{SN^2}{SP^2} = \frac{HA}{AS}, \quad \text{and} \quad SN^2 \,.\, AS = SP^2 \,.\, HA,$$

and the circle in the cylinder *placed where it is* balances the circle in the conoid *placed at* H. Summing, for all such circles, the cylinder, *placed where it is*, balances the conoid placed at H,

$$\text{i.e.} \quad \frac{\text{Cylinder}}{\text{Conoid}} = \frac{AH}{\frac{1}{2}AD} = 2.$$

† *Ibid.*, Prop. 2.

To determine the centre of gravity of the conoid the circular sections of the inscribed cone ABC (see Fig. 1.18) are similarly balanced against the conoid: let AC cut SP in F, we have,

$$\frac{DC^2}{SP^2} = \frac{AD}{AS},$$

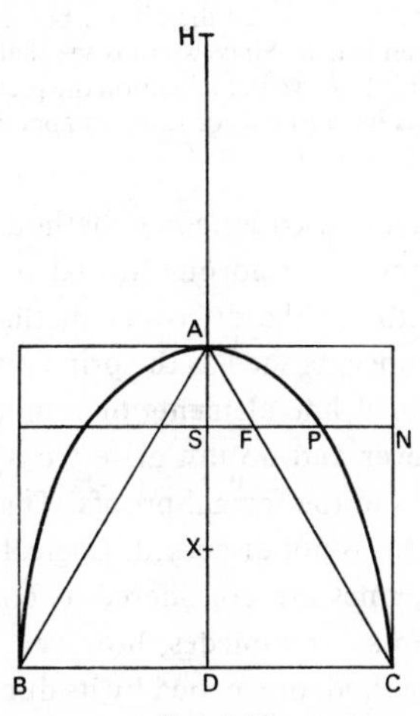

FIG. 1.18

also

$$\frac{DC^2}{SF^2} = \frac{AD^2}{AS^2} \quad \text{(from similar triangles)}$$

and

$$\frac{SF^2}{SP^2} = \frac{AS}{AD}.$$

Hence,

$$SF^2 . AD = SP^2 . AS,$$
$$SF^2 . AH = SP^2 . AS,$$

and the circle in the cone, ***placed at H***, balances the circle in the conoid, ***placed where it is***. Summing, for all such circles,

$$AH \sum SF^2 = \sum SP^2 . AS = AX \sum SP^2,$$

where AX is the distance of the centre of gravity of the conoid from A. But,

$$\frac{\sum SF^2}{\sum SP^2} = \frac{\text{Volume of cone}}{\text{Volume of conoid}} = \frac{2}{3} = \frac{AX}{AD},$$

hence

$$AX = \frac{2}{3} AD.$$

To translate these results into the notation of the integral calculus could only be misleading.† The comments of Archimedes himself are valuable:‡

> This has not therefore been proved by the above, but a certain impression has been created that the conclusion is true. Since we thus see that the conclusion has not been proved, but we suppose it is true, we shall mention the previously published geometrical proof, which we ourselves have found for it, in its appointed place.

The fact that he repeatedly used *weighing* methods in his classical treatises on statics and hydrostatics in a rigorous formal treatment indicates clearly that the experimental status of the discovery methods was not due to their dependence on physical concepts such as the principle of the lever but because of the presence of undefined line elements or infinitesimals. These elements have no breadth whatsoever and do not arise from any process of repeated subdivision such as occurs in the formal proofs. They may, in fact, represent ultimate indivisibles but this is not discussed. Logical difficulties arise immediately when areas and volumes are considered as constructed from breadthless line (or plane) elements. Archimedes, however, was not concerned with the improvement of the method, or worried by its dubious logical foundations, for he never intended to present it as a formal proof. Ingenious and interesting as the method is it nevertheless depended as much on special geometric relations as did the formal proofs.

The approach throughout is entirely static and the concept of a moving ordinate or plane section as the rate of increase of an area or volume is correspondingly and understandably absent. Without this concept and the idea of differentiation and integration as inverse processes the summation of line and plane elements to make up areas and volumes scarcely makes sense. Indeed, although it is certainly true to say that the foundations of the integral calculus were laid by the Greeks, the related problem—that of finding tangents to curved lines, which ultimately led to the differential calculus—remained comparatively unexplored.

† The account given by Heath (*Supplement*, p. 8) is in this sense misleading in that he considers the elements as infinitely narrow strips and suggests that, as the breadth is the same in each case, "it divides out". In fact the Archimedean infinitesimals have no breadth.

‡ Given from Dijksterhuis, *op. cit.*, p. 318.

1.9. Curves, Normals, Tangents and Curvature

The number of curves known to the ancients was few and this limitation inevitably inhibited the development of general methods. Of the known curves only in the case of the circle, the conic sections and the Archimedean spiral was any extensive investigation of properties undertaken.

Euclid defines the tangent to a circle† as a straight line which meets the circle and when produced does not cut it. Circles are said to touch which meet and yet do not cut one another.‡ In order to find the tangent in the case of a convex curve it is necessary to show that,

(i) a line meets the curve in a single point;
(ii) all other points on the line lie outside the curve.

In practice the Greeks used the characteristic proof by *reductio ad absurdum* to establish the tangent property. The *reductio* proof runs as follows:

(i) let a line be drawn from a point P on the curve according to a specified construction;
(ii) let the above line, when produced, meet the curve again in some other point Q;
(iii) from (i) and (ii) above a contradiction is deduced;
(iv) the alternative hypothesis is adopted, i.e. that a line constructed as above does not meet the curve in any other point, however far it be produced. Hence the line is a tangent and is said to *touch* the curve at P. In this way Euclid finds the tangent to a circle§ and Apollonius determines the tangents to the conic sections.‖ In classical Greek mathematics a tangent line was therefore taken to be a line with *one* point in common with the curve. Since every other line through this point meets the curve in a second point an important subsidiary notion is derived,†† i.e. between a tangent line and the curve no other straight line can be drawn (see Fig. 1.19).

In the fifth book of the *Conics* Apollonius applies similar considerations to the determination of the number of normals through a given point O.‡‡ He

† Euclid III, Def. 2.
‡ Euclid III, Def. 3.
§ Euclid III. 16.
‖ Apollonius, *The Conics* I. 17, 32; I. 33, 35; I. 34, 36. See Heath, T. L., *Apollonius of Perga*, Cambridge, 1896; Ver Eecke, P., *Les Coniques d'Apollonius de Perge*, Paris, 1963 (1st ed. 1924).
†† Euclid III. 16.
‡‡ Apollonius, *The Conics* V, *passim*.

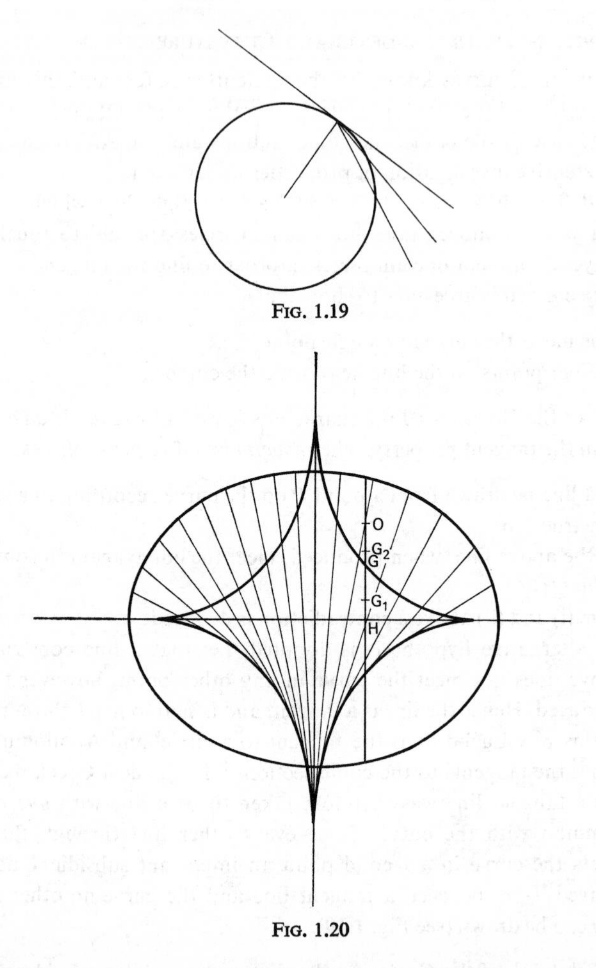

FIG. 1.19

FIG. 1.20

investigates the positions of O (see Fig. 1.20) for which there are 2, 3 or 4 normals and examines carefully the maximum and minimum properties of these lines. The *evolute* is therefore determined as the locus of points from which the number of normals drawn to the conic is 3. Apollonius does not, however, introduce this curve but restricts himself to determining GH, such that, for all points G_1 (where $G_1H < GH$), the number of normals drawn to the conic is 4, and for all points G_2, where $G_2H > GH$, the number of normals drawn to the conic is 2 (see Fig. 1.20).

Difficulties which arise in problems concerning the intersection and contact of straight lines and curves are all related to the central concept of angle. Euclid's definition of an angle† clearly extends to angles formed by curved lines. Within this category the ***rectilineal angle***, that which is formed by two straight lines, is a special case. Proclus‡ gives a long and elaborate

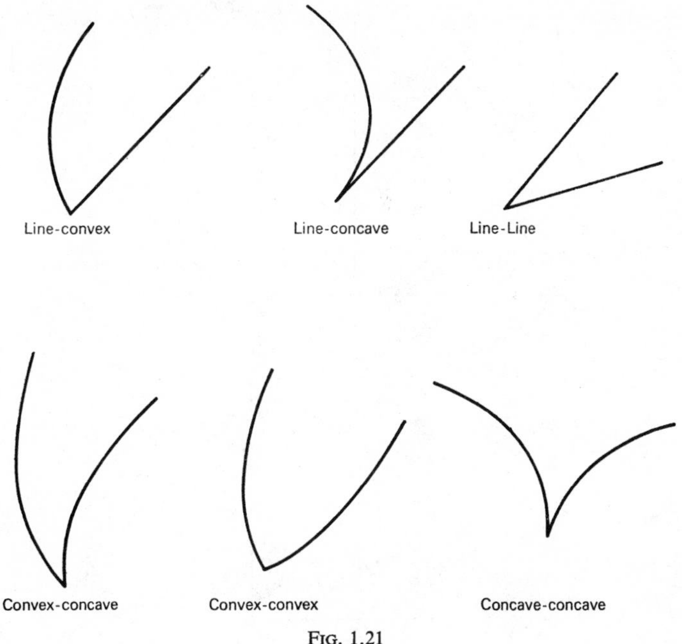

FIG. 1.21

classification of plane and solid angles which includes the following cases: line–convex, line–concave, line–line, convex–concave, concave–concave, convex–convex (see Fig. 1.21).

The above angles all arise even in the simplest consideration of the intersection and contact of circles. The title of a lost treatise of Democritus (3rd cent. B.C.) suggests that he concerned himself with the problem of the ***horn angle*** and the status of such angles continued to be a subject for mathematical controversy at least until the end of the seventeenth century.

† Euclid, Defs. 8, 9.

‡ *The Philosophical and Mathematical Commentaries of Proclus*, 2 vols. (trans. Taylor, T.), London, 1788, I, pp. 144–5.

Has a *horn angle* a magnitude in the same sense as a rectilineal angle? And if so, how is it to be measured? If two circles with differing radii touch a straight line at a point A, are the angles of contact equal or unequal? (see Fig. 1.22).

Euclid remarks[†] that no acute rectilineal angle is greater than the angle

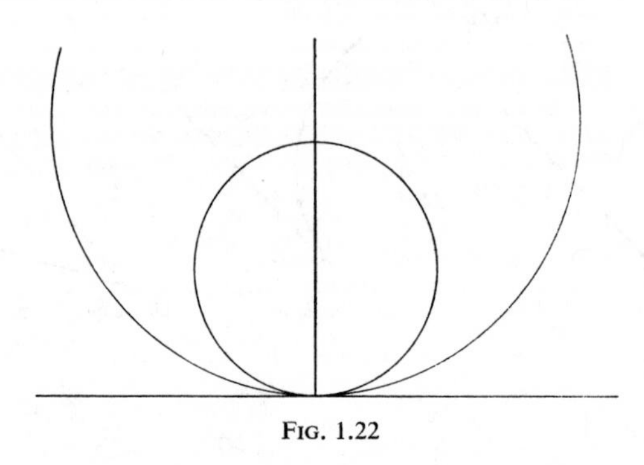

FIG. 1.22

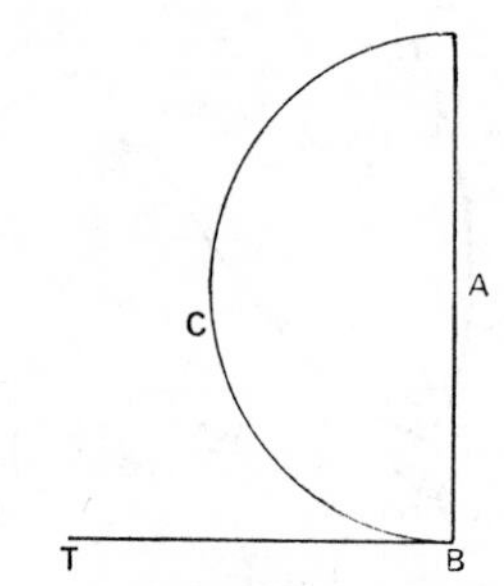

FIG. 1.23

$\widehat{CBA} >$ any acute rectilineal angle

$\widehat{CBT} <$ any acute rectilineal angle

between the diameter and the circumference and no acute rectilineal angle is less than the angle between the tangent and the circumference—a view which appears to suggest the idea of a limit to those accustomed to think in terms of limits (see Fig. 1.23).

Proclus also records[‡] that the *first distance* under the point was regarded by

† Euclid III, Prop. 16.

‡ Taylor, *op. cit.* I, p. 143. Proclus writes: "But they call it quantity, who say that it is the first interval under a point, that is immediately subsisting after a point. In the number

some as the angle; amongst those, Plutarch said that there must be some first distance, although the distance under the point being continuous, it is impossible to obtain the actual first, since every distance is divisible without limit. In this latter concept there appears to be some idea of the rate of change or divergence of two curves at a point but it is an idea so vaguely expressed that it is scarcely surprising that it remained undeveloped. The full history of the concept of curvature, linked as it clearly is with the origins of the differential calculus, has yet to be written.† It does appear, however, that the Greek concern with static geometrical form led, in this instance, to the formulation of a basically unsatisfactory concept which was only ultimately clarified by the full exploitation of kinematic methods in geometry.

The kinematic approach, however, was not by any means unknown in Greek mathematics. Aristotle reports, "they say a moving line generates a surface and a moving point a line... ".‡ Proclus§ refers to those who call a line "the flux of a point" and finds this definition highly satisfactory. Archimedes|| defines a spiral in purely kinematic terms.

> If in a plane a straight line be drawn and, while one of its extremities remains fixed, after performing any number of revolutions at a uniform rate return again to the position from which it started, while at the same time a point moves at a uniform rate along the straight line, starting from the fixed extremity, the point will describe a spiral.

The well-known tangent construction, however, is once again drawn up in such terms as to conceal all traces of the steps by which the initial approach was made: if O be the initial point; if ON, drawn perpendicular to the radius vector OP, meets the tangent at P in T, and if the circle, centre O, radius OP, meets the initial line in K, then, *either* $OT >$ arc PK, $OT <$ arc PK *or* $OT =$ arc PK. The proof is thus completed by *reductio ad absurdum* in the usual way.

It is tempting to think that Archimedes may have used the triangle ERP

of which is Plutarch, who constrains Apollonius also into the same opinion. For it is requisite (says he) there should be some first interval, under the inclination of containing line or superficies. But since the interval, which is under a point, is continuous, it is not possible that a first interval can be assumed; since every interval is divisible in infinitum. Besides, if we any how distinguish a first interval, and through it draw a right line, a triangle is produced, and not one angle."

† Coolidge, J. L., The unsatisfactory story of curvature, *Amer. Math. Monthly* 59 (1952), pp. 375–9.

‡ Aristotle, *De anima* I. 4. 409ª, 4.

§ Taylor, *op. cit.* I, p. 123.

|| Given from Dijksterhuis, *op. cit.*, p. 264.

(see Fig. 1.24) as a species of differential triangle, similar to triangle POT, and this view is supported by the terms of the original definition which gave to the point P two motions, the first produced by a rotation of the line OP about O, and the second by a linear motion along OP.

Thus, if $OP = r$, $\widehat{POK} = \theta$ and $r = a\theta$, we have, $\delta r = a\delta\theta$ and $ER/RP \simeq EO/OT$ (approximating arcs to chords).

$$\text{But} \quad \frac{ER}{RP} = \frac{\delta r}{r\delta\theta} = \frac{a\delta\theta}{r\delta\theta} = \frac{a}{r}, \quad \text{and} \quad OT = r^2/a = r\theta = \text{arc } PK.$$

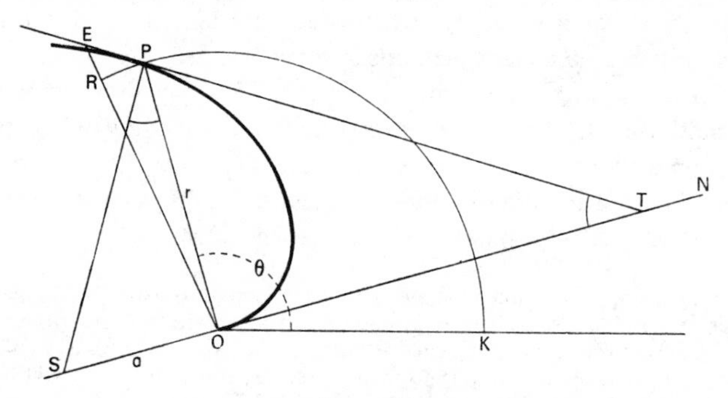

FIG. 1.24

The above result represents a rectification in the sense that the circular arc PK is stretched out along the line PT. The property of the Archimedean spiral, that the polar subnormal is constant (which provides a means of drawing the tangent to the spiral independently of the circular arc), is not, however, mentioned by Archimedes, i.e. if SP be the normal and SO the polar subnormal, then

$$\frac{SO}{OP} = \frac{SO}{r} = \frac{OP}{OT} = \frac{r}{r\theta} \quad \text{and} \quad SO = a.$$

Many other fascinating kinematic methods were evolved by the Greeks over a period of five centuries in the search for the solutions to the three famous classical problems of antiquity. Much of this material was assembled and classified by Pappus of Alexandria (*ca.* A.D. 320). To Pappus is due the division of curves into plane, solid and linear loci,† a classification which is constantly utilised in the early seventeenth century. The circle and straight

† *Pappus d'Alexandrie: La Collection Mathématique* (2 vols.) (trans. Ver Eecke, P.), Paris, 1933, I, pp. 38–9.

line were plane loci, the conic sections were solid loci and all other plane curves were classified as linear loci. Within this latter group were included all the so-called mechanical curves, i.e. those conceived as generated by some kind of motion. Of particular note here are the *Conchoid of Nicomedes* (see Fig. 1.25) developed in connection with the trisection problem, the *Cissoid of Diocles* (see Fig. 1.26) developed in connection with the Delian problem and

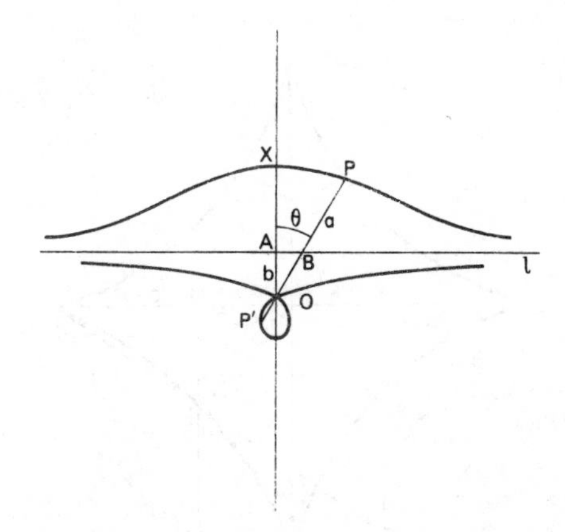

FIG. 1.25. *The conchoid of Nicomedes.* Let O be a fixed point on the line OX perpendicular at A to a fixed line l: let any line through O meet l in B: on this line mark off $BP = BP' = a$. The locus of P (and P') is a conchoid. *Polar equation* (OX is the initial line), $r = b \sec \theta \pm a$, where $OA = b$.

the *Quadratix of Hippias* (see Fig. 1.27) probably introduced for the trisection of the angle but subsequently applied by Dinostratus and Nicomedes to the squaring of the circle.

Pappus himself made many notable contributions to the study of curves: his investigations included certain skew curves derived by orthogonal projection. He obtained the quadratrix as the intersection of a cone of revolution with a right cylinder whose base is the spiral of Archimedes.† Pappus also formulated the well-known proposition on the volume of a solid of revolution which is now known as the Pappus–Guldin theorem.‡

† Ver Eecke, P., *op. cit.* I, pp. 199–201.
‡ *Ibid.* II, pp. 510–11.
$\frac{V_1}{V_2} = \frac{A_1 d_1}{A_2 d_2}$, where V_1, V_2 are the volumes of solids of revolution formed by

Apart from the cycloid† the curves known in the early seventeenth century were essentially those known to the Greeks and the development of infinitesimal methods proceeded hand in hand with the investigation of the tangent properties, arcs and areas, centroids and volumes of revolution of known curves. Any new techniques were first tried out on curves such as the parabola whose properties were already fully charted: they were then extended to more

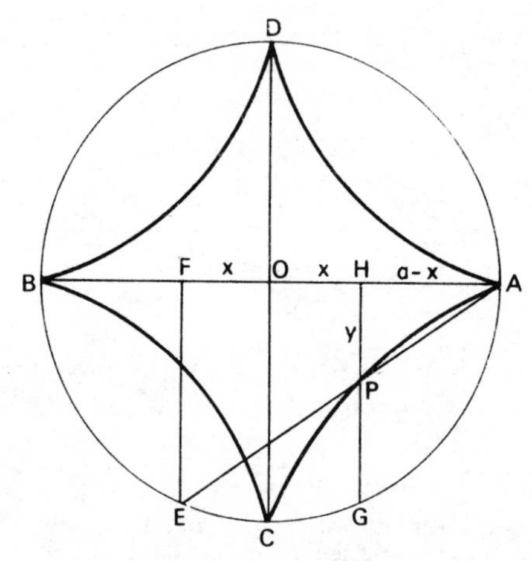

FIG. 1.26. *The cissoid of Diocles.* Let *CD* and *BA* be mutally perpendicular diameters of a circle, centre *O*, and let *FE* and *HG* be equal semi-chords perpendicular to *AB*. The cissoid is defined as the locus of the intersection of *AE* with *HG*. To obtain the *cartesian equation* let $OH = x$, $HP = y$, $OA = a$, $HG = b$,

$$\text{then} \quad \frac{y}{a-x} = \frac{a-x}{b} = \frac{b}{a+x}$$

$$\text{and} \quad y^2(a+x) = (a-x)^3.$$

HA and *HG* are mean proportionals between *HP* and *HB*.

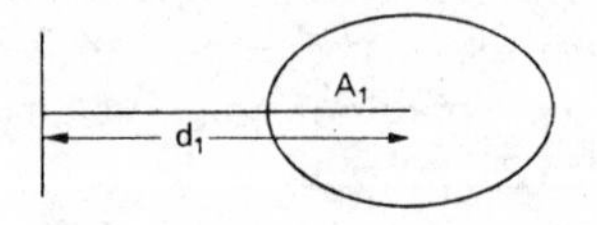

rotating A_1, A_2 about an axis distant d_1, d_2 from their respective mass centres. See Ver Eecke, P., Le théorème dit de Guldin considéré au point de vue historique, *Mathésis* 46 (1932), pp. 395–7.

† See Whitman, E. A., Some historical notes on the cycloid, *Amer. Math. Monthly* 50 (1943), pp. 309–15.

difficult curves such as the conchoid of Nicomedes with its interesting and valuable point of inflexion. The cissoid† and the quadratrix both played notable parts in seventeenth century mathematical controversies. The parabola and the Archimedean spiral were of particular importance in that they led easily to the invention of whole classes of curves‡ introduced purely in

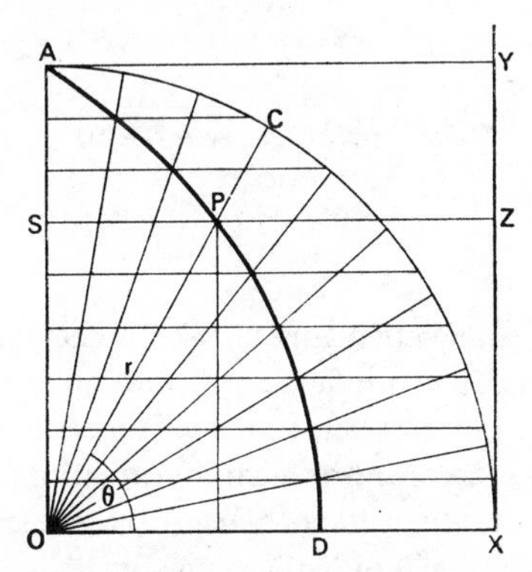

FIG. 1.27. *The quadratix of Hippias.* Let $OA = a$, the radius of the circle, centre O, rotate uniformly to the position OC in *the same time* as the straight line AY moves uniformly towards SZ. The quadratix is the locus of the intersection P of OC and SZ. *Polar equation* (OX is the initial line), $r \sin\theta = (2a/\pi)\theta$. As $\theta \to 0$, $r \to 2a/\pi$, $OD = 2a/\pi$, $\pi = 2a/OD$.

order to demonstrate the power of new techniques for the determination of tangents and quadratures.

The kinematic generation of curves was particularly valuable in its contribution to the developing concept of the tangent to a curve as the resultant motion of a point on the locus. At the beginning of the seventeenth century extension of the field of known curves depended primarily on physical intuition and curves were introduced whose generation could be understood in terms of motion. Later, the extension of algebraic notation made possible the introduction of a wide variety of curves whose properties were only studied in so far as they exhibited the new methods of analysis.

† Hofmann, J. and Hofmann, J. E., Erste Quadratur der Kissoide, *Deutsche Mathematik* 5 (1941), pp. 571–84.

‡ $(y/a)^m = (x/b)^n$, $(r/a)^m = (r\theta/b)^n$.

CHAPTER 2

The Transition to Western Europe

2.0. Introduction

It has often been suggested that later Greek mathematics foundered because it had grown so complex and difficult that few could understand it. At all events the Romans for the most part contented themselves with making compilations from Greek sources and original contributions were not substantial. These compilations for a long period constituted the sole scientific inheritance of the Latin West. In the field of mathematics Boethius (*ca.* 480–524) contributed a treatise on elementary geometry and arithmetic based on the works of Euclid and Nicomachus of Gerasa (*ca.* 100) which, from the seventh century onwards, seems to have provided the main source of mathematical knowledge. The territorial conquests of the Arabs cut off Greek learning from European scholars for many centuries and in consequence the art of demonstration seems to have been entirely lost. The circle quadrature of Francon of Liège exemplifies the depths to which geometry had sunk.†

In the twelfth century Greek science first began to filter into Western Christendom. Contacts were established with translators working on Arabic and Greek texts at Toledo and in southern Italy. In Chartres, where Roger Bacon lectured in 1245, Ptolemaic astronomy and Aristotelian physics were first welcomed. For several centuries the direction and scope of mathematical activity was, to a very considerable extent, determined by the order in which the Greek works were received.

The most important mathematical works recovered during the twelfth

† Winterberg, C., Der Traktat Franco's von Luettich *De Quadratura Circuli, Abhandlungen zur Geschichte der Mathematik* 4 (1882), pp. 135–90. Assuming the ratio of circumference to diameter to be exactly 22/7 Francon transforms the circle (diameter 14) into a rectangle (14×11). He was, however, quite unable to replace this rectangle by an equivalent square. See also Tannery, P., *Mémoires Scientifiques* 5, Paris, 1922, pp. 86–90, 233, 257–62, 294.

century[†] were the *Elements* of Euclid, the *Conics* of Apollonius, Hero's *Pneumatica*, a pseudo-Euclidean treatise, *De pondoroso et levi*, dealing with the lever and the balance, a portion of Archimedes' treatise *On Floating Bodies* and the *De mensura circuli*.[‡] In the thirteenth century the power and influence of Aristotle, already established through his logical works, was extended by translations of the *Physica* and *De caelo*. The pseudo-Aristotelian treatise, the *Mechanica* (possibly written by Strato), was also influential. Because of the recovery of these works interest tended to shift from logic and metaphysics to philosophy and science.

In the field of mathematics the Arabs transmitted much valuable material which had not been available to the Greeks. In particular, Al-Khowarizmi (*ca.* 840) contributed a work on algebra[§] and an account of the Hindu–Arabic numerals including the use of zero as a place-holder. This new material was widely publicised by Leonardo Fibonacci of Pisa (*ca.* 1175–1250) in the *Liber abaci* (1202). The extensive use of a convenient system of numeration made possible the development of standard computational techniques but it was not until the sixteenth century that systematic texts began to appear setting out these processes. It then became possible to explore the analogy between algebra and arithmetic, to generalise arithmetic processes by the use of algebraic symbols and, more important, to extend some of the calculating rules established in arithmetic to algebraic manipulation.

2.1. On Hindu Mathematics

The history of early Hindu mathematics has always presented considerable problems for the West. Nineteenth century studies, although sufficiently startling in some cases[||] as to arouse interest, were nevertheless often undertaken by Sanskrit scholars from the West whose knowledge of mathematics was not sufficiently profound to enable them to do much more than note results. Although more substantial studies in recent years by Hindu mathematicians have done a great deal towards filling some of the gaps in our knowledge it is still not possible to form a clear picture of either method or motivation in Hindu mathematics. Notwithstanding, even a straightforward ordering of

† Cf. Crombie, A. C., *Augustine to Galileo*, London, 1952, p. 215.

‡ See Clagett, M., *Archimedes in the Middle Ages* I, Madison, 1964.

§ *The Algebra of Mohammed Ben Musa* (ed. Rosen, F.), London, 1831. See also Karpinski, L. C., *Robert of Chester's Latin Translation of the Algebra of Al-Khowarizmi*, New York, 1915.

|| Whish, C. M., On the Hindu quadrature of the circle, *Trans. Royal Asiatic Society* 3 (1835), pp. 509–23.

results becomes meaningful when enough material has been collected to form a coherent pattern.

The Hindus seem to have been attracted by the computational aspect of mathematics, partly for its own sake and partly as a tool in astrological prediction. No trace has been found of any proof structure such as that established by Euclid for Greek mathematics nor is there any evidence which suggests Greek influence. Nevertheless, some of the results achieved in connection with numerical integration by means of infinite series anticipate developments in Western Europe by several centuries.

The invention of the place-value system is assigned by some authorities[†] to the first century B.C. and its use appears to have become fairly widespread by A.D. 700. The Pythagorean problem of incommensurable numbers was either unknown or not taken seriously and it was therefore possible to look forward with increasing confidence to the perfection of computational methods and the calculation of numerical magnitudes with ever-increasing accuracy.

In arithmetic zero ranked as a number and operations with zero were defined by Brahmagupta[‡] (628) as follows:

$$a-a = 0, \quad a+0 = a, \quad a-0 = a, \quad 0 \,.\, a = 0, \quad a \,.\, 0 = 0.$$

In algebra, however, zero seems to have been regarded as an infinitesimal quantity ultimately reducing to 0. Hence Brahmagupta[§] directs that $a \div 0$ shall be written $a/0$ and $0 \div a$ as $0/a$. The values of these expressions were left unspecified. Bhâskara II (1150) remarks in connection with division by zero:[||]

> In this quantity consisting of that which has cipher for its divisor, there is no alteration, though many may be inserted or extracted; as no change takes place in the infinite and immutable God, at the period of the destruction or creation of worlds, though numerous orders of beings are absorbed or put forth.

The idea expressed here may be written crudely,

$$a/0 = \infty \quad \text{and} \quad \infty + k = \infty.$$

Ganeśa (1558) is even clearer,[††] $a/0$ is "an indefinite and unlimited or infinite quantity: since it cannot be determined how great it is. It is unaltered by the addition or subtraction of finite quantities. . . ."

† Datta, B. and Singh, A. N., *History of Hindu Mathematics* (2 vols), Lahore, 1935, 1938, 2nd ed., Bombay, 1962, I, pp. 38–51.

‡ *Ibid.* I, p. 239.

§ *Ibid.* I, pp. 241–2.

|| *Ibid.*, p. 243.

†† *Ibid.*, p. 244.

According to Singh[†] arithmetic and geometric series were known in the fourth century B.C. and from A.D. 300–1350 problems are commonly quoted involving the sums of the squares and the cubes of the natural numbers as well as various arithmetico-geometric series. Pingala (200 B.C.) seems to have had some knowledge of the binomial coefficients.

In the fifteenth century computational techniques were skilfully applied[‡] in the development of expansions for the evaluation of $\sin x$, $\cos x$, $\tan^{-1} x$. There can be no question that numerical integration methods were fully understood and that many of the series developed in the West by Gregory, Leibniz and Newton were already known to the Hindus some two centuries previously. Of particular interest is the inverse tangent series, obtained by Gregory in 1671 and by Leibniz rather later. This series was known to the Hindus in the fifteenth century as *Talakulattura's Series* and is equivalent to

$$r\theta = \frac{r \sin \theta}{1 \cos \theta} - \frac{r \sin^3 \theta}{3 \cos^3 \theta} + \frac{r \sin^5 \theta}{5 \cos^5 \theta} \cdots$$

(convergent for $0 < \theta < \pi/4$).

The method of arriving at this series which is given in the *Yukti-Bhăsa*[§] can be briefly summarised in modern notation as follows; let Apq be a circular arc (see Fig. 2.1), radius OA, and let the radii Op, Oq, meet the tangent at A in P and Q respectively; let pm be the perpendicular from p to OQ. If $OA = Op = Oq = 1$, then

$$\frac{PQ}{OP} = \frac{\sin \widehat{QOP}}{\sin \widehat{OQP}} = \frac{pm/Op}{OA/OQ} = pm \,.\, OQ$$

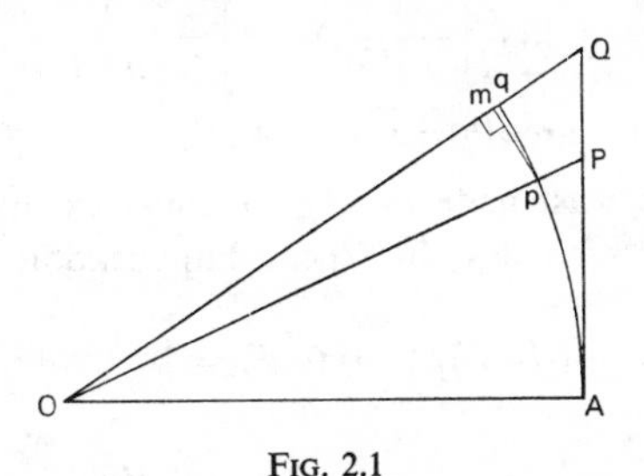

FIG. 2.1

† Singh, A. N., On the use of series in Hindu mathematics, *Osiris* 1 (1936), pp. 606–28.

‡ *Ibid.*; see also Sengupta, P. C., History of the infinitesimal calculus in ancient and mediaeval India, *Jahresbericht der Deutschen Mathematiker Vereinigung* 40 (1931), pp. 223–7.

§ The *Yukti-Bhăsa* is a Malayalam commentary on the *Tantra-samgraha* of Nilakantha (1500). See Marar, K. Mukunda and Rajagopal, C. T., On the Hindu quadrature of the circle, *Journal of the Bombay Branch of the Royal Asiatic Society* 20 (1944), pp. 65–82.

and

$$pm = \frac{PQ}{OP\,.\,OQ}.$$

If PQ be small, then $pm = \text{arc}\; pq = \dfrac{PQ}{OP^2} = \dfrac{PQ}{1+AP^2}$.

Let $B\widehat{O}A < \pi/4$, $\tan A\widehat{O}B = t$; divide AB into n equal parts at points P_0, $P_1, P_2, P_3, \ldots, P_n$, and let OB meet the circle in b (see Fig. 2.2), then

$$\text{arc}\; Ab = \lim_{n\to\infty} \sum_0^{n-1} \frac{P_rP_{r+1}}{1+AP_r^2} = \lim_{n\to\infty} \sum_0^{n-1} \frac{t/n}{1+(rt/n)^2}$$

$$= \lim_{n\to\infty} \sum_0^{n-1} \{(t/n)\,(1-(rt/n)^2+(rt/n)^4-\ldots)\}.$$

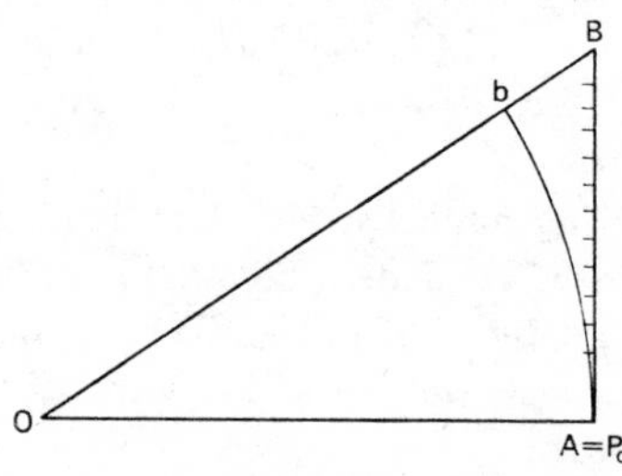

FIG. 2.2

Hence,

$$\text{if} \lim_{n\to\infty} \left[\frac{1}{n^{p+1}} \sum_0^{n-1} r^p\right] = \frac{1}{p+1},$$

$$\text{arc}\; Ab = t-t^3/3+t^5/5-\;\ldots$$

Apparently no use was made of the binomial expansion, the series for $(1+(rt/n)^2)^{-1}$ being arrived at by repeated application of the multiplicative identity

$$(1+(rt/n)^2)\,(1-(rt/n)^2) = 1-(rt/n)^4.$$

Hence,

$$\frac{1}{1+(rt/n)^2} = 1-(rt/n)^2+\frac{(rt/n)^4}{1+(rt/n)^2}$$

and

$$(1+(rt/n)^2)\,(1-(rt/n)^2+(rt/n)^4) = 1+(rt/n)^6,$$

$$1-(rt/n)^2+(rt/n)^4 = \frac{1}{1+(rt/n)^2}+\frac{(rt/n)^6}{1+(rt/n)^6}.$$

The series for $(1+(rt/n)^2)^{-1}$ is thus constructed term by term.

Of particular interest is the application of the general result,

$$\lim_{n\to\infty} \frac{\sum\limits_{0}^{n-1} r^p}{n^{p+1}} = \frac{1}{p+1}.$$

In the West, Archimedes had made use of $\sum r$, $\sum r^2$ and the Arabs (*ca.* 1000) developed expressions for $\sum r^3$, $\sum r^4$. Beyond this no further extensions occurred until the seventeenth century.

The concept of differentiation was also explored by the Hindus in connection with the variable motion of planets. Examples are quoted from Bhâskara and from Brahmagupta† in which numerical methods are used to calculate the apparent motion of planets. These results are formally equivalent to the determination of the rates of change of trigonometric functions.

Nevertheless, the customary presentation of results gave very little indication of the means by which they were derived; this, together with the absence of any formalised proof structure, militated against continuous mathematical development. Under the circumstances it was inevitable that results should frequently be found, lost and rediscovered under different circumstances. Despite the extraordinary degree of success which the Hindus achieved in numerical integration by means of infinite series no attempt was made to relate the processes of differentiation and integration on a theoretical basis and no specialised language evolved in which these operations could be expressed. The infinitesimal calculus originated in the seventeenth century in Western Europe as a result of prolonged and sustained effort to achieve these ends. The fact that the Leibniz–Newton controversy hinged as much on priority in the development of certain infinite series as on the generalisation of the operational processes of integration and differentiation and their expression in terms of a specialised notation does not justify the belief that the development and use of series for numerical integration establishes a claim to the invention of the infinitesimal calculus.

Much of the Hindu attitude and approach to mathematics was certainly conveyed to Western Europe through the Arabs. The algebraic methods formerly considered to have been invented by Al-Khowarizmi can now be seen to stem from Hindu sources. The place-value system involving the use of nine numerals and a zero as place-holder is undoubtedly of Hindu origin and its transmission to the West had a profound influence on the whole course of mathematics.

† See Sengupta, P. C., *op. cit.*

2.2. The Arabs

The original Arab contribution was more important in optics and perspective than in pure mathematics. Advances made in trigonometry arose in connection with observational astronomy and had no immediate impact on mathematics as such. Nevertheless, although the Arabs had no special feeling for the rigorous methods of the Greek geometers they understood the Exhaustion proofs in Euclid's *Elements* and knew how to use an argument by *reductio ad absurdum*.

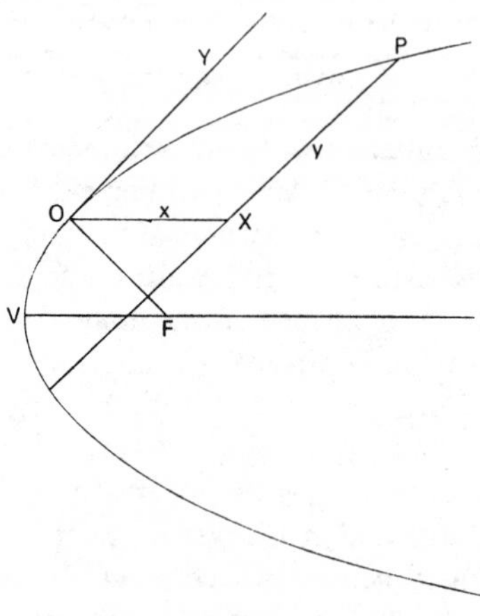

Fig. 2.3

Ibn-al-Haitham† (Alhazen) in particular made striking advances in applying such methods in the calculation of the volumes of solids of revolution. Whereas Archimedes had concerned himself in the case of the parabola with rotation about the axis only, Ibn-al-Haitham extended and developed this work by considering the volumes of solids formed by the rotation of parabolic segments about lines other than the axis.

If O be any point on a parabola, vertex V and focus F, then OX, drawn parallel to VF, is a *diameter* and XP, drawn parallel to the tangent OY to meet the parabola at P, is an *ordinate* (see Fig. 2.3). By a standard theorem in geometrical conics, $XP^2 = 4OF \,.\, OX$, so that the equation of the parabola referred to oblique axes OX and OY can in general be written in the form

† Suter, H., Die Abhandlung über die Ausmessung des Paraboloides von el. Hasan b. el-Hasan b. el-Haitham, *Bibl. Math.* (3f.) 12 (1911–12), pp. 289–332.

$y^2 = kx$. In the special case where O is the vertex the axes are rectangular. Ibn-al-Haitham makes use of this relation to determine the volumes of solids formed by rotating parabolic segments about any diameter or ordinate.

For rotation about OX (see Fig. 2.4) let OX be divided into n equal parts, each equal to h and let $Op = rh$, $Oq = (r+1)h$, where r is an integer less than n. Complete the parallelogram $OXPY$. In each case, $PX^2 = k.OX = k.nh$. Let ps, the ordinate at p, meet PY at m: draw ql parallel and equal to pm and st parallel and equal to pq. We now have, $ps^2 = k.Op = k.rh$ and

$$\frac{\text{Volume of solid formed by rotating parallelogram } pqts}{\text{Volume of solid formed by rotating parallelogram } pqlm} = \frac{ps^2}{pm^2}$$

$$= \frac{ps^2}{PX^2}$$

$$= \frac{Op}{OX}$$

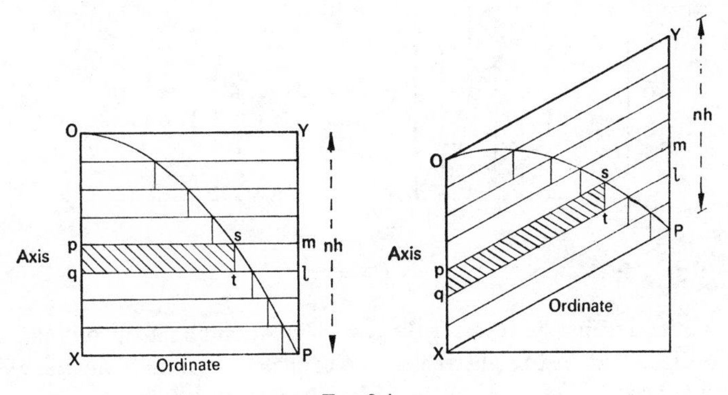

FIG. 2.4

Hence, by summation,

$$\frac{\text{Volume of inscribed solid}}{\text{Volume of solid formed by rotating parallelogram } OXPY}$$

$$= \frac{\sum_{1}^{n-1} r}{\sum_{1}^{n} n} = \frac{n(n-1)}{2n^2} < \frac{1}{2}.$$

Similarly,

$$\frac{\text{Volume of circumscribed solid}}{\text{Volume of solid formed by rotating parallelogram } OXPY} = \frac{\sum_1^n r}{\sum_1^n n} = \frac{n(n+1)}{2n^2} > \frac{1}{2}.$$

The proof can thus be completed by *reductio ad absurdum:* the volume of the solid formed by rotating the segment OPX about the diameter OX is equal to one half of the solid formed by rotating the parallelogram $XPYO$ about OX.

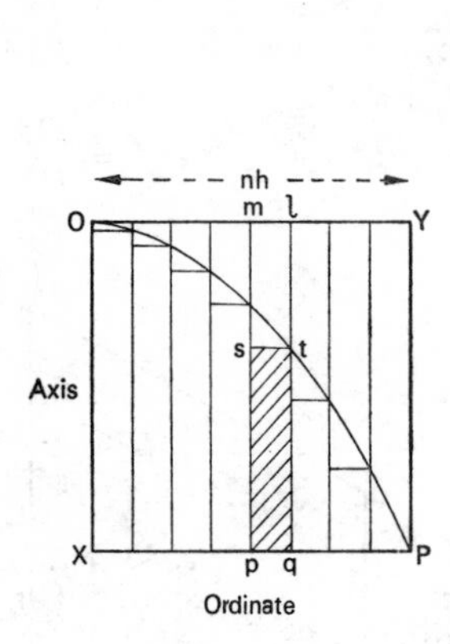

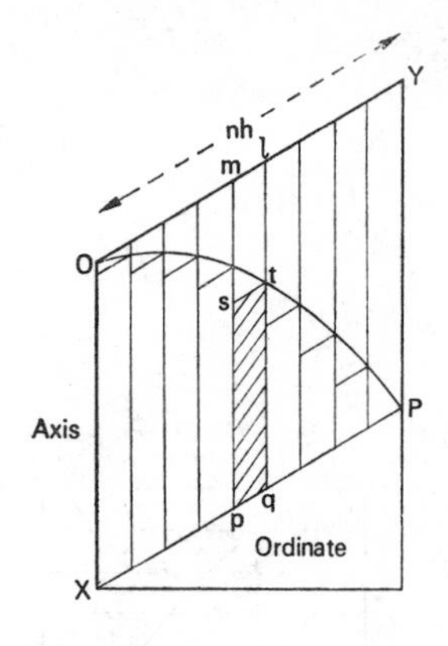

FIG. 2.5

More interesting, however, is the case of rotation about the ordinate PX.† The ordinate can be at right angles to a diameter, or at an oblique angle (see Fig. 2.5). Let PX be divided into n equal parts, each equal to h, and let $Xp = (r-1)h$, $Xq = rh$ (where r is an integer less than, or equal to, n). As before, complete the parallelogram $OXPY$: draw ql parallel to PY to meet the curve in t and OY in l: complete the parallelogram $pqts$. We have, $PX^2 = k\,.\,OX = k\,.\,PY$, $PY = (n^2h^2)/k$; also, $Xq^2 = k\,.\,lt$, $qt = ql-lt = PY-lt = (n^2h^2)/k-(r^2h^2)/k$, and

$$\frac{\text{Volume of solid formed by rotating } pqst \text{ about } pq}{\text{Volume of solid formed by rotating } pqlm \text{ about } pq} = \frac{(n^2-r^2)^2}{(n^2)^2}.$$

† Kepler called this a *parabolic spindle* (*Nov. ster.* I. 26). Cavalieri correctly found its volume (*Ex. geom. sex.* IV, Prop. XXIV).

Hence, by summation,

$$\frac{\text{Volume of inscribed solid}}{\text{Volume of solid formed by rotating } XPOY \text{ about } XP} = \frac{\sum_0^{n-1} (n^2-r^2)^2}{n^5} < 8/15.$$

By a similar argument the volume of the circumscribed solid is derived and the conclusion follows by *reductio*.† To arrive at the above inequalities Al-Haitham required to make use of expressions for $\sum r^p$ up to and including $p = 4$. He uses an interesting geometrical method to sum these series (see Fig. 2.6) and, although he only takes the results as far as $n = 4$, they can be readily extended to other values of n. The methods are completely different from those of Archimedes:‡ the expression for $\sum r^4$ had not, so far as is known, been given before, at any rate in the West.§

2.3. The Influence of Aristotle

Until the end of the twelfth century mathematical activity in the West remained at a low ebb and progress was impeded by lack of technical proficiency. The writings of Aristotle, however, promoted a general interest in the nature of mathematics and much consideration was given to the question of its relationship to natural science. Although Aristotle's philosophy was not accepted uncritically his interpretation of the respective roles of mathematics and physics in the study of natural phenomena was fairly generally adopted.

† $$\lim_{n\to\infty} \sum_1^n \frac{(n^2-r^2)^2}{n^4} = 1-2 \lim_{n\to\infty} \sum_1^n \frac{r^2}{n^2} + \lim_{n\to\infty} \sum_1^n \frac{r^4}{n^4} = 1-\frac{2}{3}+\frac{1}{5} = \frac{8}{15}.$$

In calculus notation, let $y^2 = kx$, $y_0^2 = kx_0$,

then,

$$\begin{aligned} V &= \pi \int_0^{y_0} (x_0-x)^2\, dy \\ &= \pi \int_0^{y_0} (x_0^2 - 2x_0 y^2/k + y^4/k^2)\, dy \\ &= \pi \left(x_0^2 y_0 - \frac{2}{3} x_0^2 y_0 + \frac{x_0^2 y_0}{5}\right) \\ &= \pi x_0^2 y_0 \left(1-\frac{2}{3}+\frac{1}{5}\right) = \frac{8}{15}\pi x_0^2 y_0. \end{aligned}$$

‡ Archimedes, *On Spirals*, Prop. X.

§ It is by no means clear that this work was original with Ibn-al-Haitham. It is possible that he had access to unknown sources. No reference is made to Archimedes, *On Conoids*, although Al-Haitham was certainly familiar with the *De mensura circuli*. Cf. Wieleitner, H., Das Fortleben der Archimedischen Infinitesimalmethoden bis zum Beginn des 17 Jahrhundert insbesondere über Schwerpunktbestimmungen, *Quellen und Studien* (B) 1 (1931), pp. 201–20.

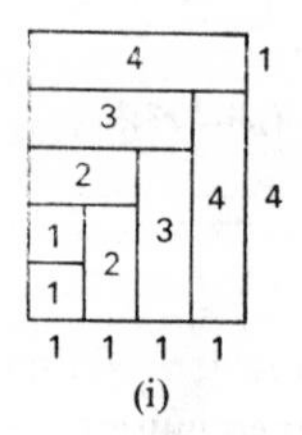

(i)

(iii)

(ii)

(iv)

(i) *Sums of natural numbers*

$$\frac{n(n+1)}{2} = \sum_1^n r$$

(ii) *Sums of squares of natural numbers*

$$(n+1)\sum_1^n r = \sum_1^n r^2 + \sum_1^n \sum_1^r r$$

$$(n+1)\sum_1^n r = \sum_1^n r^2 + \sum_1^n \frac{r(r+1)}{2}$$

$$(n+\tfrac{1}{2})\sum_1^n r = \tfrac{3}{2}\sum_1^n r^2$$

$$\frac{(2n+1)n(n+1)}{6} = \sum_1^n r^2$$

(iii) *Sums of cubes of natural numbers*

$$(n+1)\sum_1^n r^2 = \sum_1^n r^3 + \sum_1^n \sum_1^r r^2$$

Hence, substituting for $\sum r^2$, $\dfrac{n^2(n+1)^2}{4} = \sum_1^n r^3$

(iv) *Sums of fourth powers of natural numbers*

$$(n+1)\sum_1^n r^3 = \sum_1^n r^4 + \sum_1^n \sum_1^r r^3 = \sum_1^n r^4 + \sum_1^n \frac{r^2(r+1)^2}{4}$$

and, substituting for $\sum r^3$, $\sum r^2$,

$$\sum_1^n r^4 = \frac{n(n+1)\,(2n+1)\,(3n^2+3n-1)}{30}$$

FIG. 2.6

The investigation of the physical world was considered as a part of a single philosophical activity which concerned the search for reality and truth. Physics was the science concerned with defining the substance underlying and causing change and motion: mathematics, being concerned only with the abstract quantitative aspects of material things, could provide no direct knowledge of causal relations but could only describe them. Aristotle's definitions of cause and substance, however, were both so wide that there was little prospect of disentangling the physical aspects of those things susceptible to mathematical investigation from their more general aspects.

The history of the exact sciences in Europe from the thirteenth to the eighteenth century is a history of the gradual reversal of Aristotelian philosophy.† This took place as a slow process of emancipation completed in two major phases, the first by the end of the fourteenth century and the second by the end of the seventeenth century. Throughout the whole period the increasing power of mathematics and its use in investigating and interpreting the physical world played a major part in determining new attitudes to scientific explanation. The gradual penetration of mathematics into the preserve of physics was naturally accompanied by a separation of the aspects of the physical world susceptible to quantitative measurement from those, which, by virtue of their nature, defied quantitative description. The possibility, or otherwise, of applying methods of mathematical analysis was therefore a major factor in determining and delimiting the area of scientific investigation. The extended use of mathematics as an instrument of scientific enquiry was, however, not only important for the progress of science but also in the development of mathematics itself: it provided a spur and an impetus towards further work and made available new concepts and intuitions from the physical world which could be utilised in the study of form, pattern and structure in the field of pure mathematics.

2.4. The Continuum, Indivisibles and Infinitesimals

Philosophers, speculating on the nature of the continuum have encountered grave conceptual difficulties from the time of Zeno and Democritus. Each age has produced its own crop of paradoxes of the infinite and those of the thirteenth and fourteenth centuries were no less ingenious than those of any other.

Although logically problems concerning the mathematical continuum are entirely distinct from those of the physical world‡ it was inevitable, both in

† Maier, A., *Die Vorläufer Galileis im 14 Jahrhundert*, Rome, 1949, pp. 1–2.

‡ Cf. Russell, B., *op. cit.*, pp. 346–7.

antiquity and in the Middle Ages, that no such distinction should be recognised. Any consideration of the division or multiplication of physical magnitudes was invariably associated with the numerical divisors and multipliers by which these operations were conducted. If, for example, division is possible without end then there is no last divisor and the number of such divisors (i.e. positive integers) is infinite in the sense that there is no divisor so great that a greater cannot be found.

The central issue in all mediaeval discussion was the question of the existence of the infinite and controversy arose as a result of attempts to support or refute the views of Aristotle as expressed in the *Physics*.† Aristotle distinguishes carefully between the *actual* and the *potential* infinite. He argues that, since the world is finite, no physical magnitude can by multiplication become *actually* infinite. The process of division, on the other hand, can be continued without end and there is no stage at which it is not possible to go further. From this two conclusions emerge: first, there are no indivisible magnitudes and, second, number can be increased without limit. Since there is no physical magnitude so small that it is not possible to produce a smaller so also there is no integer so great that it is not possible to produce a greater. Hence the number of positive integers is *potentially* rather than *actually* infinite. Aristotle rejects the notion of a continuum composed either of mathematical points or of any other kind of indivisibles.‡ Although he never denied the reality of points he considered them as possessing neither dimension nor magnitude§ just as zero was not considered to exist as a number.||

The philosophers of the thirteenth and fourteenth centuries recognised, for the most part, the existence of points, lines and planes, not, of course in the sense that these could exist singly as separate entities in space, but as the boundaries of lines, planes and solids.†† A solid, being everywhere divisible into plane sections, contained infinitely many planes, a plane infinitely many lines and a line infinitely many points. The question then arose as to whether, in fact, the continuum might be conceived as made up entirely of points and conversely whether any finite continuum might be so divided as to produce nothing but points.

On the whole Aristotle's views on the nature of the continuum found fairly general acceptance in the thirteenth and fourteenth centuries: Roger Bacon,

† Aristotle, *Physics* III. 5. 6. 204^a–207^a.
‡ *Phys.* III. 6. 206^a, 17.
§ *Phys.* VI. 231^a, 25.
|| *Phys.* III. 7. 207^b, 7.
†† Maier, *op. cit.*, p. 158.

Albertus Magnus, Thomas Aquinas, Duns Scotus, William of Ockham, Jean Buridan and his school, Nicole Oresme, Albert of Saxony and many others all rejected indivisibles of any shape or form.† Alternative points of view were developed, however, which aroused interest and controversy. The defence and refutation of Platonic and Democritian theories led to some profitable exploration of the logical foundations for operations on infinite numbers and many contradictions emerged which were not finally resolved until the twentieth century.‡ Although few arguments were entirely free from metaphysical considerations the idea gradually emerged of an omnipotent God whose powers were only limited to the extent that it was not possible for him to create a self-contradictory universe.

The Platonists, or Pythagoreans,§ amongst whom were Henry of Harclay and Nicholas of Autrecourt,|| constructed solids from planes, planes from lines and lines from points. Ultimately therefore the material continuum was considered to be made up of indivisible points. Many difficulties arose in connection with this theory. If a point has no dimensions how is it possible that by multiplication or addition of points a finite segment be produced? If, on the other hand, a point has dimensions in what sense is it to be regarded as indivisible?

Difficulties also arose in connection with the correspondence of points on concentric circles,†† or indeed on any system of lines or curves produced by parallel or central projection.

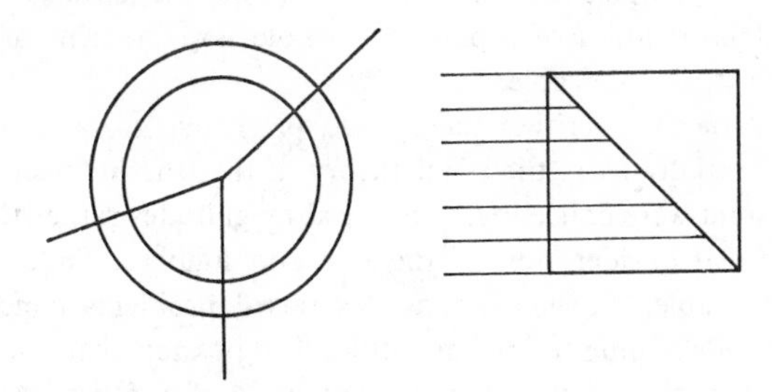

FIG. 2.7

† Cf. Lasswitz, K., *op. cit.* I, pp. 186–201.

‡ Russell, B., *op. cit.*, p. 353.

§ The theories of the Platonists have come to be associated with the Pythagorean ideas expressed in Plato's *Timaeus*.

|| Cf. Maier, A., *op. cit.*, pp. 161–2.

†† Drabkin, I. E., Aristotle's wheel: notes on the history of a paradox, *Osiris* 9 (1950), pp. 162–98.

Every radius cuts the circumference of each circle in one, and only one, point: every line perpendicular to the side of a square cuts that side and the diagonal in one, and only one, point (see Fig. 2.7). Hence, there exists a one-one correspondence between points on concentric circles and also between points on the side and diagonal of the square. In both of these cases how is it possible that the number of such points can be other than equal? Since, however, the lines are unequal in the Euclidean sense, how is it possible that the number of points should be equal? This paradox constituted a strong argument against the point structure of the continuum.† The principal difficulty lay, therefore, not in the notion of a continuum constructed from points, but in the tacit assumption that a finite line or space contains a definite number of points and that a greater line or space contains a greater number of points. Since it is inconceivable that a finite line should be made up of a finite number of dimensionless points the number of such points must be infinite. The supporters of the point theory were therefore led to conclude that any two infinite numbers could not be regarded as equal and they were further obliged to postulate different orders of infinite magnitudes. Examples were not hard to find. If the world has existed from eternity then the number of days, of months and of years is infinite; yet the number of days is certainly greater than the number of months and the number of years.‡ Towards the end of the fourteenth century a clearer view of these matters emerged. In particular, Albert of Saxony,§ who himself rejected indivisibles, demonstrated by a number of examples that between two infinite magnitudes standing to each other in the relation of a part to a whole no statement of the form, $A > B$ or $B < A$, is applicable.

The atomist or Democritian theory was presented in various forms all differing substantially from the point theory in that the indivisibles making up the continuum were conceived as physical magnitudes rather than dimensionless points.|| Consideration of the size and number, finite or infinite, constant or variable, of such indivisibles raised problems concerning the relative magnitude of infinitesimal quantities. Is it possible that one indefinitely small quantity can be regarded as a multiple of another? Alternatively are all such quantities to be regarded as equal in size? Is the magnitude of an indivisible to be regarded in any sense as dependent on the process of division

† Cf. Maier, A., *op. cit.*, p. 165.
‡ *Ibid.*, p. 169.
§ *Ibid.*, pp. 170–2.
|| *Ibid.*, pp. 172–7.

through which it is arrived at? Gregory of Rimini† maintained that the infinite *number* of indivisibles making up a finite continuum was independent of the *size* of the continuum. An individual element or indivisible was, accordingly, proportional to the continuum from which it was drawn.

All these issues so closely involve the relations between infinitely small quantities that there is a temptation to regard the philosophical discussions of the thirteenth and fourteenth centuries as part of a tentative exploration of the logical foundations of the differential and integral calculus.‡ It should be emphasised, however, that the discussions were severely limited, both in scope and content and that, except in the field of mechanics, very little original work was done in mathematics until the treatises of Archimedes had been recovered and fully mastered. The questions asked were directed primarily towards the establishment of certain truths regarding the reality and substance of the physical world and any mathematical insights gained were in a sense incidental to this purpose.

2.5. The Growth of Kinematics in the West

Whereas in dynamics movement is studied in relation to the forces associated with it, in kinematics only its spatial and temporal aspects are considered. It follows that in kinematics we are concerned with the description of movement and not with its causes and it can therefore properly be regarded as the geometry of movement.

If a particle be known to trace a given curve the geometric properties of that curve can be used to predict the subsequent positions of the particle; conversely, if a curve be defined as the path of a point moving under specified conditions, then the laws of kinematics can be utilised to provide information as to certain geometric properties of the curve. For example, knowledge of the instantaneous motion at a given point on the curve enables us to draw the tangent to the curve at that point. The development, in the fourteenth century, of certain important concepts of motion including instantaneous velocity can therefore be seen to have direct and immediate bearing on the study of the tangent properties of curves. Furthermore, the introduction of graphical methods of representation led to the establishment of a link between the velocity–time graph, the total distance covered and the area under the curve, and this in turn is closely connected with the integral calculus (see Figs. 2.8 and 2.9).

† *Ibid.*, pp. 172–7.
‡ See Crombie, A. C., *op. cit.*, p. 242; Maier, A., *op. cit.*, p. 177.

The imaginative insights gained by the use of kinematic concepts in geometry were responsible for some of the more powerful methods developed during the seventeenth century for the study of curves. The work of Isaac Barrow, for instance, which certainly influenced Newton, is dominated by the idea of curves generated by moving points and lines.

The oldest extant discussion of kinematic statements is found in Aristotle's *Physics*.† Here Aristotle uses a simple straight-line representation to denote motions which are equal, slower or quicker respectively. He had no means of defining velocities other than by comparing the spaces covered in various times. Let A move through the distance s_1 with velocity v_1 in time t_1, and let

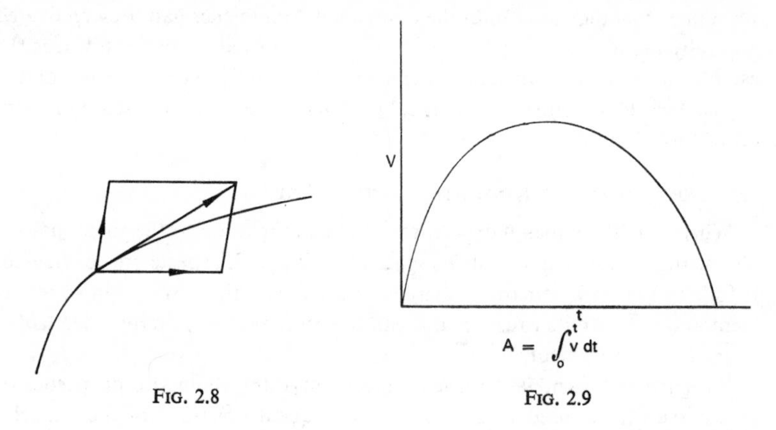

FIG. 2.8 FIG. 2.9

B move through the distance s_2 with velocity v_2 in time t_2. Then, Aristotle says,

(i) $v_1 = v_2$, provided $s_1 = s_2$, when $t_1 = t_2$;

(ii) $v_1 > v_2$, provided $s_1 > s_2$, when $t_1 = t_2$, also, if $v_1 > v_2$, then $s_1 > s_2$, when $t_1 = t_2$. It is also possible, if $v_1 > v_2$, that $s_1 > s_2$ when $t_1 < t_2$.

In modern notation all these relations can be expressed in terms of the single formula, $v = s/t$.

Aristotle goes on to discuss circular motion.‡ Although he is prepared to consider that the length of arc traversed by a moving point in a given time might be considered as a measure of rotatory motion he rejects the idea of

† *Physics* VI. 2. 232ᵃ–233ᵃ. See also Schramm, M., *op. cit.* Discussed by Mau, J., in *History of Science* 3 (1964), pp. 127–31.

‡ *Physics* VII. 4. 248ᵃ⁻ᵇ.

comparing it with an equivalent linear path on the grounds of the incommensurability of the circle and the straight line. Since with Aristotle velocities (or motions) could only be compared by considering the actual spaces moved through he felt it necessary to restrict these comparisons to cases where the actual figures traversed were of the same geometrical form.

The pseudo-Aristotelian treatise, the *Mechanica*,[†] extends further the discussion of circular motion and includes the helpful comment[‡] that "of two points on the same radius that which is further from the centre is quicker". The discussion which follows includes the now familiar figure of the parallelogram of velocities and a statement of the theorem, "if the two displacements

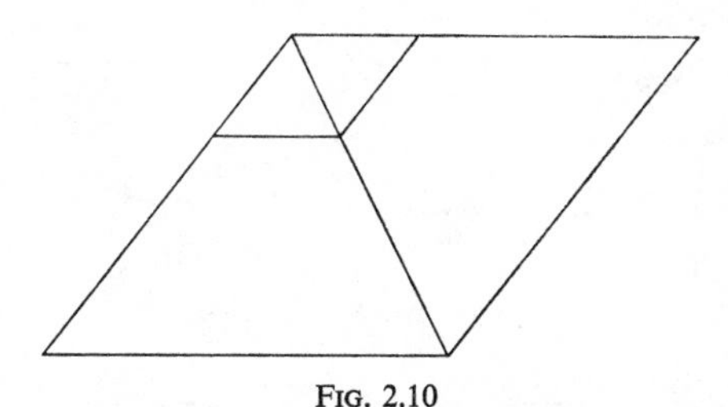

Fig. 2.10

of a body are in a fixed proportion, the resulting displacement must necessarily be a straight line, and this line is the diagonal of the figure, made by the lines drawn in this proportion" (see Fig. 2.10).

The author of the treatise goes on to explain that if the ratio between the two displacements is not fixed then the resulting path will not be a straight line but a curve. In the case of circular motion he gives to the moving point two displacements, one tangential and the other towards the centre. A constant relation between these would, he says, produce motion along the chord. Since the relation varies, motion is along the arc. Although this analysis of circular motion is confused it nevertheless represents the first attempt to explain curvilinear motion in terms of component linear velocities changing in direction and magnitude (see Fig. 2.11).

The parallelogram of velocities (or displacements) is also treated in the *Mechanica*[§] of Hero of Alexandria (*ca.* A.D. 100). Let a point A move uni-

† Attributed to Strato.

‡ *The Mechanica* (trans. Forster, E. S.), Oxford, 1913.

Given from Clagett, M., *The Science of Mechanics in the Middle Ages*, Madison, 1959, p. 41.

formly along a line AB, whilst in the same time the line AB moves parallel to itself into a new position CD. Since, at any given instant, the ratios of the distances moved along and parallel to AB are as $AB : AC$, the resultant path of the moving point is the line AD. Motion along the diagonal is *compounded* of two *simple* motions along the lines AC and AB (see Fig. 2.12).

In the twelfth century a number of important mechanical treatises circulated which have subsequently become associated with the school of Jordanus Nemorarius. Certain ideas contained in these works were widely discussed in the fourteenth century and later influenced the mechanics of Stevin and Galileo. In the *Liber de motu* Gerard of Brussels† considers the motion of

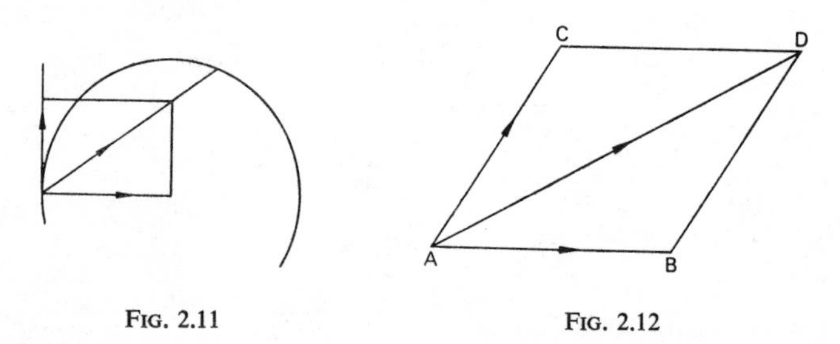

FIG. 2.11

FIG. 2.12

bodies by translation and rotation. He says of a body that it is moved *equally* or *uniformly* when it is moved "equally in all its parts". Circular motion is accordingly, by definition, non-uniform and Gerard conceives the valuable notion of reducing this motion to uniformity by replacing a rotating line (or surface) by a translation in which the same line (or surface) covers an equal area (or volume) in an equal time.

Let the line AB, length r, rotate about the end-point A through an area $\pi r^2 \lambda$. This rotation is *made uniform* by a translation of the line AB through a distance $\pi r \lambda$ in the same time. But $\pi r \lambda$ is the *actual* distance moved through by a point on AB, distant $r/2$ from A. Hence the rotation is made uniform by a punctual rectilinear motion equal to that of the mid-point of the rotating line (see Fig. 2.13).

The conclusion applies equally to any finite segment of a rotating line. For the rotating segment ab (see Fig. 2.14) let the circular area moved through be $\pi\lambda(r_1^2 - r_2^2) = (r_1 - r_2) \times \pi\lambda(r_1 + r_2)$, where $r_1 = Ab$, $r_2 = Aa$ and $\pi\lambda(r_1 + r_2)$ is the

† Clagett, M., The *Liber de motu* of Gerard of Brussels and the Origins of Kinematics in the West, *Osiris* 12 (1956), pp. 73–175.

motion of the mid-point of the segment ab; thus, since, $r_1 - r_2 = ab$ the conclusion follows immediately.

Gerard's approach is wholly geometrical. He replaces the circular sector traversed by the rotating line AB by an equivalent triangle, with altitude AB

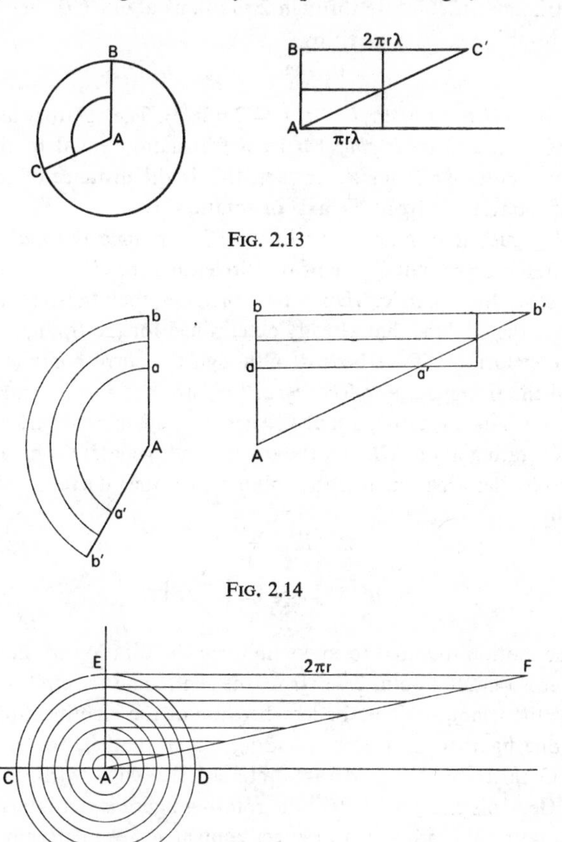

FIG. 2.13

FIG. 2.14

FIG. 2.15

and base BC' equal in length to the circular arc BC. In making this geometric transformation he explicitly recognises the correspondence between the *lines of the triangle* drawn parallel to the base, BC', and the circular paths of the points on the line AB. This correspondence he exploits more fully when he discusses how to *make uniform* the rotatory motion of a circular area about a diameter CD. He replaces the *lines of the circle* by the *lines of a triangle* (see Fig. 2.15) and then goes on to determine correctly the punctual rectilinear velocity necessary to *make uniform* the rotatory motion of the triangle about the axis CD.

If the triangle AEF rotate through 2π radians about CD then the volume V of the resulting solid is given by

$$V = \tfrac{2}{3}\pi r^2 \times 2\pi r.$$

Let $V = $ area $AEF \times d$, then $d = \tfrac{4}{3}\pi r = 2\pi(2r/3)$. The rotatory motion of the triangle is therefore *made uniform* by a translation equal to the punctual motion of a point $\tfrac{2}{3}r$ from A. In fact, this is the distance of the centre of gravity of the triangle from the axis of rotation, CD.

Gerard argues that since every line of the triangle is *equal* and *equally moved* as the corresponding line of the circle then the circle is *equal* and *equally moved* as the triangle. Hence he concludes that the required uniform velocity for the circle is that already determined for the triangle. Unfortunately this conclusion is false because although the corresponding lines of the circle and the triangle are indeed *equal* they are not *equally moved* since the lines of the triangle rotating about CD generate cylindrical surfaces; those of the circle rotating about CD, on the other hand, generate spherical surfaces.

Let the circular area generate the volume $\tfrac{4}{3}\pi r^3$, then the required translation is given by

$$\pi r^2\, d' = \tfrac{4}{3}\pi r^3,$$

$$d' = \tfrac{4}{3} r = 2\pi\left(\frac{2}{3\pi}\, r\right).$$

Hence the motion required to make uniform the rotation of the circle about a diameter is that of a point $2r/3\pi$ from the centre and the circle is not equally moved as the triangle. None the less, because of the identification of a set of lines in one figure with a corresponding set in another figure the methods used by Gerard are of considerable interest. It is also worth noting that in those of Gerard's examples where the rotating figure is not intersected by the axis of rotation the process is in effect equivalent to a determination of the centre of gravity of the rotating figure by the Pappus–Guldin theorem.†

If an area A rotate through 2π radians about an axis OY (which does not intersect the area) then the volume of the resulting solid is given by

$$V = 2\pi \int_0^a xy\, dx,$$

† Gerard seems to have been unaware of this theorem or of the connection of his results with the determination of centres of gravity.

but, by definition, $\int_0^a xy\,dx = \bar{x}A$, where $\bar{x}$ is the distance of the centre of gravity of A from OY (see Fig. 2.16).

Hence, $V = 2\pi\bar{x}A$

$= A\times$ length of path of the centre of gravity of the area A as it rotates through 2π radians about the axis OY.

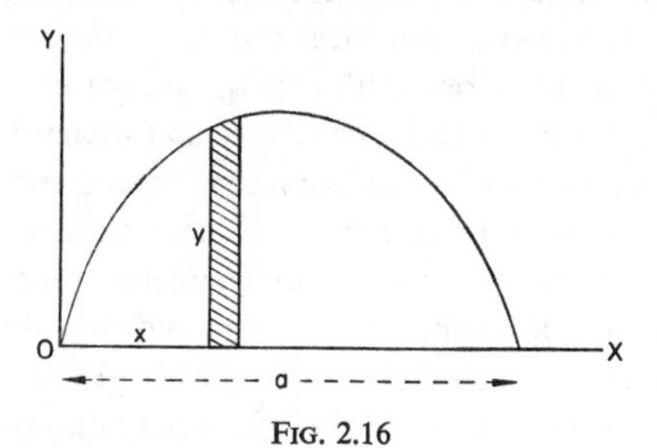

FIG. 2.16

2.6. The Latitude of Forms

At Merton College, Oxford, between the years 1328 and 1350, the distinction between kinematics and dynamics was made explicit. In the work of Thomas Bradwardine, William Heytesbury, Richard Swineshead and John Dumbleton the foundations for further study in this field were laid through the clarification and formalisation of a number of important concepts including the notion of instantaneous velocity (*velocitas instantanea*).†

The study of space and motion at Merton College arose from the mediaeval discussion of the intension and remission of forms, i.e. the increase and decrease of the intensity of qualities. The distinction between intension and extension is exemplified in the case of heat by the difference between temperature, or degree of heat, and quantity of heat; in the case of weight between density, or weight per unit volume and total weight. For local motion the distinction is between velocity (or motion) at a given instant (instantaneous velocity) and total motion over a period of time, i.e. distance covered.

William Heytesbury distinguishes between uniform and difform (non-uniform) motion. In the case of difform motion he says:‡

> In nonuniform motion, however, the velocity at any given instant will be measured (*attendetur*) by the path which *would* be described by the most rapidly moving point

† Clagett, M., *The Science of Mechanics*, pp. 205, 210.
‡ *Ibid.*, p. 236.

> if, in a period of time, it were moved uniformly at the same degree of velocity (*uniformiter illo gradu velocitatis*) with which it is moved in that given instant, whatever [instant] be assigned.

The most notable single achievement of the Merton School was the establishment of the mean speed theorem for uniformly accelerated (uniformly difform) motion, i.e. $s = (v_1+v_2)t/2$, where s is the distance covered in time t and v_1 and v_2 are the initial and final velocities. Various proofs of this theorem were presented, some purely arithmetical and others depending on some skilful manipulation of infinite series. Although in some cases velocities (or intensions) were represented by single straight lines and Richard Swineshead actually uses a geometric analogy[†] to explain intension and remission of qualities it was not until knowledge of the Merton College work reached France and Italy that it received the full benefits of geometric representation and so linked kinematics with the geometry of straight and curved lines.

Nicole Oresme first expounded the technique of two-dimensional representation some time between the years 1348 and 1362 although it seems from a treatise of Giovanni di Casali that the latter was already using graphing techniques in 1346.[‡] It is not my purpose here to discuss the question of priorities nor even to consider how far such methods constituted an anticipation of the invention of coordinate geometry by Descartes and Fermat.[§] There are, however, certain aspects of this form of geometric representation such as the relation between ordinate, gradient and area, which concern us deeply.

Oresme describes the geometric representation of all kinds of qualities in the following terms:[||]

> Every measurable thing except numbers is conceived in the manner of continuous quantity. Hence it is necessary for the measure of such a thing to imagine points, lines, and surfaces—or their properties—in which, as Aristotle wishes, measure or proportion is originally found.... Hence every intension which can be acquired successively is to be imagined by means of a straight line erected perpendicularly on some point or points of the (extensible) space or subject of the intensible thing.

The extension in time of successive things is to be called *longitude* and represented by a horizontal line. Lines of intension can be imagined "going in any direction you wish" but in practise it is more convenient to erect them perpendicular to the lines of extension and to call them lines of *latitude*. The *quan-*

Clagett, M., *The Science of Mechanics*, pp. 306, 335–6.

‡ *Ibid.*, pp. 332–3.

§ Cf. Wieleitner, H., Über den Funktionsbegriff und die graphische Darstellung bei Oresme, *Bibl. Math.* (3) 14 (1914), pp. 193–248.

|| Given from Clagett, *op cit.*, p. 347.

tity of any linear quality is to be imagined by a surface the base of which is the extension or longitude of the quality and the altitude of which is the latitude or intension of the quality (see Fig. 2.17). Examples are then given of the geometric *forms* of various qualities. Qualities which are *uniform* and

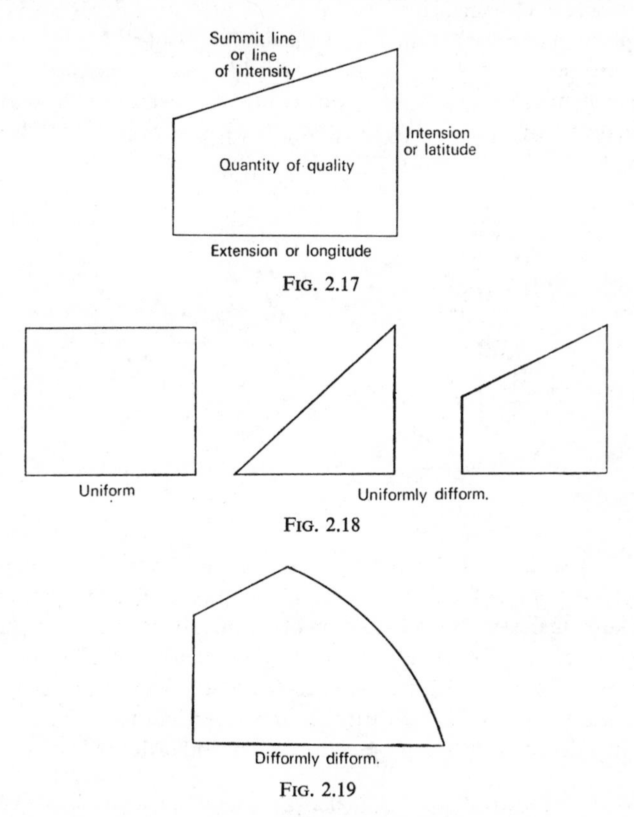

FIG. 2.17

FIG. 2.18

FIG. 2.19

uniformly difform are represented by rectangles, triangles and trapezia (see Fig. 2.18). If the *line of intension* or *summit line* is a curve or is made up of more than one line the quality is *difformly difform* (see Fig. 2.19).

In an interesting passage† Oresme discusses the possibility of using a three-dimensional coordinate system to represent a quality having extension in space as well as in time. These two extensions are represented by perpendicular

† *Ibid.*, p. 358.

axes in a horizontal plane and the corresponding intensities erected at each point in the plane (see Fig. 2.20).

Most important of all, perhaps, is the section in which Oresme describes the "acquisition and measure of quality and velocity". The acquisition of a linear quality is to be imagined as the motion of a point flowing over a line, the acquisition of a surface quality by the motion of a line "dividing the part of the surface altered from the other part which has not yet been altered". Similarly the acquisition of a corporeal quality is to be imagined by the motion of a surface "dividing the part altered from the part not yet altered".†

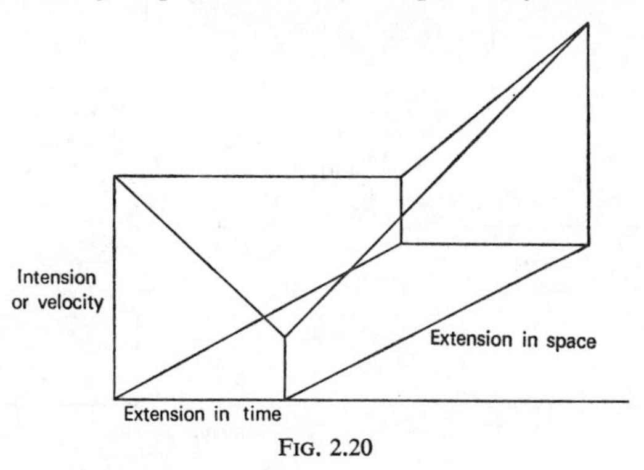

Fig. 2.20

Oresme was already using some of the concepts and terminology subsequently to become so familiar as Newton's *fluxions*.‡ Oresme, however, was much more interested in the final *form* of the quality, triangular, rectangular or trapezoidal, and its relation to the varying intensity than in considering either area or ordinate in terms of extension or abscissa. A rectangle corresponds to a uniform velocity, a triangle or a trapezium to uniform acceleration. From the latter the mean speed theorem is readily derived (see Fig. 2.21).

† Given from Clagett, M., *The Science of Mechanics*, p. 378. "Acquisitio autem intensiva qualitatis punctualis ymaginanda est per motum puncti continue ascendentis super punctum subiectivum. Acquisitio vero intensiva qualitatis linearis ymaginanda est per motum linee perpendiculariter ascendentis super lineam subiectivam, et suo fluxu vel ascensu derelinquentis superficiem, per quam designatur qualitas acquisita." Furthermore, the intensive acquisition of a punctual quality is to be imagined by the motion of a point continually ascending above the subject point, while the intensive acquisition of a linear quality is to be imagined by the motion of a line perpendicularly ascending above the subject line; and by its flux or ascent it describes the surface by which the acquired quality is designated. (Given from Clagett's translation, p. 358.)

‡ The Merton Calculators also used the terms *fluxus* and *fluens*. With Oresme, however, the terms were first related to specific geometric representation.

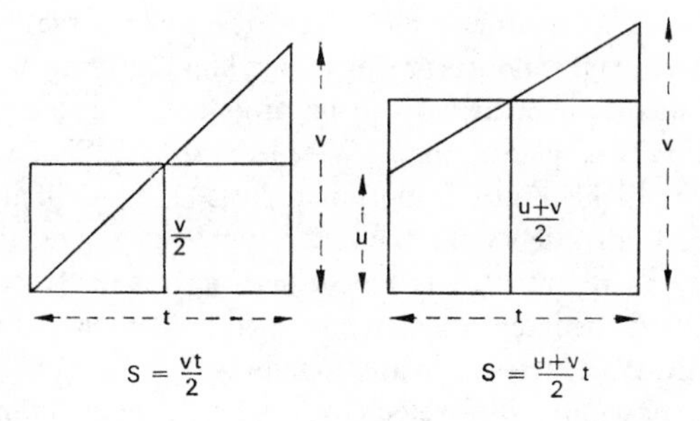

FIG. 2.21

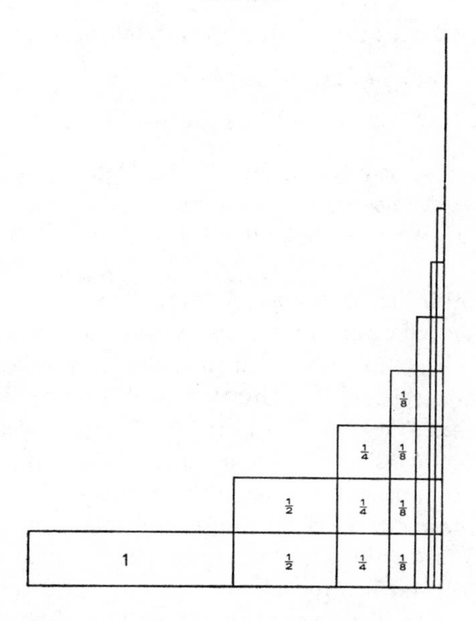

Summing by vertical columns we have

$$4 = 1+2\cdot\tfrac{1}{2}+3\cdot\tfrac{1}{4}+4\cdot\tfrac{1}{8}+5\cdot\tfrac{1}{16}+\ldots$$

Summing by horizontal rows we have

$$4 = 2+1+\tfrac{1}{2}+(\tfrac{1}{2})^2+(\tfrac{1}{2})^3+(\tfrac{1}{2})^4+\ldots$$

FIG. 2.22

This concern with the surface quality (or area under curve) is so pronounced with Oresme that it is not surprising to find him enquiring under what conditions equal qualities (or areas) will be produced. A rectangle, altitude 1 unit, length 2 units, is placed upon a base line AB (see Fig. 2.22). An equal rectangle is then divided into proportional parts ($1, \frac{1}{2}, \frac{1}{4}, \frac{1}{8}$, and so on) and the parts erected separately upon the first rectangle in a vertical column. The surface quality so represented is then equal in area to twice the original rectangle although the intensity, as represented by the increasing altitudes of successive vertical rectangles, increases to infinity.

Hence, if a body move with velocity v, $2v$, $3v$, $4v$, and so on, in successive time intervals, t, $t/2$, $t/4$, $t/8$, . . . , the total distance covered will not exceed $4vt$, even though the velocity ultimately becomes infinite,

i.e. $$4vt = vt+2vt/2+3vt/2^2+4vt/2^3+\ldots$$

or, $$4 = 1+2.1/2+3.(1/2)^2+4.(1/2)^3+5.(1/2)^4+\ldots$$

This type of series had also figured in the Merton College demonstrations but Oresme's geometric summation was a new development and led to further exploration of infinite series in the fifteenth century by Alvarus Thomas† and others.

Most significant in Oresme's approach is the linking of two fundamental notions: firstly, the representation of a physical quality by a surface; secondly, the concept of a surface as the *flux* or motion of a line parallel to itself. There seems no reason to suppose that Oresme regarded a line as the sum of its points or a surface as the sum of its lines. This is a static idea and the whole emphasis with Oresme is on the description of lines, surfaces and solids by motion. It was not necessary for him to postulate an infinity of lines drawn upon a surface; he argues, in fact, that if an ordinate be erected at *any* point on the base line proportional to the intensity of the quality then the total area under the summit line (or line of intensity) will correspond with the quantity of the quality. Since he concerns himself in detail only with rectilinear figures the areas can be calculated directly without the necessity for infinitesimal techniques. In fact, by establishing this correspondence between certain properties of physical qualities and rectilinear geometric forms Oresme is in a sense developing a means of avoiding integration rather than a means of integrating. The linking, however, of a whole range of physical phenomena with the determination of a surface area gave a new emphasis to the role of

† See Wieleitner, H., Zur Geschichte der unendlichen Reihen im christlichen Mittelalter, *Bibl. Math.* (3) 14 (1914), pp. 150–68.

geometry and particularly to the central problem of quadrature in the field of natural science.

If Oresme himself did not find it necessary to draw lines of intensity everywhere on the base line others who followed him may well have done so; for most of the sixteenth-century editions of his work and that of the Merton Calculators are illustrated with innumerable diagrams exhibiting an endless variety of geometric *forms* covered with arrays of parallel ordinates.† Although no substantial advances appear to have been made some exploration of curvilinear *forms* inevitably took place. Jacobus de Sancto Martino‡ in discussing a curvilinear *form* (a circular segment) makes some useful comments (see Fig. 2.23).

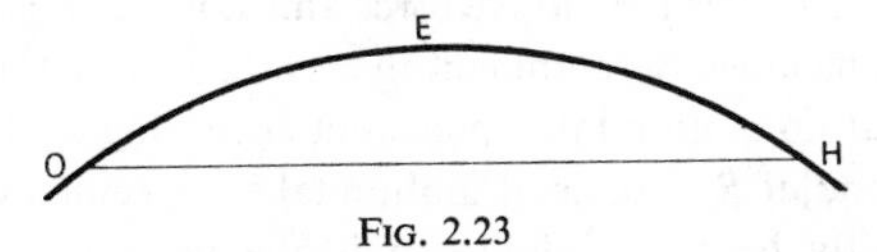

FIG. 2.23

He notes that the velocity is greatest at the highest point E, that acceleration (intension) ceases at E and that deceleration (remission) begins at E, that acceleration decreases from O to E and that deceleration increases from E to H. In fact he is drawing conclusions from the form of the curve and the variation of its gradient and applying these conclusions to the velocity–time relation which the curve represents. Jacobus also discussed a quality which he calls uniformly difformly difform in which the lines of intension vary exponentially. For some reason he appears to have thought that a circular arc satisfied this condition.

The idea of a function varying exponentially was not unknown in the fifteenth century and arose in the first instance from a particular relation developed by Thomas Bradwardine.

2.7. The Function Concept in the Fourteenth and Fifteenth Centuries

Although kinematics is the branch of mechanics most clearly related to geometry its development cannot be considered in isolation and apart from the general discussion of functional relationships in the physical world. Much of the background of mediaeval discussion remained purely speculative and

† *Suiseth, Ricardus; Calculator* (ed. Trinchavellus, V.), Venice, 1520. See also *De latitudinibus Nicholai Horen*, Venice, 1520.

‡ Clagett, M., *The Science of Mechanics*, pp. 392–401.

unsupported by any kind of empirical investigation. The fact that no experiments were ever done was not, however, any particular disadvantage mathematically. Mathematics, unlike physics, is not concerned with the applicability of any particular theory but rather with the possibilities of any new relations suggested by it.

Perhaps the most influential contribution of this kind was that made by Thomas Bradwardine in his *Treatise on Proportions*† (1328). In this treatise Bradwardine reviews in considerable detail Aristotle's views on the nature of forces causing motion. The basic idea running through all of Aristotle's scattered statements on motion‡ is that speed is proportional to motive force and inversely proportional to resistance, i.e. in modern notation, $v \propto F/R$, where v is the velocity ($\propto s/t$), F is the motive force and R is the resistance.

Two critical difficulties were implicit in this theory: first, if $F = R$, experience suggests that no motion takes place but according to the theory motion must occur; second, if $R = 0$, i.e. if motion takes place in a vacuum, then the velocity apparently becomes infinite. Aristotle was not unaware of these difficulties and dealt with them by specifying that $v = 0$, when $F = R$ and denying the possibility of the existence of a void.

In the sixth century John Philoponus criticised this view and put forward an alternative theory in which the time taken to move through a given distance was determined in part by the motive force and in part by the resistance R. This was later interpreted as implying, $v \propto (F-R)$. Despite the fact that this afforded a not unreasonable interpretation of experience ($R = 0$, $v \propto F$, $R = F$, $v = 0$), Bradwardine rejected it in favour of his own theory in which he asserted that what Aristotle had in mind was the idea that velocity increases arithmetically as the ratio F/R increases geometrically.

Thus, if u_0 be the velocity corresponding to a particular ratio λ of force to resistance, then a ratio of λ^2 will be necessary in order to produce a velocity of $2u_0$ and ratios of λ^3, λ^4, λ^5 to produce velocities of $3u_0$, $4u_0$, $5u_0$, respectively.

Let $F_0/R_0 = \lambda$, and let $F/R = \lambda^n$, then $v = nu_0$ and $F/R = (F_0/R_0)^{v/u_0}$.
Hence, $v/u_0 = \log_a(F/R)$, where $a = F_0/R_0$.

Whether sound or otherwise in its application to the physical world the concept of a function varying exponentially was of value to mathematics. In this sense Bradwardine's *Tractatus* was of major importance. Very many works appeared in which the theory of proportions was developed in rela-

† Bradwardine, T., *Tractatus de proportionibus* (ed. Crosby, H. L.), Madison, 1955.
‡ *Phys.* VII. 1. 241[b], VII. 2. 243[a], VII. 5. 249[b].

tion to this function. In particular Nicole Oresme in his *Algorismus proportionum* explored the rules for manipulating functions exponentially.[†] Although he had no satisfactory way of writing powers, he shows, by the examples he gives, that he fully understands the rules for handling both fractional and integral exponents.

2.8. Conclusion

The works of Archimedes constituted the principal source and inspiration for much of the seventeenth-century geometry which played such an important role in the development of the infinitesimal calculus. In consequence it is technically possible to unravel this complex network of processes without any direct reference to the mediaeval period. Nevertheless, in order to achieve more than a superficial understanding of the contributions of Stevin, Galileo, Kepler and Cavalieri[‡] it is necessary to take into consideration all the important currents of thought which developed during the Middle Ages and which ultimately led to the clarification of a number of important notions. For many centuries the works of Archimedes were completely unknown in the Latin West[§] and it was for this reason that other ideas had a chance to establish themselves and so to pervade mediaeval thinking that their eventual application in the field of mathematics became inevitable. The development in the seventeenth century of new and original ideas and techniques and the discovery of results beyond those known to Archimedes were, at least in some measure, due to the introduction and application of these new concepts.

† Curtze, M., *Der "Algorismus proportionum" des Nikolaus Oresme*, Berlin, 1868.

‡ Cf. Goldbeck, E., Galileis Atomistik und ihre Quellen, *Bibl. Math.* (3) 3 (1902), pp. 84–112. Crombie, A. C., *op. cit.*, pp. 23–30. Wiener, P. P., The tradition behind Galileo's methodology, *Osiris* 1 (1936), pp. 733–46.

§ Clagett, M., Archimedes in the Middle Ages: the *De mensura circuli*, *Osiris* 10 (1952), pp. 587–618.

CHAPTER 3

Some Centre of Gravity Determinations in the Later Sixteenth Century

3.0. Introduction

During the sixteenth century Greek mathematical texts became more widely available and the publication of Latin translations of Euclid's *Elements* and the *Conics* of Apollonius accompanied by critical commentaries[†] made it possible for mathematicians to acquire the basic skills and techniques essential for complete mastery of the works of Archimedes. Such was the respect and admiration generated by these works that a desire arose to extend and develop this field of study by emulating the methods of the great Greek geometer. This was accompanied by a corresponding tendency to decry the influence of mediaeval thought, to deplore unprofitable speculation on the nature of the infinite and to concentrate on purely technical manipulations based on the *exhaustion* method of Archimedes.

In particular, the Archimedean treatise *On the Equilibrium of Planes* seems to have excited the admiration of sixteenth-century scholars and suggested an extension, by means of the same methods, to the determination of the centres of gravity of solids. Although Archimedes makes use of various results for the centres of gravity of solids[‡] there was no systematic work available in the sixteenth century which included any proofs. Francesco Maurolico was among the first to apply himself to repair the deficiency.[§]

† Cf. Hofmann, J. E., *Geschichte der Mathematik*, Berlin, 1953–7, I, pp. 107–10.

‡ See Heath, *Archimedes*, p. 265.

§ Maurolico, F., *Admirandi Archimedis Syracusani monumenta omnia mathematica, quae extant*, Palermo, 1685. *De momentis aequalibus, liber quartus*, pp. 156–80. This work, written in 1548, was not published until 1685 but seems to have been accessible to other mathematicians including Commandino. See also Napoli, F., Intorno alla vita ed ai lavori di Francesco Maurolico, *Bull. Boncamp.* 9 (1876), pp. 1–156.

3.1. Francesco Maurolico (1494–1575)

The application of the method of moments to centre of gravity determinations was important for the development of the calculus in that it provided a valuable field for the deployment of infinitesimal methods through which the concept could be approached both arithmetically and geometrically. It linked volumetric and area determinations, thereby providing a basis for the geometric transformations which played such a fundamental role in integration before the development of any general concept of function.† An important paper of Leibniz, the *Analysis tetragonistica ex centrobarycis*,‡ shows clearly that these processes were not only important as anticipations of the calculus but also that they played a significant role in its actual invention.

Maurolico's determination of the centre of gravity of a conoid§ is of interest in that, through the method of moments, the volume of a conoid is brought into association with the area of a triangle.

Maurolico first proves|| that if a system of weights, I, K, M, N, be suspended from a balance arm AB at points A, E, F, B, respectively, and another set of weights, O, P, Q, R, be suspended from the arm CD at points C, G, H, D, and if,

$$\frac{I}{O} = \frac{K}{P} = \frac{M}{Q} = \frac{N}{R},$$

and also, $AE = CG$, $EF = GH$, $BF = DH$ (see Fig. 3.1), then the centres of gravity V and Y are equidistant from B and D, i.e. $VB = YD$.

The axis AN $(= nh)$ of a conoid†† is then divided into n parts, each equal to h, and cylinders inscribed and circumscribed in the usual way (see Figs. 3.2 and 3.3). Let the volumes of the inscribed cylinders be denoted by $i_1, i_2, i_3, i_4, \ldots, i_{n-1}$, and let their respective radii be $r_1, r_2, r_3, \ldots, r_{n-1}$. Since the cylinders are of equal altitude their volumes are proportional to the squares of their radii, i.e.

$$\frac{i_1}{r_1^2} = \frac{i_2}{r_2^2} = \frac{i_3}{r_3^2} = \ldots \frac{i_{n-1}}{r_{n-1}^2}.$$

† Wieleitner, H., Das Fortleben der Archimedischen Infinitesimal methoden bis zum Beginn des 17 Jahrh., *Quellen und Studien*, Studien (B), 1 (1931), pp. 201–20. See also Stein, W., Der Begriff des Schwerpunktes bei Archimedes, *Quellen und Studien*, Studien (B) 1 (1931), pp. 221–4.

‡ *Cat. crit.*, Nos. 1086, 1089, 1090, 1092, 1106.

§ *De momentis aequalibus*, pp. 172–80.

|| *Ibid.*, Prop. XXI, p. 175.

†† *Ibid.*, Prop. XXII, pp. 175–6.

Since, however, BNC is a paraboloid, $r_1^2 \propto h$, $r_2^2 \propto 2h$, $r_3^2 \propto 3h$, ..., then $i_1/1 = i_2/2 = i_3/3 = \ldots$.

Maurolico then shows that the areas of the rectangles inscribed in a triangle BNC with altitude AN (divided into n parts each equal to h) also increase in

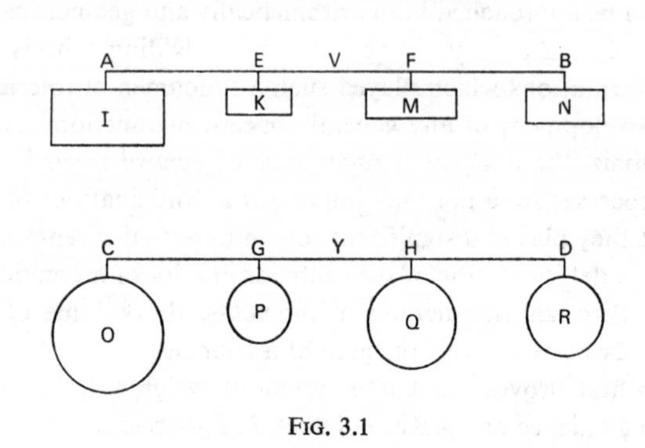

FIG. 3.1

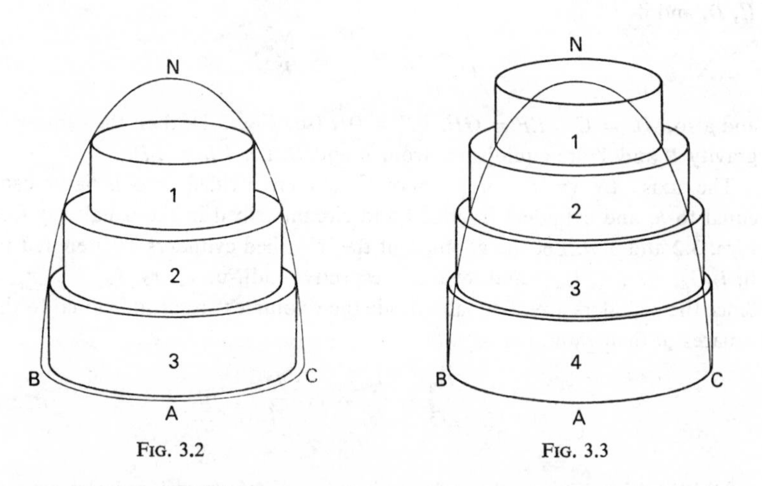

FIG. 3.2 FIG. 3.3

the ratios, 1 : 2 : 3 : 4 : ... (see Fig. 3.4). Hence the weights of the inscribed cylinders balanced on the lever arm AN can be replaced by the corresponding rectangles balanced on the arm AN. Similar considerations apply for the circumscribed figures and therefore any results established for the triangle apply also for the conoid. Maurolico then goes on to investigate the centres of gravity of the figures inscribed and circumscribed to the triangle by a somewhat novel method.†

First, consider a row of n small triangles, each of height b and base h, suspended on a lever arm AN ($= nh$) (see Fig. 3.5). From considerations of

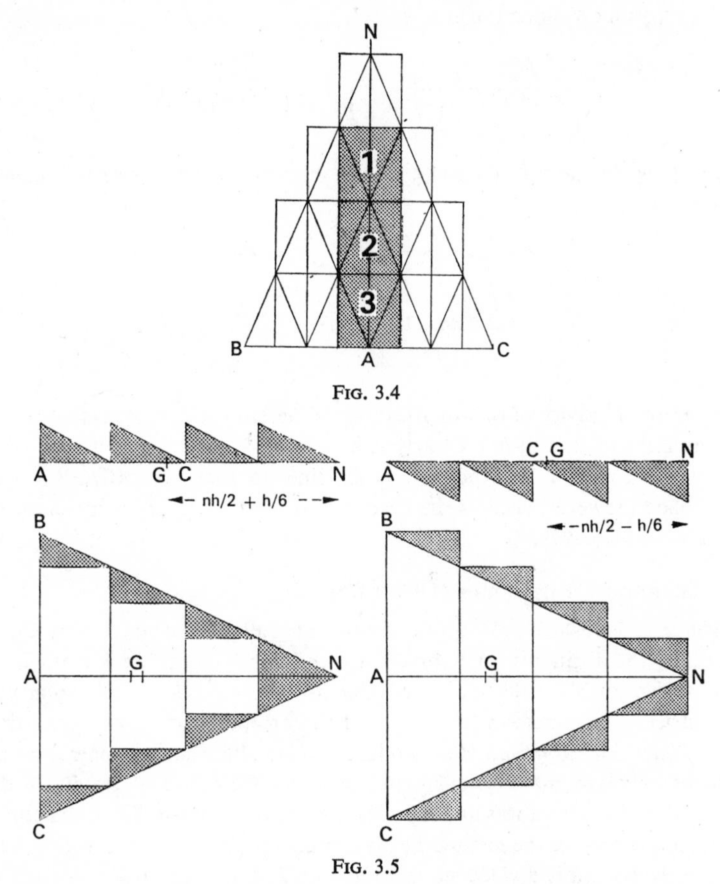

Fig. 3.4

Fig. 3.5

symmetry it is immediately clear that the centre of gravity is at a distance $h/6$ from the mid-point C of AN, i.e. if G be the centre of gravity, then $GN = nh/2 \pm h/6$, according to the arrangement of the triangles. This result remains unaltered in any vertical displacement. Now, in the triangle BNC,

† *De momentis aequalibus*, Props. XIX, XX, pp. 172–4.

altitude $AN(=nh)$, the inscribed and circumscribed figures can be considered as formed by the subtraction and addition of small triangles from the original triangle. If Δ denote the area of ΔBNC then the areas of the inscribed and circumscribed figures are respectively, $\Delta(1-1/n)$ and $\Delta(1+1/n)$. Taking moments about N along the arm AN, we have,

$$\Delta \times \frac{2}{3}nh \mp \frac{\Delta}{n}\left(\frac{nh}{2} \pm \frac{h}{6}\right) = \Delta\left(\frac{n\mp 1}{n}\right) \times GN,$$

where G is the centre of gravity of the inscribed or circumscribed figure. Hence,

$$h\left(\frac{2}{3}n \mp \frac{1}{2} - \frac{1}{6n}\right) = \frac{n\mp 1}{n} \times GN,$$

and

$$GN = \frac{4n^2 \mp 3n - 1}{6(n\mp 1)}h = \frac{4n \pm 1}{6}h = \frac{2}{3}nh \pm \frac{h}{6}.$$

The centre of gravity of the inscribed figure therefore differs from that of the circumscribed figure by $h/3$. Clearly as $n \to \infty$, $h \to 0$ and the centre of gravity of the conoid is at a point G in the line AN such that $AG/GN = 1/2$. Maurolico, however, accepts the necessity for the customary completion by *reductio ad absurdum.*

3.2. Federigo Commandino (1509–1575)

The first published work on the centres of gravity of solids was by Commandino, the distinguished commentator and translator of Greek mathematical works. In his little book the *Liber de centro gravitatis solidorum*† he uses strictly Archimedean methods and the orthodox proof by exhaustion and *reductio ad absurdum.* He introduces no simplification or generalisation and the only new material of interest concerns the centre of gravity of the conoid.‡ Although he was undoubtedly influenced by Maurolico to whom he refers in the preface the method he gives differs from that of Maurolico. The axis of the conoid is divided successively into 2, 4, 8, ..., equal parts and at each stage the positions of the centres of gravity of the inscribed and circumscribed solids are determined. Let these be denoted by $g_2, g_4, g_8, \ldots$ and $G_2, G_4, G_8, \ldots$, respectively. Commandino shows that $g_2g_4 = G_2G_4$, $g_4g_8 =$

† Commandino, F., *Liber de centro gravitatis solidorum*, Bologna, 1565.
‡ *Ibid.*, pp. 40–45.

$= G_4G_8$, and so on. Thus, if the centre of gravity of the conoid lies between g_2 and G_2, g_4 and G_4, g_8 and G_8, . . ., it lies at the mid-point of g_2G_2, g_4G_4, g_8G_8 (see Figs. 3.6 and 3.7).

The sixteenth-century use of the method of moments was a clumsy affair in which the system was reduced by manipulation of the weights in pairs according to the rule, $M_1/M_2 = d_2/d_1$. The full implication of Commandino's method becomes clear when moments are taken in the usual way about the vertex, V (see Fig. 3.6).

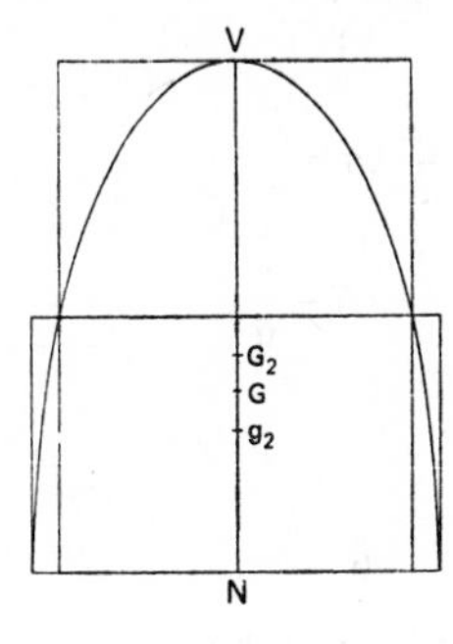

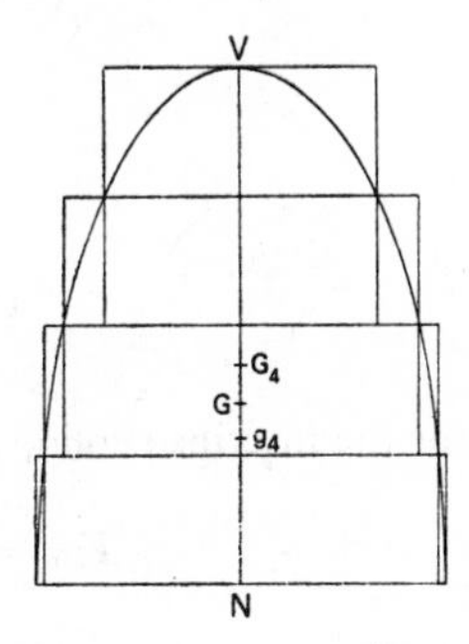

FIG. 3.6

G

N g2 g4 G4 G2 V

FIG. 3.7

Let n be the number of equal divisions of the axis VN, let h be the altitude of each cylinder and let,

$$I_n = i_1+i_2+i_3+\ldots i_{n-1},$$

$$C_n = c_1+c_2+c_3+\ldots c_n,$$

where, as previously, $$\frac{i_1}{1} = \frac{i_2}{2} = \frac{i_3}{3} = \ldots \frac{i_{n-1}}{n-1}$$

and $$\frac{c_1}{1} = \frac{c_2}{2} = \frac{c_3}{3} = \ldots \frac{c_n}{n}.$$

The centres of gravity of i_r and c_r are at distances $(rh+h/2)$ and $(rh-h/2)$ respectively from the vertex V.

Hence, taking moments about V, we have, for the circumscribed solid,

$$\sum_1^n r(rh-h/2) = VG_n \sum_1^n r,$$

$$\frac{h\sum_1^n r^2}{\sum_1^n r} - \frac{h}{2} = VG_n$$

$$= \frac{h(2n+1)}{3} - \frac{h}{2}$$

$$= \frac{2}{3}nh - \frac{h}{6},$$

and
$$VG_n = \frac{2}{3}VN - \frac{h}{6}.$$

Similarly for the inscribed solid, we have,

$$Vg_n = \frac{2}{3}VN + \frac{h}{6}.$$

Hence, the centre of gravity of the conoid is given by

$$VG = \frac{Vg_n + VG_n}{2} = \frac{2}{3}VN.$$

Commandino, however, was satisfied to determine the positions of the centres of gravity only in a limited number of cases: he did not concern himself with convergence and completed the proof by *reductio ad absurdum*. It was left to the Flemish mathematician, Simon Stevin, to take the next step forward and to stipulate conditions for the direct passage to the limit.

3.3. Simon Stevin (1548–1620)

Simon Stevin of Bruges was perhaps the most original mathematician of the second half of the sixteenth century† and his contributions both to pure and applied mathematics were notable. His work was first systematically studied in the present century by Bosmans.‡ As the founder of scientific and technical

† Sarton, G., Simon Stevin of Bruges, *Isis* 21 (1934), pp. 241–303.

‡ Bosmans, H., Le calcul infinitésimal chez Simon Stevin, *Mathésis* 37 (1923), pp. 12–18, 55–62, 105–9. Sur quelques exemples de la méthode des limites chez Simon Stevin, *Annales de la Société Scientifique de Bruxelles* 37 (1913), pp. 171–99.

Dutch Stevin was opposed to the exclusive use of Latin in scientific writing and after 1583 he published only in Dutch. In consequence the *Mechanics* and the *Hydrostatics*† were little known outside the Low Countries until the preparation of French and Latin editions by Girard and Snell‡ respectively. By this time Stevin's methods were half a century old but even so, they were not without influence on Kepler, Cavalieri and Grégoire de Saint-Vincent.

Although Stevin was essentially concerned to present mathematics as a practical tool in a form readily comprehensible to the engineer and scientist he was nevertheless always careful to distinguish between proof, example and illustration. He was well acquainted with the works of his predecessors and was undoubtedly fully conscious of the significant methodological changes he was making. The second book of the *Mechanics* is devoted to centre of gravity determinations and the methods he used indicate a deliberate and systematic effort to modify and simplify the Archimedean proof structure. Probably the most radical step which he took was the abandonment as a general routine of the completion of all propositions by a *reductio* proof. Unlike the Greeks Stevin was prepared to make use of the general consideration that any two magnitudes, the difference between which can be shown to be less than any assigned magnitude, are equal to one another.

The direct way in which he applied this principle can best be illustrated by looking at one or two simple examples. To prove that the centre of gravity of a triangle *ABC* (see Fig. 3.8) lies in the median *AD* he draws the inscribed figure in the usual way. Since the centre of gravity of each of the inscribed parallelograms lies in *AD* then the centre of gravity of the inscribed figure lies in *AD*. He continues as follows:§

> Now as here three quadrilaterals have been inscribed in the triangle, so an infinite number of such quadrilaterals can be inscribed therein, and the centre of gravity of the inscribed figure will always be (for the reasons mentioned above) in the line *AD*. But the more such quadrilaterals there are, the less the triangle *ABC* will differ from the inscribed figure of the quadrilaterals. For if we draw lines parallel to *BC* through the middle point of *AN, NM, ML, LD*, the difference of the last figure will be exactly half the difference of the previous figure. We can therefore, by infinite approximation, place within the triangle a figure such that the difference between the latter and the triangle shall be less than any given plane figure, however small. From which it follows that,

† *De Beghinselen der Weeghconst*, Leyden, 1586. *De Beghinselen des Waterwichts*, Leyden, 1586. All references are given from *The Principal Works of Simon Stevin* I (ed. Dijksterhuis, E. J.), Amsterdam, 1955.

‡ *Les Oeuvres mathématiques de Simon Stevin de Bruges* (ed. Girard, A.), Leyden, 1634. *Hypomnemata mathematica* (ed. Snell, W.), Leyden, 1608.

§ *The Principal Works of Simon Stevin*, I, pp. 229–31.

taking AD to be centre line of gravity, the apparent weight of the part ADC will differ less from the apparent weight of the part ADB than any plane figure that might be given, however small....

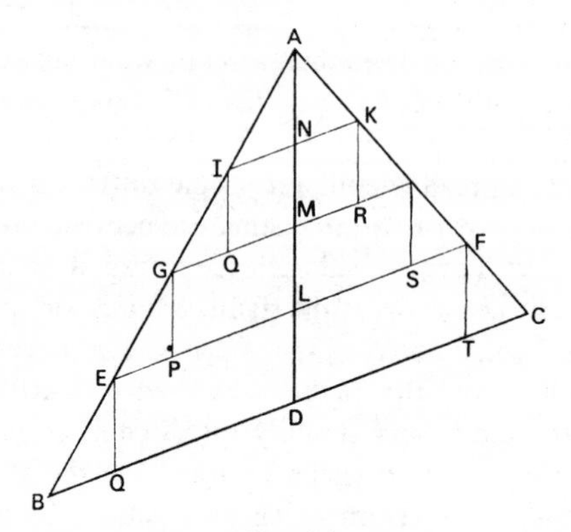

FIG. 3.8

If the area of the triangle ABC be denoted by Δ and that of the inscribed figure by I_n, where n is the number of inscribed quadrilaterals, then $\Delta - I_n = \Delta/n$. By Euclid X. 1, this can be made less than any assigned quantity, however small, i.e.

$$\lim_{n \to \infty} \Delta/n = 0.$$

Now, if Δ_1, Δ_2, denote the areas of triangles ABD and ADC, respectively, we have

$$\Delta - I_n = \Delta/n,$$

$$\Delta_1 - \frac{I_n}{2} < \Delta/n,$$

$$\Delta_2 - \frac{I_n}{2} < \Delta/n,$$

and $|\Delta_1 - \Delta_2| < \Delta/n$, so that, given any plane area δ, however small, it is possible by suitable choice of n to make $|\Delta_1 - \Delta_2| < \delta$.

Instead of the usual *reductio* proof Stevin submits the following:

A. Beside any different apparent gravities there may be placed a gravity less than their difference;
O. Beside the present apparent gravities ADC and ADB there cannot be placed any gravity less than their difference;
O. Therefore the present apparent gravities ADC and ADB do not differ.

Stevin is utilising the symbolism of ancient formal logic to emphasise the importance of the step he is taking. The argument can be rephrased as follows:

1. If two quantities differ they differ by a finite quantity.
2. These quantities differ by less than any finite quantity.
3. These quantities do not differ.†

Unfortunately Stevin was not prepared to present this important innovation in the form of a general proposition so that it might be cited whenever subsequently required. He repeated it *in extenso* time and time again‡ just as Archimedes did with the *reductio* proof. Even so, Stevin should be given credit for the insight and understanding he showed in emphasising by the use of the syllogistic form the concept of a direct passage to the limit and the significant divergence from the Archimedean proof structure. Although the importance of his work was recognised in the seventeenth century, by the time it became known the influence of Cavalieri had become so great that a mathematically rigorous approach to the limit had ceased to be regarded as practicable.

Less important but nevertheless of interest are Stevin's arithmetical illustrations which he calls *proofs by numbers*. An easy example of such a proof is found in the *Hydrostatics*§ when he sets out to find the total pressure on a vertical wall. Consider a square *ACDE* placed vertically with *AC* in the surface and divided into n equal horizontal strips by lines parallel to *AC* (see Fig. 3.9). Let $AC = CD = 1$ foot, then the area of $ACDE = 1$ ft^2 and the area of each strip is $1/n$ ft^2. Now, if such a strip were to be placed horizontally at a depth of h ft below the surface a weight of $(h\times 1/n)$ ft^3 of water would rest upon it. Placed vertically, however, we have

$h_1\times 1/n <$ pressure on strip $< h_2\times 1/n$, where h_1 is the distance of the top of the strip and h_2 is the distance of the bottom of the strip, below the surface.‖

† $(p \rightarrow q) \rightarrow (\sim q \rightarrow \sim p)$.

‡ See *De Beghinselen der Weeghconst*, Props. X, XV, XVI, XVIII, XXIX.

§ *De Beghinselen des Waterwichts*, Prop. XI, Ex. IV. Stevin says: "We have given above three examples with a mathematical proof, which indeed explains the cause more perfectly than any other, but since there is no harm in giving the proof by means of numbers in order to make everything clearer, we shall give this 4th example by means of numbers." *The Principal Works of Simon Stevin* I, p. 433.

‖ The *pressure* is measured in terms of the weight (or volume) of water resting (or pressing) against the strip.

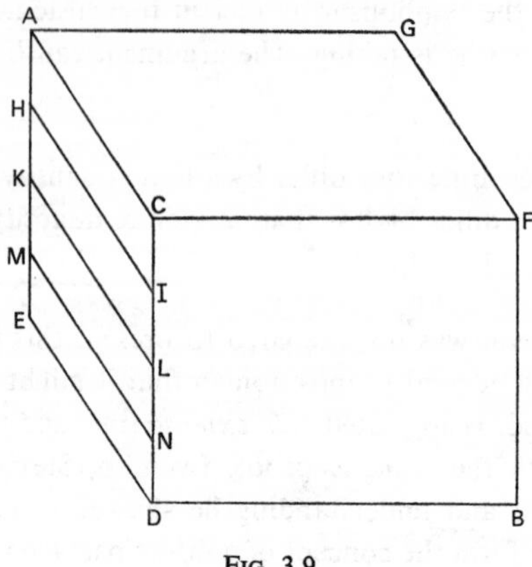

FIG. 3.9

For the rth strip we have, in general, $h_1 = (r-1)/n$, $h_2 = r/n$. Hence, adding for all such strips,

$$\frac{1}{n^2}(0+1+2+\ \dots\ +(n-1)) < \text{total pressure on } ACDE$$
$$< \frac{1}{n^2}(1+2+3+\ \dots\ +n),$$

and
$$\frac{n(n-1)}{2n^2} < \text{total pressure} < \frac{n(n+1)}{2n^2},$$
$$\frac{1}{2}-\frac{1}{2n} < \text{total pressure} < \frac{1}{2}+\frac{1}{2n}.$$

Stevin works out numerical results for 4, 10 and 1000 divisions and notes that the greater the number of parts the nearer the result is to $\frac{1}{2}$. Further, he observes, "such a thing can be proved with regard to any given part, however small".† Hence, he is able to conclude that, since the difference between the total pressure and $\frac{1}{2}$ can be made as small as we please and consequently smaller than any finite quantity, there exists no difference between the total pressure and $\frac{1}{2}$, i.e. $\lim_{n\to\infty} (\frac{1}{2n}) = 0$.

This arithmetic is refreshingly direct and simple compared with the usual

† *The Principal Works of Simon Stevin* I, p. 437.

type of proof by proportions but Stevin did not intend it to replace the formal proof. Although he was always concerned to provide enough understanding to make of mathematics a serviceable tool his formal proofs were intended to comply with the utmost standards of mathematical rigour.†

3.4. Luca Valerio (1552–1618)

Valerio was yet another mathematician who, perceiving that Archimedes had neglected the centres of gravity of solids and that Commandino had considered only a limited number of simple cases, set himself to repair this deficiency whilst at the same time attempting a significant modification in method. A skilful geometer and a virtuoso in the theory of proportions, Valerio worked entirely within the framework of Euclidean method and in consequence was widely read and greatly admired in the seventeenth century. As a Neapolitan who taught in Rome Valerio is unlikely to have had the opportunity to read the works of Stevin before the publication of his own *De centro gravitatis solidorum* in 1604.‡

Because Valerio's work is couched entirely in geometric language the generalisations he makes appear at first sight less striking than those of Stevin. Nevertheless the attempt he makes to dispose of the necessity for constant repetition of *reductio* proofs is, in some respects, more thorough and far-reaching than that of Stevin. The standard *exhaustion* proof contained two major steps, both of which Valerio attempted to generalise in a series of important theorems so that for a whole class of convex curves and solids it was no longer necessary to establish results by special methods.

The first generalisation occurs in Proposition VI when Valerio considers a convex plane figure enclosed by a curve, in which parallelograms have been inscribed and circumscribed in the usual way, and establishes that, in general, the difference between the areas of the inscribed and circumscribed figures is equal to the area of the parallelogram on the base (see Fig. 3.10). Let BAC be any plane figure in which AD is a diameter, BC is a chord, and in which chords parallel to BC decrease towards A.§ For the inscribed and circum-

† Bosmans (see Le calcul infinitésimal, *Mathésis* 37 (1923), p. 58) compares Stevin with Cavalieri. "Mais, y apporter ce perfectionnement essentiel, c'est en revenir purement et simplement à la méthode de Stevin! On regarde parfois, et non sans raison, Cavalieri comme le père de l'analyse infinitésimale moderne. Stevin fut à tout le moins un ancêtre, qui sut joindre la simplicité à la rigeur."

‡ Valerio, L., *De centro gravitatis solidorum*, libri tres, Rome, 1604; see Preface, I.

§ *Ibid.*, Prop. VI, Book I, p. 14: "Omni figurae circa diametrum in alteram partem deficienti..."

scribed figures we have

$$I_n = i_1+i_2+i_3+ \ldots i_n,$$
$$C_n = c_1+c_2+c_3+ \ldots c_n,$$

and

$$C_n-I_n = (c_1-i_1)+(c_2-i_2)+ \ldots (c_n-i_n)$$
$$= c_n.$$

Hence, by suitable choice of *n* this difference can be made less than any given quantity, however small.

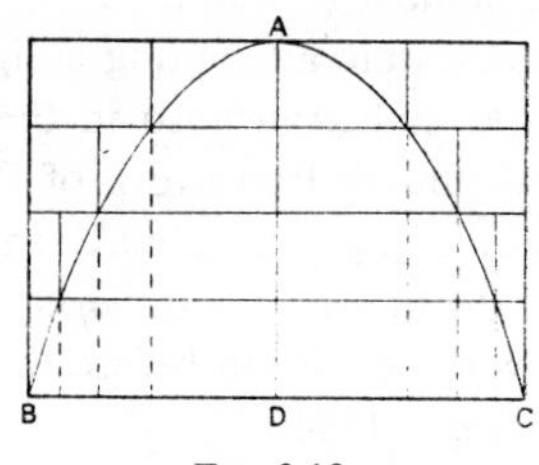

FIG. 3.10

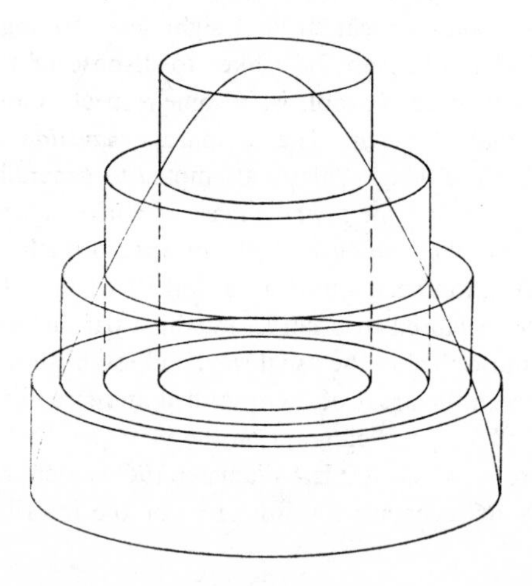

FIG. 3.11

In Proposition XI, Valerio extends this theorem to the volumes of solids with circular or elliptic bases by taking the differences (c_1-i_1), (c_2-i_2), (c_3-i_3), . . ., and fitting them inside each other like a set of nesting boxes (see Fig. 3.11). Although Valerio makes no attempt to define the areas and volumes arrived at in this way in terms of the limit process, the use which he makes of these propositions indicates that having established that, by suitable choice of n, $|C_n - I_n| < \epsilon$, where ϵ is any arbitrarily chosen quantity, however small, he is prepared to conclude that the area (or volume) enclosed by the curved line (or surface) is, in general, equal to the common limit,

i.e. $$\lim_{n\to\infty} (C_n) = \lim_{n\to\infty} (I_n) = A \text{ (or } V\text{)}.$$

He ceases therefore to regard the construction of both circumscribed and inscribed figures as necessary and is prepared to use only one sequence.†

In Book II, Valerio sets out with the intention of constructing a geometric theory of limits which will enable him to dispense altogether with *reductio* proofs. Without the benefit of any sort of functional notation Valerio could achieve little in this direction but there is no doubt that he was seeking primarily to lay down conditions which would enable him to conclude that the limit of the quotient of two variables is equal to the quotient of their limits.‡ In modern notation, if

$$\frac{\varphi_1(n)}{\varphi_2(n)} = \frac{\varphi_3(n)}{\varphi_4(n)},$$

and if, by suitable choice of n, it is possible to find l_1. l_2, l_3, and l_4, such that

$$|\varphi_1(n) - l_1| < \epsilon, \quad |\varphi_2(n) - l_2| < \epsilon, \quad |\varphi_3(n) - l_3| < \epsilon, \quad |\varphi_4(n) - l_4| < \epsilon,$$

where ϵ is any arbitrarily chosen quantity, however small. then,

$$\frac{l_1}{l_2} = \frac{l_3}{l_4},$$

i.e. $$\lim_{n\to\infty} \frac{\varphi_1(n)}{\varphi_2(n)} = \frac{\lim\limits_{n\to\infty} \varphi_1(n)}{\lim\limits_{n\to\infty} \varphi_2(n)} = \lim_{n\to\infty} \frac{\varphi_3(n)}{\varphi_4(n)} = \frac{\lim\limits_{n\to\infty} \varphi_3(n)}{\lim\limits_{\to\infty} \varphi_4(n)}.$$

If, therefore, a relation of this form be established between the areas (or volumes) of certain inscribed and circumscribed figures this relation also holds between the areas (or volumes) of the figures themselves. It is interesting to examine the effect of these propositions on the presentation of a standard theorem and it will be useful to take for this purpose the determination of

† See Valerio, Book I, Props. XXXIX, XLI; Book II, Props. XII, XVIII.
‡ *Ibid.*, Book II, Prop. III, p. 6.

the volume of a conoid so that comparison can be made with the Archimedean proofs already discussed in Chapter I.†

Let BAC be a conoid with a circular base‡ (diameter BC) and axis AD (see Fig. 3.12). Join BA, AC, to form the triangle BAC. Let the axis AD be divided into n equal parts and let figures be circumscribed to the conoid and to the triangle in the usual way. Let V be the volume of the conoid and V that of

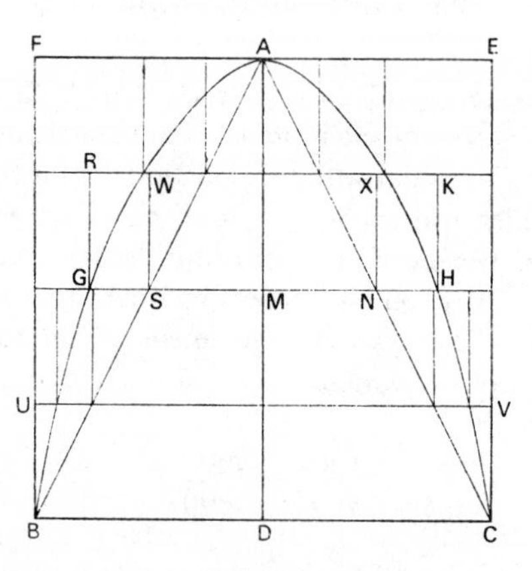

FIG. 3.12

the circumscribed figure: similarly, let Δ be the area of the triangle and Δ_n the area of the circumscribed figure. Now, by suitable choice of n we have,§

$$|V-V_n| < \epsilon, \quad \text{and} \quad |\Delta-\Delta_n| < \epsilon$$

where ϵ is any finite magnitude, however small.

Now, if M be any point of division on the axis AD and if GMH be a diameter which cuts the conoid in G, H, and the sides AB, AC, of the triangle BAC in S and N, we have

$$\frac{AM}{AD} = \frac{MH^2}{DC^2} = \frac{GH^2}{BC^2} = \frac{\text{circle on } GH}{\text{circle on } BC} = \frac{\text{cylinder } GRKH}{\text{cylinder } BUVC},$$

also, by similar triangles (see Fig. 3.12),

$$\frac{AM}{AD} = \frac{MN}{DC} = \frac{SN}{BC} = \frac{\text{rectangle } SWXN}{\text{rectangle } BUVC}.$$

† See pp. 42–3 above.
‡ Valerio, Book II, Prop. XVIII, pp. 28–31.
§ From Prop. III and Prop. VI, Book I.

Hence, $$\frac{\text{cylinder } GRKH}{\text{cylinder } BUVC} = \frac{\text{rectangle } SWXN}{\text{rectangle } BUVC}.$$

and, adding for all such cylinders,

$$\frac{V_n}{\text{volume of enveloping cylinder}} = \frac{\Delta_n}{\text{area of rectangle } BFEC}.$$

Therefore, by Prop. III, Book II,

$$\lim_{n \to \infty} \frac{V_n}{\text{volume of enveloping cylinder}} = \lim_{n \to \infty} \frac{\Delta_n}{\text{area of rectangle } BFEC} = \frac{1}{2},$$

and $V = \frac{1}{2}$ volume of cylinder on the same base and with the same altitude.

Further insight is gained into Valerio's use of limits from an ingenious method he gives for determining the volume of a hemisphere.† *BEFC* is a cylinder, *BAC* is a hemisphere, both on a circular base, diameter *BC*, centre *D*. *FDE* is a cone, vertex *D*, on a circular base, diameter *FE* (see Fig. 3.13). Let the common axis *AD* be divided into *n* equal parts and let figures be circumscribed to the hemisphere and inscribed to the cone in the usual way.

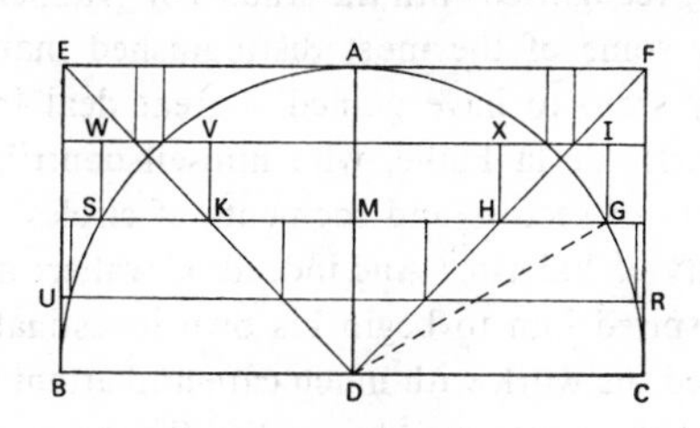

FIG. 3.13

Let *M* be a point of division on the axis *AD* and let a diameter through *M* cut the hemisphere in *S*, *G* and the cone in *K*, *H*.

In the triangle *DMG*,

$$MG^2 = DG^2 - MD^2 = DG^2 - MH^2$$
$$= DC^2 - MH^2,$$
$$\frac{MG^2}{DC^2} = 1 - \frac{MH^2}{DC^2},$$
$$\text{and} \quad \frac{SG^2}{BC^2} = 1 - \frac{KH^2}{BC^2}.$$

† Book II, Prop. XII, pp. 18–20.

Hence,

$$\frac{\text{cylinder } SWIG}{\text{cylinder } BURC} = 1 - \frac{\text{cylinder } KVXH}{\text{cylinder } BURC},$$

and, adding for all such cylinders,

$$\frac{\sum \text{cylinder } SWIG}{\sum \text{cylinder } BURC} = 1 - \frac{\sum \text{cylinder } KVXH}{\sum \text{cylinder } BURC}.$$

But, $$\lim_{n \to \infty} \frac{\sum \text{cylinder } SWIG}{\sum \text{cylinder } BURC} = \frac{\text{volume of hemisphere}}{\text{volume of cylinder}},$$

and $$\lim_{n \to \infty} \frac{\sum \text{cylinder } KVXH}{\sum \text{cylinder } BURC} = \frac{\text{volume of cone}}{\text{volume of cylinder}},$$

therefore, $$\frac{\text{volume of hemisphere}}{\text{volume of cylinder}} = 1 - \frac{1}{3} = \frac{2}{3}.$$

Although the contributions made by Valerio to the development of mathematics were not fully recognised until the studies of Wallner† and Bosmans‡ in the present century some of the most distinguished mathematicans of the seventeenth century seem to have gained a great deal from the *De centro gravitatis*. Jean-Charles de la Faille, who himself contributed a treatise on the centres of gravity of sectors and segments of circles and ellipses,§ often invokes the authority of Valerio.|| And indeed, Cavalieri mentions him as an author who had inspired him to begin his own investigations; Grégoire de Saint-Vincent studied the work with much care and attention.

Notable as are all these attempts to modify the proof structure of Archimedes another stream of development was already in motion which was to lead to the temporary eclipse of the *exhaustion* method. In 1644 Mersenne†† includes the names of Maurolico, Commandino, Stevin and Valerio with those of Archimedes, Apollonius and Pappus among those who "have en-

† Wallner, C., Über die Entstehung des Grenzbegriffes, *Bibl. Math.* (3) 4 (1903), pp. 246–59.

‡ Bosmans, H., Les démonstrations par l'analyse infinitésimale chez Luc Valerio, *Annales de la Société Scientifique de Bruxelles* 37 (1913), pp. 211–28.

§ Faille, Jean-Charles de la, *De centro gravitatis partium circuli et ellipsis*, Antwerp, 1663. Cf. Bosmans, H., Le mathématicien Anversois Jean-Charles della Faille de la Compagnie de Jésus, *Mathésis* 41 (1927), pp. 5–11.

|| Cf. Bosmans, H., Les démonstrations par l'analyse infinitésimale chez Luc Valerio, *Annales de la Société Scientifique de Bruxelles* 37 (1913), p. 228.

†† Mersenne, M., *Universae geometriae mixtaeque mathematicae synopsis*, Paris, 1644: preface.

riched geometry", and indeed all this sixteenth-century work may be more appropriately considered as a postscript to Greek mathematics rather than as a part of the mainstream of seventeenth-century development. The search for quick results and simplified techniques led rapidly to a concentration on methods of discovery rather than on methods of demonstration and in consequence the impetus to improve and modify the Archimedean proof structure was lost.

CHAPTER 4

Infinitesimals and Indivisibles in the Early Seventeenth Century

4.1. Johann Kepler (1571–1630)

Among the first of the seventeenth-century mathematicians to abandon the formal proof structure of Archimedes and to introduce the free use of infinitesimals in the determination of areas and volumes was Johann Kepler. As an astronomer Kepler was primarily concerned to interpret the universe in terms of ordered mathematical harmony and throughout his entire life he pursued this goal with tireless energy and unremitting determination. In so doing he was willing to go to any lengths to develop from a mathematical hypothesis sufficient data to test a theory but he was nevertheless prepared at any time to abandon the theory and start afresh should it fail to fit the observed facts.†

As a mathematician Kepler sought to demonstrate in the simplest possible way the existence of mathematical form and structure in the external world and, upheld by his Platonic–Pythagorean philosophy, he often allowed faith, analogy and intuition to guide him when traditional methods failed. Many of the mathematical problems which Kepler attempted to resolve in the interests of astronomy were extremely difficult by any standards and well beyond the scope of existing seventeenth-century techniques. Although Kepler became conversant with the whole range of Greek mathematics he was more interested in results than in proofs. If an Archimedean demonstration supplied him with a required property he was glad to make use of it, but when formal methods failed to provide the answers to his problems the practical necessities of the situation drove him onwards to devise new techniques and processes.

† Cf. Burtt, E. A., *The Metaphysical Foundations of Modern Physical Science*, London, 1925, pp. 44–60.

Many of the astronomical problems with which he was faced required the evaluation of trigonometric integrals and in some of these cases, finding no theoretical method available, Kepler was obliged to resort to numerical summation. One such problem arose in connection with the theory of planetary magnetism† in which the entire sphere of the planet was regarded as a magnet, one pole being attracted and the other repelled by the sun. After many attempts to determine the relation between the attractive and repulsive forces he finally concluded that it was necessary to divide off the arc into infinitely many small and equal parts and to find the sum of the sines (or chords) of all the arcs. This was duly accomplished and he arrived (presumably by incomplete induction) at a conclusion formally equivalent to the definite integral

$$\int_0^{\alpha} \sin \theta \, d\theta = 1 - \cos \alpha.$$

He was later able to confirm this result by reference to a theorem of Pappus.‡ On another occasion he was defeated by an elliptic integral of the form $\int_0^{\alpha} r \, d\theta$, where $r = k\sqrt{(1+2r \cos \theta + e^2)}$. This he was unable to resolve other than by dividing the arc into 1^0 intervals, calculating values of r and summing.§

When he argued by analogy he sometimes made mistakes. Unable to determine theoretically the length of an elliptic arc he argues‖ that, since the area of an ellipse is equal to that of a circle, the radius of which is the geometric mean of the major and minor axes (area $= \pi ab = \pi r^2$, where $a/r = r/b$), then the circumference of the ellipse should correspondingly be equal to the circumference of a circle the radius of which is the *arithmetic* mean of the semi-axes, i.e. $\pi(a+b)$. Although arguments of this kind could lead him into difficulties a happy instinct often saved the situation. He was not afraid to admit defeat and the unresolved problems he left behind him acted as a stimulus to further efforts throughout the century.

† Kepler, J., *Opera Omnia* (ed. Frisch, C.), Frankfort-on-Main, 1858–71. See Vol. III, p. 390; Vol. V, p. 80; Vol. VI, p. 407. See also Günther, S., Über eine merkwürdige Beziehung zwischen Pappus und Kepler, *Bibl. Math.* N.S. 2 (1888), pp. 81–7.

‡ See Eneström, G., Sur un théorème de Kepler équivalent à l'intégration d'une fonction trigonométrique, *Bibl. Math.* N.S. 3 (1889), pp. 65–6. Also Eneström, G., Über die angebliche Integration einer trigonometrischen Funktion bei Kepler, *Bibl. Math.* (3) 13 (1915), pp. 229–41. See also Chap. I, pp. 38–41, and *Pappus d'Alexandrie: La Collection Mathématique* I, pp. 281–5.

§ *Opera omnia* III, p. 466.

‖ *Ibid.* III, p. 401.

The *Nova stereometria*,† published in 1615, contained a fascinating collection of infinitesimal methods for estimating the volumes of solids of revolution. Like Stevin, Kepler had practical reasons for preferring a simplified presentation, for the book was written as a handbook to enable wine gaugers to estimate more efficiently the volumes of wine casks. The Latin edition of 1615 was followed a year later by a popular edition in the German language that established the beginnings of German mathematical terminology.‡

In this work Kepler uses a wide range of methods including visual imagery, geometric transformation, analogy and tabulation but most of all the cutting of small sections varying in size and shape, parallel to no given direction and chosen at will in the most convenient form to meet the needs of a particular problem. Probably no one since Kepler has used infinitesimals quite so freely.§

Of the circle he says that the circumference has as many parts as it has points, each part forming the base of an isosceles triangle with vertex at the centre of the circle.|| The circle is accordingly made up of an infinity of small triangles, each with its base on the circumference and with altitude equal to the radius of the circle. Replacing these triangles by a single triangle with the circumference as base the area of the circle is thus given in terms of the circumference and radius (see Fig. 4.1). The properties of circular cylinders, cones and prisms can all be derived by similar considerations.†† The sphere can be regarded as made up of infinitely many small cones, each with its base on

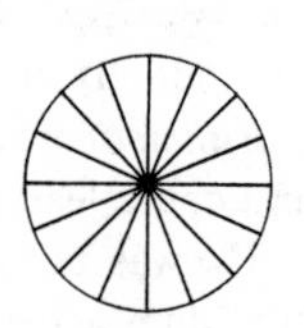

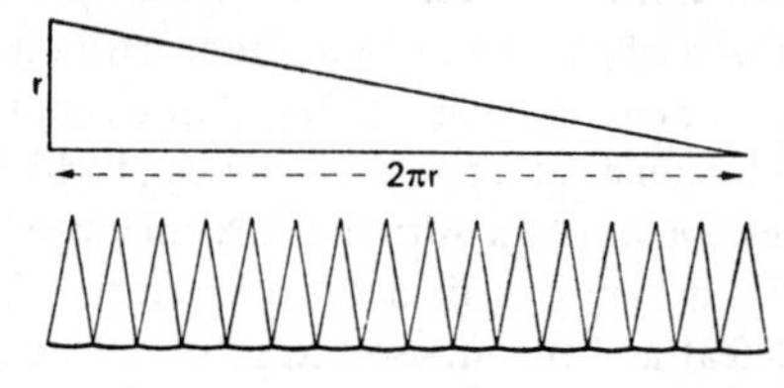

FIG. 4.1

† Kepler, J., *Nova stereometria doliorum vinariorum*, Linz, 1615; *Opera omnia* IV, pp. 551–646.

‡ *Opera omnia* V, pp. 497–610; *Ausszug aus der Uralten Messe-Kunst Archimedis*, Linz, 1616.

§ Cf. Struik, D. J., Kepler as a Mathematician, in *Johann Kepler, 1571–1630*, Baltimore, 1931.

|| *Opera omnia* IV, pp. 557–8. This concept of the circle is not, of course, new with Kepler and is found with Nicholas of Cusa, Viète and Stifel among others.

†† *Ibid.* IV, pp. 559–60.

the surface of the sphere and its vertex at the centre[†] (see Fig. 4.2). Hence,

$$\text{volume} = \tfrac{1}{3}\,\text{surface area} \times \text{radius}.$$

Passing on to solids of revolution in general Kepler considers the rotation of certain closed plane figures about axes in the plane. Whereas Archimedes had confined himself to rotations about the principal axes, Kepler generates a variety of different solids by rotating about lines in the plane other than the principal axes.

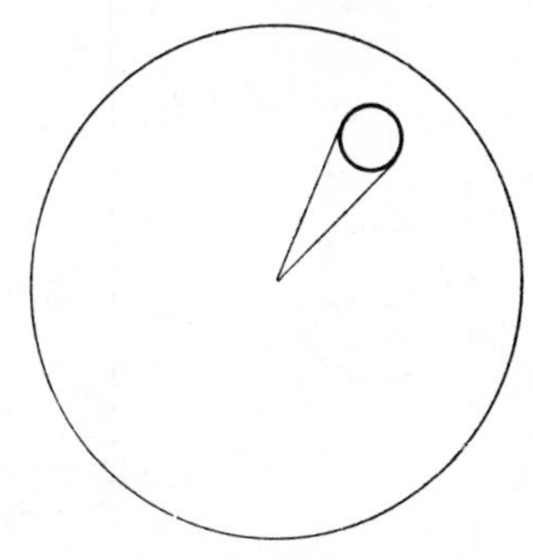

FIG. 4.2

A solid formed by rotating a plane figure about a line in the plane which does not meet the curve is called a *ring*. If the line is parallel to an axis of symmetry then the volume of the ring is equal to that of a cylinder with height equal to the circumference of the circle described by the centre of the figure and with base equal to the section itself.[‡] For, if the ring be cut (by planes passing through the axis) into infinitely many thin slices, then the part nearest the axis will be thinner than the outer part and the average thickness will be given by $t = (t_1+t_2)/2$, where t_1 and t_2 represent the thicknesses at equal distances from the centre of the section. The section can thus be replaced by a disc with volume At, where A is the area of the plane figure: the volume of the entire ring is thus given by Al, where l is the length of the circular path of the centre of the section (see Fig. 4.3). This is, of course, a simplified version of the Pappus–Guldin theorem. Kepler was well aware that, in this form, the result did not extend to non-symmetrical sections: he applied it in a form which could be readily checked to a rotating square (or rectangle)[§] (Fig. 4.4).

† *Ibid.* IV, p. 563.
‡ *Ibid.* IV, pp. 582–3.
§ *Ibid.* IV, p. 583.

The rotation of circular segments produces a variety of solids† (Fig. 4.5): if the major segment *MPN* (Fig. 4.6) rotate about the chord *MN*, we have an *apple*; if the semicircle *CPD* rotate about a diameter *CD*, we have a *sphere*; if the minor segment *IPK* rotate about a chord *MN* we have an *apple ring* (Fig. 4.7); if *IPK* rotate about a diameter *CD* we have a *spherical ring* (Fig.

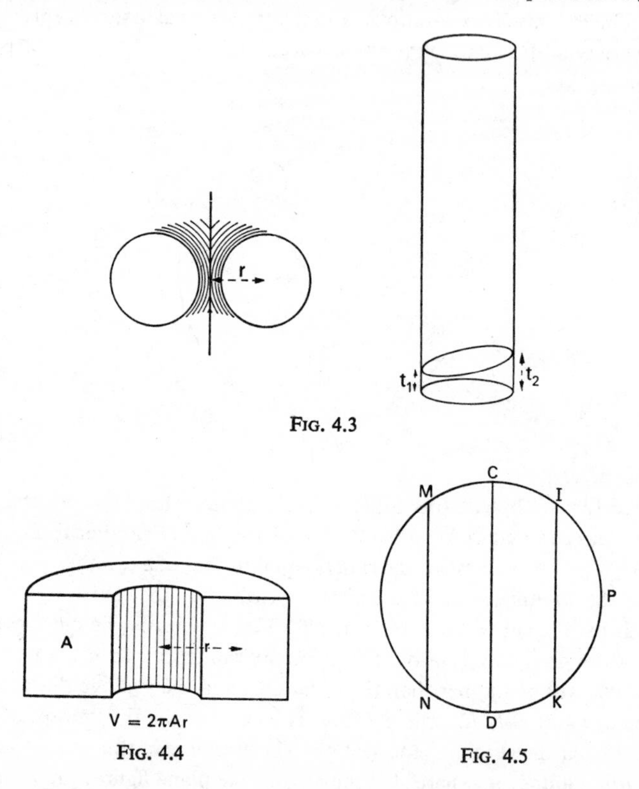

FIG. 4.3

FIG. 4.4

FIG. 4.5

4.8); if *IPK* rotate about the chord *IK* we have a *lemon* (see Fig. 4.9). In order to investigate the relations between the volumes of these various solids Kepler introduced a form of geometric representation which not only served him well for this particular purpose but which also operated as a fruitful source of inspiration for many of the geometric integration methods developed throughout the century.

† *Opera omnia* IV, pp. 584–5.

Kepler conceived the segment *MPN* (see Fig. 4.10) as made up of a set of chords parallel to *MN*. As the segment rotates about *MN* each chord describes the surface of a cylinder with radius equal to the perpendicular distance of the chord from *MN* taken along the axis *AP*. The volume of the solid generated (the *apple*) can accordingly be represented as the sum of the sur-

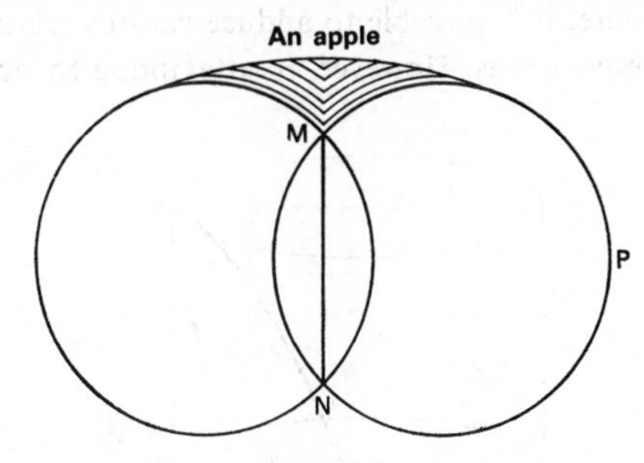

FIG. 4.6

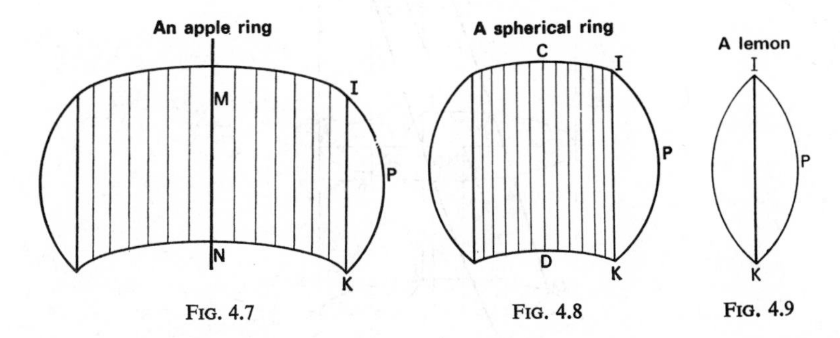

FIG. 4.7 FIG. 4.8 FIG. 4.9

faces of all such cylinders.[†] Hence, if a rectangle be constructed on each chord parallel to *MN* (in the segment *MPN*) with altitude equal to the length of the circular path of that chord rotating about the axis *MN*, the resulting solid, a cylinder on the base *MPN* with oblique section, *MI″SK″N*, will be equal in volume to the solid formed by rotating *MPN* about *MN*, i.e.

$$\text{volume of } apple = \sum IK \,.\, KK'' = \text{volume of solid } MI''SK''NP.$$

On the same diagram the volumes of the other solids can readily be distinguished as follows.

† *Ibid.* IV, p. 584.

volume of *apple ring* (*IPK* rotates about *MN*) = *I''SK''IPK*,

volume of *spherical ring* (*IPK* rotates about *CD*) = *I''SK''I'P'K'*,

volume of *lemon* (*IPK* rotates about *IK*) = *I''SK''P''*,

volume of *sphere* (*CPD* rotates about *CD*) = *C'SD'P'*.

Thus, from the figure, it is possible to adduce various relations between the volumes of the above solids. However, in attempting to draw further con-

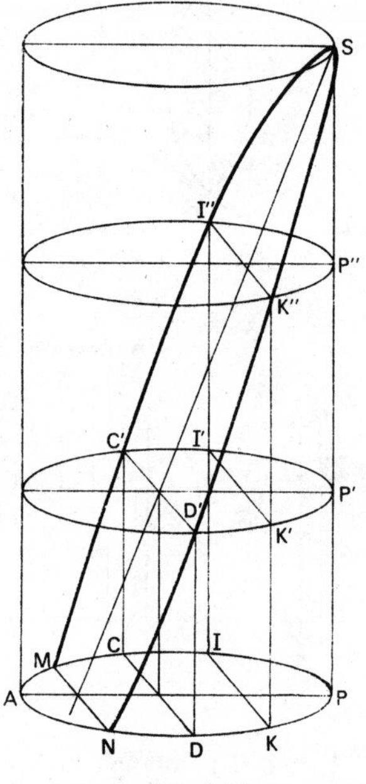

FIG. 4.10

clusions as to the volume of the *lemon*† Kepler compares it with the volume of a spherical cap formed by rotating the segment *IPK* about its axis of symmetry, *AP* (see Fig. 4.11), and gets into some interesting difficulties. The tentative solution which he puts forward is based on the mediaeval thesis, "What is true for the greatest and the smallest is also true for the intermediate values" (Fig. 4.11).

† *Opera omnia* IV, p. 594.

For the *greatest*, if $AI = AP = r$, the solids formed by rotating about IA and AP are both hemispheres. When the segment is small then the volumes of the solids approximate to the volumes of their inscribed cones. Thus, for the *greatest* and for the *smallest* we have

$$\frac{\text{volume of solid formed by rotation about } AP}{\text{volume of solid formed by rotation about } IK} = \frac{IA^2\, AP}{AP^2\, IA} = \frac{IA}{AP}.$$

At the minimum, he says, it is not possible to speak with any degree of certainty.†

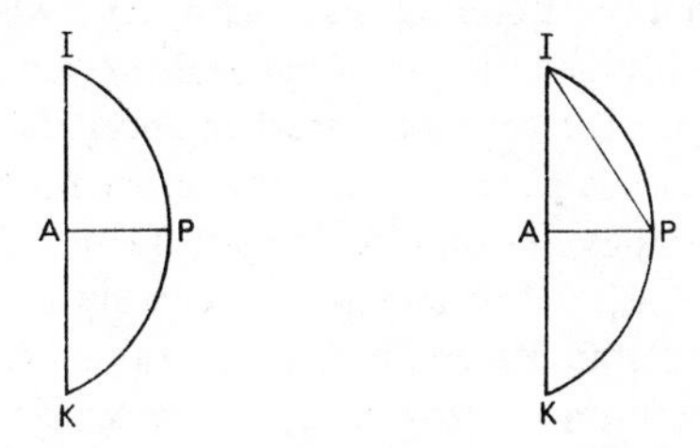

FIG. 4.11

In all Kepler considers the volumes of ninety-six different solids obtained by rotating segments of conic sections about various axes. For the consideration of other mathematicians he formulated a further problem: to determine the volumes of solids formed by the rotation about an ordinate of segments of conic sections cut off by lines parallel to the axis.‡ This problem was subsequently solved by Cavalieri. Kepler also has the distinction of being the first to formulate an inverse tangent problem, i.e. given a relation between subtangent and abscissa, to determine the curve.§ The cases he considers, however, are of no special interest since they involve only the well-known properties of tangents to conic sections.

In investigating the best dimensions for wine casks Kepler encountered problems concerning maxima and minima. By considering small variations about the maximum he was able to establish that the cube is the greatest parallelepiped which can be inscribed in a sphere|| and that amongst all circular cylinders with equal diagonals†† the greatest is that in which the ratio

† *Ibid.* IV, p. 594.
‡ *Ibid.* IV, p. 601.
§ *Ibid.* IV, p. 598.
|| *Ibid.* IV, pp. 607–9.
†† *Ibid.* IV, p. 610.

of diameter to altitude is $\sqrt{2}:1$. In the latter problem he tabulated values obtained by calculation to reinforce his comment that the volumes of such cylinders change very little in the neighbourhood of a maximum.

The influence of Kepler's *Stereometria* on the growth of integration methods has perhaps been exaggerated by some historians. Montucla says† of Kepler: "Il osa, le premier, introduire dans le langage ordinaire le nom et l'idée de l'infini", whilst Cantor‡ refers to the *Stereometria* as, "die Quelle aller späteren Kubaturen". Although many of the ideas Kepler was using have now been traced to mediaeval sources the importance of the publication of a work of this kind must still be recognised. And whereas the ostensible purpose of the work justified the use of plausible arguments and a departure from Archimedean proof structure the mathematical content so far exceeded the practical needs of wine gaugers that there can be no question that Kepler intended to make a serious contribution to mathematics. The book was widely read, stimulated intense interest and was vigorously denounced by Guldin, Anderson and others as an attack on the authority of Archimedes. How far Galileo and Cavalieri in Italy were directly influenced by Kepler in their own approach to indivisible methods is another matter.

4.2. Indivisibles in Italy

Although there is no record of any direct exchange of views between Kepler and Galileo on the subject of infinitesimals the publication of the *Stereometria* may well have acted as a spur, both to Galileo and to his pupil Cavalieri, in extending and developing their own work in this field. Although Galileo never actually published a work on indivisible methods as such§ the whole trend of the discussion in the *Two New Sciences*‖ throws light not only on his own approach to such matters but also on the background of ideas which he transmitted to Cavalieri.

Galileo had a profound respect for the Greeks and in consequence greatly admired the achievements of those mathematicians such as Valerio who had expanded and developed the work of Archimedes.†† Nevertheless it is clear

† Montucla, J. E., *Histoire des mathématiques*, Paris, 1799–1802, II, p. 29.

‡ Cantor, M., *Vorlesungen über Geschichte der Mathematik* (2nd ed.), Leipzig, 1894–1908, II, p, 750.

§ Cavalieri wrote twice to Galileo announcing his intention of publishing a work on indivisibles. The purpose of these letters was apparently to give warning in case Galileo had in mind to publish anything himself.

‖ Galilei, Galileo, *Discorsi e dimostrazioni matematiche*, Leyden, 1638. *Dialogues Concerning Two New Sciences* (trans. Crew, H. and De Salvio, A.), New York, 1914 (refs. given from the Dover edition of this work).

†† *Ibid.*, p. 30. See also pp. 101–7 above.

that for the most part his own ideas led him along entirely different paths. The influence of Scholastic thought on Galileo is, of course, most pronounced in his early works but even in *Two New Sciences* it is very much to the fore when he discusses the nature of indivisibles and the infinite. In this work Galileo reviews most of the mediaeval paradoxes and at the same time provides some new ones of his own. Most of these paradoxes are based on the concept of indivisible quantities and Galileo makes no distinction between physical atoms and line and surface elements: in fact he finds support for his belief in the efficacy of mathematical indivisibles from his desire to demonstrate the possibility of the existence of a vacuum.†

In discussing the *Wheel of Aristotle*,‡ still a popular subject in the seventeenth century, Galileo first allows the hexagon, *ABCDEF*, to roll upon the line *AS* (see Fig. 4.12) carrying with it a concentric hexagon, *HIKLMN*,

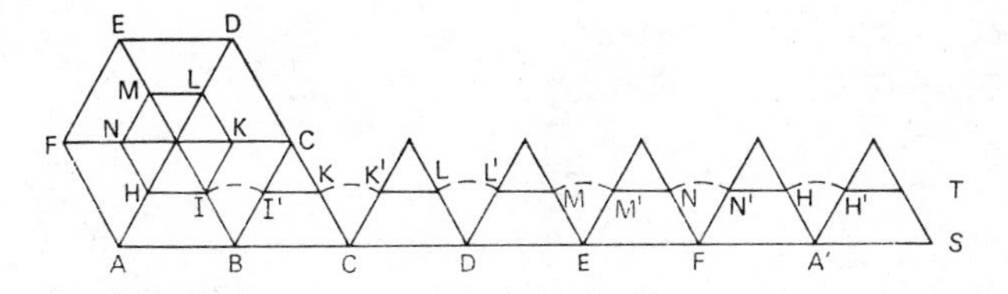

Fig. 4.12

inscribed within it. As the sides *AB, BC, CD, DE, EF, FA* take up successive positions on the line *AB*, the sides of the concentric hexagon fall into positions *HI, I'K, K'L, . . ., N'H* on the line *HT*, parallel to *AS*. Hence, in one revolution of the wheel, the total distance *HH'* moved by the point *H* along the line *HT* equals the distance *AA'* moved by the point *A* along the line *AS*. The distance *HH'*, however, is greater than the perimeter of the lesser polygon by the length of the spaces *II', KK', LL', . . ., HH'*. Now, if the number of sides of the polygon be increased to 1000, then in one complete revolution the lesser polygon will lay off 1000 sides and the same number of spaces along the line *HT*. Finally, if the rolling polygons be replaced by concentric circles, rigidly connected, the point *C* on the lesser circle will move through the same distance, *CC'*, as the point *B* on the greater circle (see Fig. 4.13). The space *CC'*, however, will, in this case, be made up of an infinite number of infinitely small and indivisible parts with an infinite number of infinitely

† *Two New Sciences*, p. 51.
‡ *Ibid.*, pp. 20–3.

small indivisible spaces interposed between the parts. Thus Galileo lent his authority to the concept of lines, solids and surfaces, made up of infinitely many indivisible elements variously distributed.[†] He issues constant warnings, however, against the very serious error of attempting to discuss the infinite by assigning to it properties which we employ for the finite.[‡] He is careful to distinguish between the *potentially* infinite, exemplified in the process of

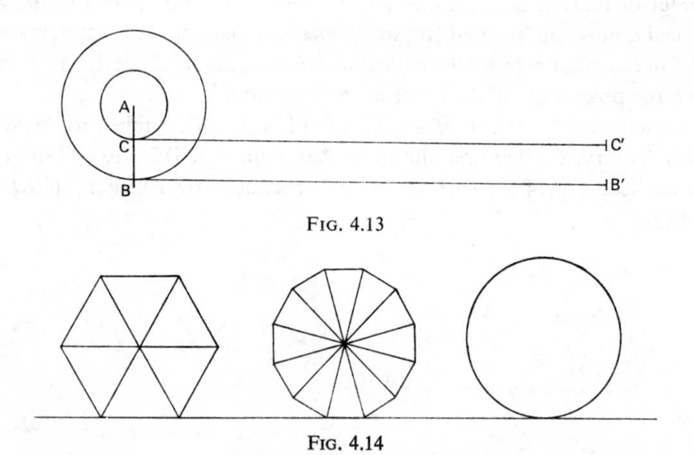

FIG. 4.13

FIG. 4.14

subdivision without end, and the *actually* infinite, arrived at by some direct process. If, for example, a sequence of polygons be constructed by repeated bisection with an ever-increasing number of sides, this process can be continued indefinitely and every polygon in the sequence lies with one of its sides on a given straight line (see Fig. 4.14). But it is possible (by means of a compass construction) to construct immediately a circle, which is a polygon with an *infinite* number of sides and which touches the straight line with one of its sides which is a single point.[§] The distinction which Galileo makes is therefore between the limiting case which he regards as a complete resolution into primary elements or indivisibles, and the approach to the limit which he regards as an endless process of continuous subdivision. Another example which he quotes relates to the difference between a finite and an infinite circle.[||] Whereas the radius of a finite circle can be indefinitely increased

† *Two New Sciences*, p. 25.
‡ *Ibid.*, pp. 34–6.
§ *Ibid.*, pp. 47–8.
|| *Ibid.*, pp. 38–9.

without the circle changing its form, if the radius becomes *actually* infinite then the circle becomes a straight line. It thereby changes its character in such a manner that it loses "not only its existence but also its possibility of existence" in that any point O which moves along it moves along an infinite straight line and is unable to return to its starting point (see Fig. 4.15).

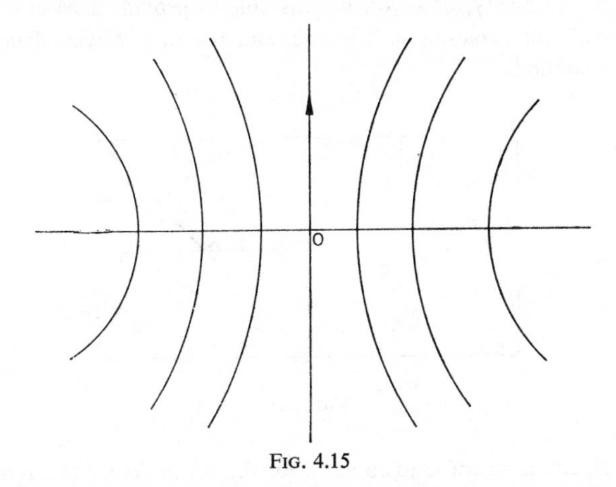

FIG. 4.15

Galileo was fascinated by the configuration of cone, hemisphere and cylinder which Valerio introduced to establish the relations between their respective volumes.† The use he made of it, however, was entirely different from that for which it was originally devised.

Let a hemisphere, AFB, diameter AB, centre C, and a cone DCE, base DE, axis FC, be inscribed in a cylinder $ABED$ and let a plane parallel to DE cut the axis CF in P, the cone in H, L, the hemisphere in I, O, and the cylinder in G, N (see Fig. 4.16). We have, accordingly, $CO^2 = PN^2 = CP^2+PO^2 = PL^2+PO^2$, and $PN^2-PO^2 = PL^2$.

Hence the circular band formed by rotating ON about the axis CF is equal in area to the circle, radius PL. Now, as the cutting plane GN moves parallel to the base DE towards the vertex C, this relation continues to hold and, at the "extreme and final end of their diminution" the band and the circle remain equal so that the circumference of the circle, radius CB, becomes equal to the point C. Appealing for the acceptance—apparently contrary to common sense and reason—of this odd concept of the equality of a finite straight

† *Ibid.*, pp. 27–30.

line and a point Galileo says: "Shall we not call them equal seeing that they are the last traces and remnants of equal magnitudes?"

Indeed, acceptance of the inevitability of contradiction in such matters was inherent in Galileo's mediaeval approach and it can scarcely be claimed that he did very much to clarify the conceptual basis for indivisible methods. Through his authority, however, he was able to provide a basis of support for the use of indivisibles in mathematics and also to give some detailed guidance as to methods.

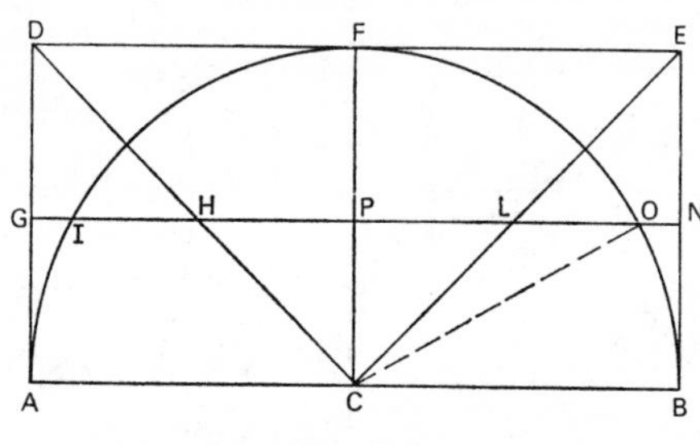

FIG. 4.16

The geometrical proof Galileo gives for the *Mean Speed* theorem† is very similar to that of Oresme. If a uniformly accelerated body moves from C to D in time AB (see Fig. 4.17), and if the final speed be represented by a line BE at right angles to AB then, in general, any line PQ, drawn parallel to BE in the triangle ABE, corresponds to the speed in the time interval AP. Since *the sum of all the parallels* in the triangle AEB equals the sum of the parallels in the rectangle $FGAB$, where $BF = \frac{1}{2}BE$, then the distance covered in the time AB with uniformly accelerated motion is equal to the distance covered in the same time with constant velocity equal to one half of the final velocity.‡

† *Two New Sciences*, pp. 173–4.

‡ *Ibid.*, pp. 173–4. Since each and every instant of time in the time-interval AB has its corresponding point on the line AB, from which points parallels drawn in and limited by the triangle AEB represent the increasing values of the growing velocity, and since parallels contained within the rectangle represent the values of a speed which is not increasing, but constant, it appears, in like manner, that the momenta (*momenta*) assumed by the moving body may also be represented, in the case of the accelerated motion, by the increasing parallels of the triangle AEB, and in the case of the uniform motion, by the parallels of the rectangle GB. For, what the momenta may lack in the first part of the accelerated motion (the deficiency of the momenta being represented by the parallels of the triangle AGI) is made up by the momenta represented by the parallels of the triangle IEF.

This concept of *the sum of the lines* as a measure of the space contained within a plane figure was to become the foundation for a great many seventeenth-century integration methods. It appears, of course, most notably in the works of Cavalieri. In establishing that the path of a projectile is a parabola† Galileo also introduced some techniques which proved particularly

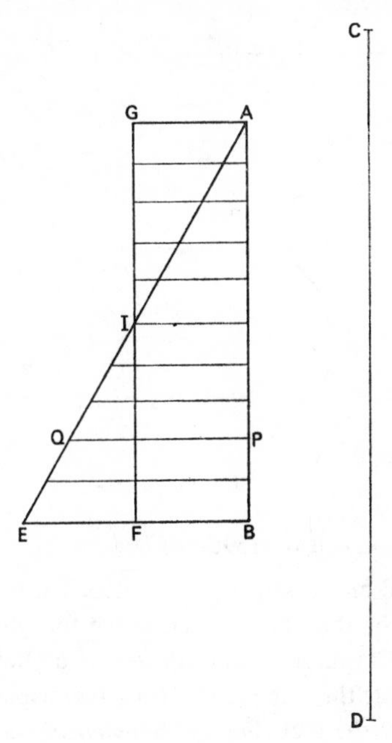

FIG. 4.17

fruitful in the study of curves. A body is conceived as starting from a given point with two motions, one horizontal and the other vertical, with constant acceleration, i.e. the distance varies as the square of the time. If, therefore, the horizontal axis represent the "flow or measure of time", if the segments *bc, cd, de*, . . ., represent equal intervals of time, and if *ci, df, eh*, . . . be drawn perpendicular to *bcde* so that $ci \propto bc^2$, $df \propto bd^2$, $eh \propto be^2$, then the body will traverse the parabola *bifh* (see Fig. 4.18). Although Galileo was here considering the actual motion of a particle describing a curve under prescribed physical

† *Ibid.*, pp. 248–50.

conditions the concept of a curve as the path of a point moving with constant velocity parallel to one axis and with variable velocity in a direction at right angles to the first was of great value to Torricelli, Barrow and later Newton, in providing a direct link between tangent and area properties and ultimately in establishing the inverse nature of differentiation and integration.

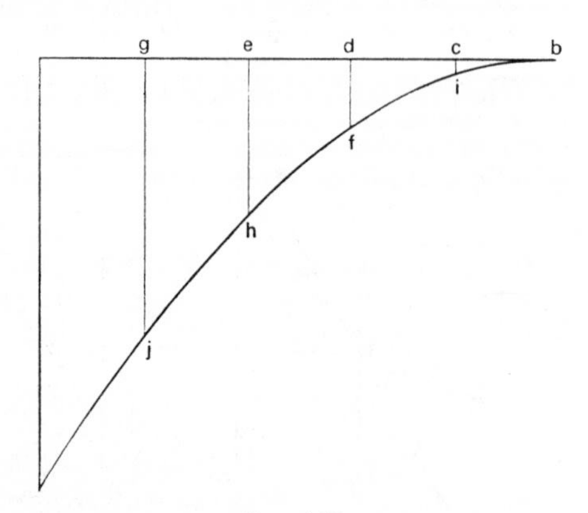

FIG. 4.18

4.3. BONAVENTURA CAVALIERI (1598–1647)

On Cavalieri, pupil and associate of Galileo, fell the task of shaping the vague concept of indivisibles into a serviceable tool in the determination of areas and volumes. Cavalieri became Professor at Bologna in 1629 and in canvassing for office at this time is said to have displayed a manuscript on indivisibles.† Whether the fact that the *Geometria*‡ was not published until 1635 was due to the expectation that Galileo himself was writing on the subject we do not know.§ At all events the *Geometria* was the first complete work entirely devoted to indivisible methods and it received wide publicity, was extensively read and its contents hotly debated. Guldin attacked it bitterly in the *Centrobaryca* and, besides denouncing the form and content of the proofs, accused Cavalieri of plagiarism.‖ In reply Cavalieri wrote the *Exercitationes geometricae sex* (1647) in which he made further advances in

† See Cavalieri, B., *Exercitationes geometricae sex*, Bologna, 1647, p. 183.

‡ Cavalieri, B., *Geometria indivisibilibus continuorum nova quadam ratione promota*; all refs. are given from the 2nd ed. (1653).

§ See p. 116 above.

‖ See *Ex. geom. sex*, p. 183.

the method of indivisibles and launched a savage counter-attack on Guldin which appeared, unfortunately, after his death. These two works of Cavalieri rapidly became the most quoted sources (save for Archimedes) on geometric integration in the seventeenth century, and Cavalieri himself is frequently named as the founder of the new analysis. The original works did not make easy reading but a great many authors took upon themselves the task of interpreting the methods, both mathematically and philosophically, and these second-hand accounts were sometimes misleading and false. For example, the account of Cavalierian indivisibles given by Torricelli which John Wallis was fortunate enough to read[†] was probably more illuminating than the original, whilst that of the philosopher Thomas Hobbes which was read by Leibniz[‡] was certainly less helpful.

Although in the history of mathematics the method of indivisibles has always been associated with the name of Cavalieri, the term indivisible is mediaeval in origin, had been familiar since Bradwardine and was used extensively by Galileo. Cavalieri, somewhat uncritically, was prepared to take over the mediaeval concept of indivisibles, on the basis of geometric intuition alone, but in order to forestall criticism he allowed himself to embark on much more detailed discussion of the nature of line and surface indivisibles than he personally felt to be either necessary or desirable. In consequence he made many obscure, difficult and even contradictory statements drawing upon himself considerable ridicule which continued unabated throughout the seventeenth century and has persisted to the present day. Criticism of Cavalieri, based on his failure to distinguish between mathematical indivisibles and physical atoms[§] and on the use he makes of a set of elements and operations not defined in a manner satisfactory in terms of twentieth-century standards, is historically unsound and often leads to a failure to appreciate the powerful way in which he used consistent indivisible techniques to deduce whole sequences of results. Cavalieri did explain his indivisibles in the only terms he knew—the language of Oresme and the mediaeval Calculators.

Although Cavalieri was persuaded to introduce a little algebraic notation into the *Exercitationes* in order to generalise some of his results by the use of exponents,[||] for the most part he presents his results entirely in verbal and

† See *Opere di Evangelista Torricelli* (ed. Loria, G. and Vassura, G.), Faenza, 1919, I (Pt. 1), pp. 138–62: *Quadratura parabolae per novam indivisibilium geometricam pluribus modis absoluta.*

‡ See Hofmann, J. E., *Die Entwicklungsgeschichte*, p. 4.

§ See Boyer, C. B., *The Concepts of the Calculus*, pp. 117–23.

|| See *Ex. geom. sex*, pp. 283–5.

geometrical form. The *Geometria*, in particular, is a long and difficult book,† although the methods which emerge after much tedious preparation are interesting and original.

For Cavalieri a surface consists of an indefinite number of equidistant parallel straight lines and a solid of a set of equidistant parallel planes.‡ These constitute the line and surface indivisibles respectively. For plane figures (or solids) a *regula*, that is, a line (or plane) drawn through the vertex, is the starting point. The *regula* moves parallel to itself until it comes into coincidence with a second line (or plane) termed the base or *tangens opposita*.§ The intercepts (lines or plane sections) of the *regula* with the original plane (or solid) figure are the elements or indivisibles making up the totality of lines (or planes)‖ of the figure (see Fig. 4.19).

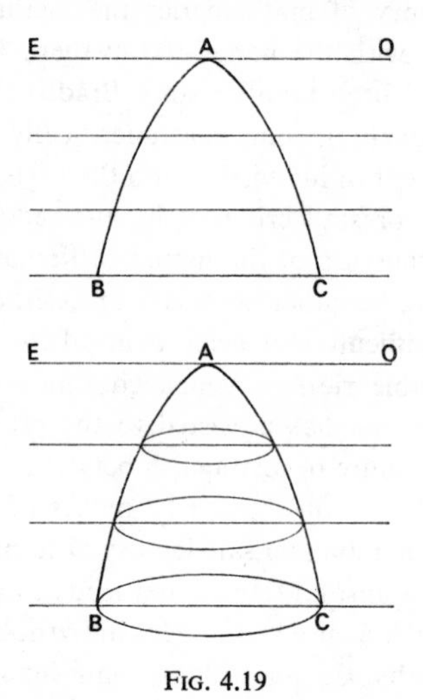

FIG. 4.19

† Marie, M., *Histoire des sciences mathèmatiques et physiques* (12 vols.), Paris, 1883–8. Marie says the *Geometria* should have the prize for dullness (see IV, p. 90).

‡ See *Geom. ind.*, pp. 104–5.

§ The generation of lines, surfaces and solids by moving points, lines and planes is to be found with Napier and Galileo and was subsequently developed by Torricelli, Barrow and Newton among others.

‖ *Geom. ind.*, p. 104 (see Fig. 4.19). "Sit figura plana quaecunque, *ABC*, duae eiusdem oppositae tangentes utcunque ductae, *EO*, *BC*, intelligantur autem per *EO*, *BC*, indefinitè extensa duae plana invicem parallela, quorum quod transit per, *EO*, ex. gr. moveatur versus planum per, *BC*, semper illi aequidistans, donec illi congruat, igitur communes sectiones talis moti, sive fluen is plani & figurae, *ABC*, quae in toto motu fiunt, simul collectae à me vocantur:"

In the techniques developed by Cavalieri the indivisibles of two or more configurations are associated together in the form of ratios and from these ratios the relations between the areas (or volumes) of the figures themselves are derived. In moving from a relation between the sums of the indivisibles and thus to a relation between the spaces the infinite is employed, but purely in an auxiliary role.†

Cavalieri's defence of indivisible methods was based primarily on the idea of a device or *artificium* which works rather than on any definite or dogmatic views as to the nature of indivisibles and the spaces which they occupy.‡ Nevertheless the classic problem of the nature of the continuum imposed itself upon him and, from the outset, he felt himself obliged to try to meet some of the arguments which he felt might be directed against his methods. He admits that some might well doubt the possibility of comparing an indefinite number of lines, or planes *(indefinitae numero lineae, vel plana)*.§ When such lines (or planes) are compared, he says, it is not the numbers of such lines which are considered but the spaces which they occupy. Since each space is enclosed it is bounded and one can add to it or take away from it without knowing the actual number of lines or planes. Whether indeed the continuum consists of indivisibles or of something else neither the space nor the continuum is directly measurable. The totalities, or sums, of the indivisibles making up such spaces are, however, always comparable. To establish a relation between the areas of plane figures, or the volumes of solid bodies, it is therefore sufficient to compare the sums of the lines, or planes, developed by any *regula*.||

The foundations for Cavalieri's indivisible techniques rest upon two distinct and complementary approaches which he designates by the terms *collective* and *distributive* respectively.†† In the first, the *collective sums*, $\sum l_1$ and $\sum l_2$, of the line (or surface) indivisibles for two figures P_1 and P_2 are obtained separately and then used to establish the ratio of the areas (or volumes) of the figures themselves. If, for example, $\sum l_1/\sum l_2 = \alpha/\beta$, then $P_1/P_2 = \alpha/\beta$. This approach was given the most extensive application by Cavalieri‡‡ and he exploited it with skill and ingenuity to obtain a fascinating collection of new results which he exhibited in the *Exercitationes*. The *distributive* theory,§§

† See Brunschvicg, L., *Les Étapes de la philosophie mathématique*, pp. 162–6.

‡ See *Ex. geom. sex*, p. 241.

§ See *Geom. ind.*, pp. 111–12.

|| *Ibid.*, pp. 111–12.

†† *Ibid.*, p. 483.

‡‡ *Ibid.*, Books I–VI; *Ex. geom. sex*, Books II ff.

§§ *Geom. ind.*, Book VII; *Ex. geom. sex*, Book I.

on the other hand, was developed primarily in order to meet the philosophic objections which Cavalieri felt might be raised against the comparison of indefinite numbers of lines and planes. Fundamental here is *Cavalieri's Theorem*:† the spaces (areas or volumes) of two enclosed figures (plane or solid) are equal provided that any system of parallel lines (or planes) cuts off equal intercepts on each. In brief, if for every pair of corresponding intercepts l_1 and l_2, $l_1 = l_2$, then $P_1 = P_2$. An immediate extension follows: if $l_1/l_2 = \alpha/\beta$, then $P_1/P_2 = \alpha/\beta$. Cavalieri himself only made use of this method in a limited number of cases where α/β is constant for all such pairs of intercepts. This technique, however, in the hands of Grégoire de Saint-Vincent and others in the seventeenth century became a valuable means of integration by geometric transformation. Ultimately, whichever of the two methods was applied, Cavalieri was prepared to concede that absolute rigour required in each case an Archimedean proof with completion by *reductio ad absurdum.*

In the first group of theorems Cavalieri develops the concept of equal and similar figures.‡ If l and p represent the line and surface elements parallel to any *regula* of the plane and solid figures, P and V respectively, then in some sense, $\sum l = P$ and $\sum p = V$. For equal plane figures, P_1 and P_2, $\sum l_1 = \sum l_2$; similarly, for equal solids, V_1 and V_2, $\sum p_1 = \sum p_2$. If $P_1 \neq P_2$, then $P_1/P_2 = \sum l_1/\sum l_2$, and if $V_1 \neq V_2$, then $V_1/V_2 = \sum p_1/\sum p_2$.

For *similar* parallelograms, P_1 and P_2, we have§ $l_1/l_2 = a/b = c/d$ (see Fig. 4.20). Hence, $P_1/P_2 = \sum l_1/\sum l_2 = \sum a/\sum b = ac/bd = a^2/b^2$.

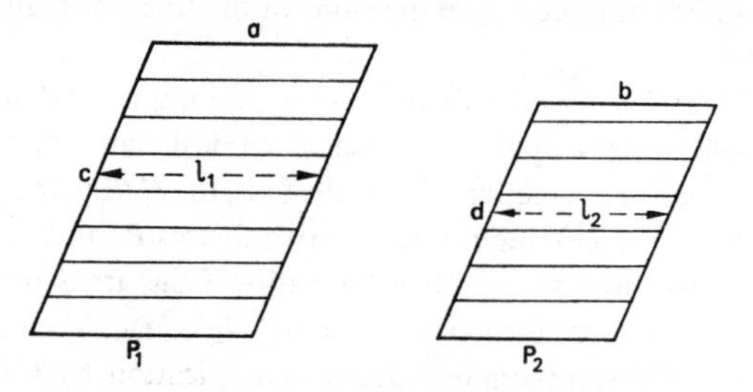

FIG. 4.20

For any pair of similar figures,|| S_1 and S_2 inscribed in the parallelograms P_1 and P_2 (see Fig. 4.21), such that, for corresponding line elements l_1 and l_2,

† *Geom ind.*, Book VII, Prop. I, pp. 484–8. See also Evans, G. W., Cavalieri's theorem in his own words, *Amer. Math. Monthly* 24 (1917), pp. 447–51.

‡ *Geom. ind.*, Book I.

§ *Ibid.*, Book II, p. 120.

|| *Ibid.*, Book II, pp. 127–31.

we have, $l_1/l_2 = a/b = c/d$, then, $\sum l_1/\sum l_2 = \sum a/\sum b = ac/bd = a^2/b^2 = (l_1/l_2)^2$.

For similar solids,† we have, $l_1/l_2 = a/d = b/e = c/f$ (see Fig. 4.22) so that, $V_1/V_2 = \sum p_1/\sum p_2$, where p_1 and p_2 are similar plane sections of V_1 and V_2. But, for similar plane figures, $p_1/p_2 = \sum l_1/\sum l_2 = ab/de$. Hence, $V_1/V_2 = \sum ab/\sum de = abc/def = a^3/d^3 = (l_1/l_2)^3$.

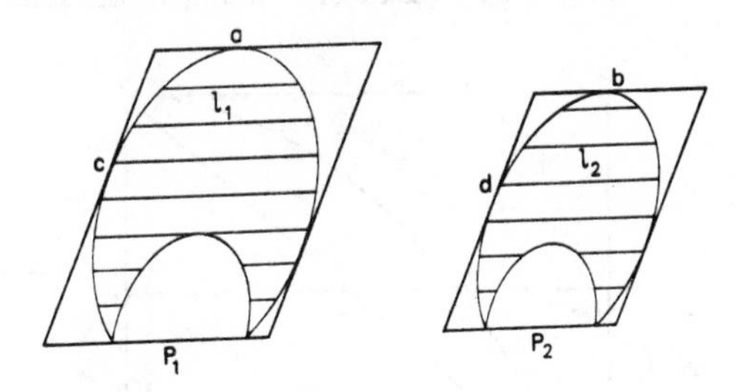

FIG. 4.21

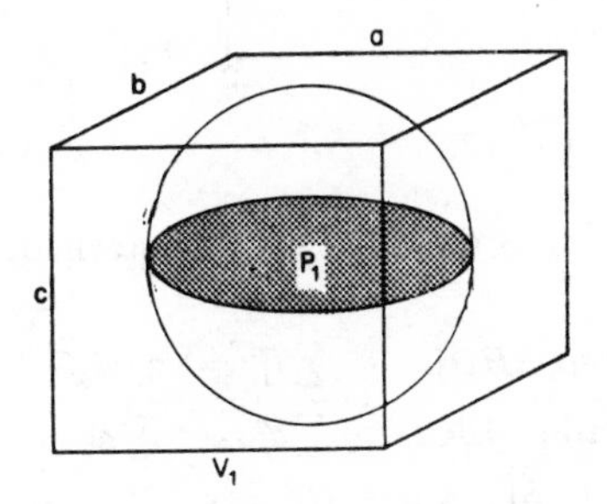

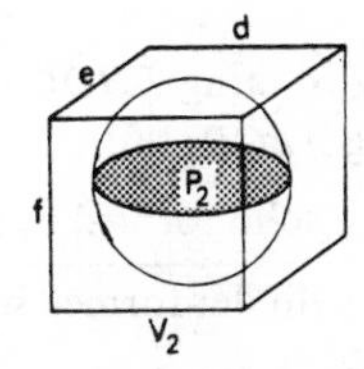

FIG. 4.22

In the systematic integration methods which follow, the concept of the *powers* of line elements plays an important role. Consider a plane curve, *DHB*, inscribed in the rectangle *ABCD* (see Fig. 4.23). Taking *AB* as *regula* let *EF*, drawn parallel to *AB*, cut the curve in *H* and the diagonal *BD* in *G*. Now, if $HF = l_1$, and $GF = l_2$, then

$$\frac{\text{space } CBHD}{\text{space } CBAD} = \frac{\sum l_1}{\sum a} \qquad \text{(where } AB = CD = a\text{)}.$$

† *Ibid.*, Book II, pp. 133–45.

If, for a particular curve *DHB*, it is possible to express l_1 in terms of some power of l_2, i.e. $l_1 = f(l_2)$, then

$$\frac{\sum l_1}{\sum a} = \frac{\sum f(l_2)}{\sum a}.$$

The space *BHDC* can thus be obtained by considering *the sums of the powers of the lines* of the triangle *BDC*. The volume of the solid of revolution

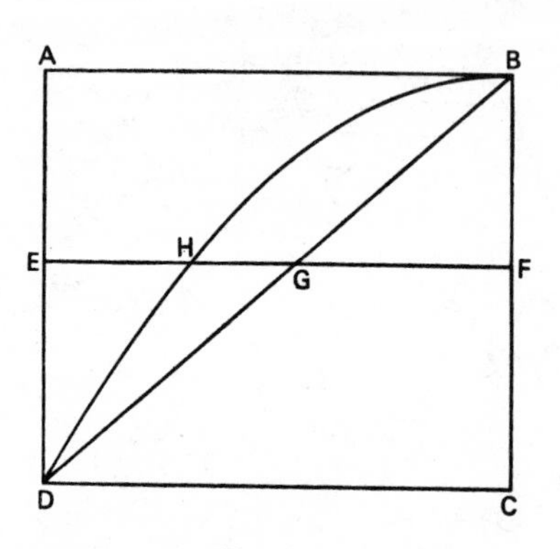

FIG. 4.23

formed by rotating *BHDC* about *BC* is given in terms of the cylinder formed by rotating *ABCD*; thus,

$$\frac{\text{solid formed by rotating } BHDC}{\text{cylinder formed by rotating } ABCD} = \frac{\sum l_1^2}{\sum a^2} = \frac{\sum [f(l_2)]^2}{\sum a^2}.$$

For curves of the form, $y/a = (x/b)^n$, where $HF = y$, $BF = x$, $AD = b$, $AB = a$, we have

$$\frac{\text{space } BHDC}{\text{space } ABCD} = \frac{\sum l_1}{\sum a} = \frac{\sum y}{\sum a}.$$

Since, $l_1/a = (x/b)^n$ and $l_2/a = x/b$, then $l_1/a = (l_2/a)^n$. Hence,

$$\frac{\text{space } BHDC}{\text{space } ABCD} = \frac{\sum l_1}{\sum a} = \frac{\sum (l_2)^n}{\sum a^n}.$$

For the volume of the solid of revolution formed by rotation about *BC*, we have

$$\frac{\text{solid formed by rotating } BHDC}{\text{solid formed by rotating } ABCD} = \frac{\sum (l_1)^2}{\sum a^2} = \frac{\sum (l_2)^{2n}}{\sum a^{2n}}.$$

In this way a whole range of quadratures and cubatures can be reduced to the determination of *the sums of the powers of the lines of a triangle* taken parallel to a given *regula.* In the *Geometria* Cavalieri considers only $\sum l$ and $\sum l^2$ but even here he is able to demonstrate the efficacy of the method by applying it to the ellipse, parabola, hyperbola, pyramid, cone, various solids of revolution and the Archimedean spiral. In the *Exercitationes*, stimulated to further efforts by suggestions received from France,† he launched a further

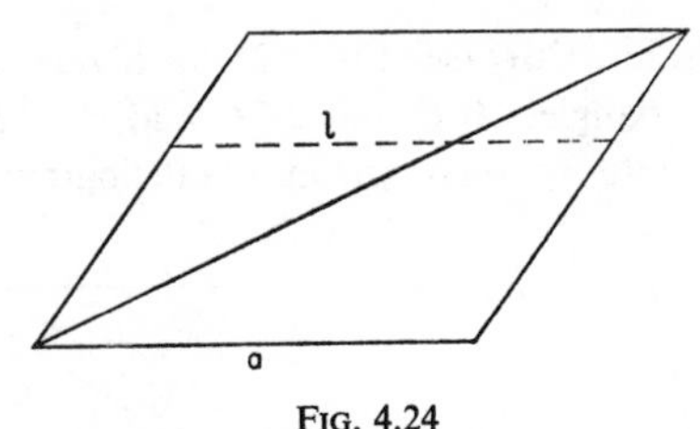

FIG. 4.24

attack on the problem and, using a consistent method throughout, determined successively the sums of the powers of the lines of the triangle up to and including $\sum l^9$,

$$\text{i.e.}\quad \frac{\sum l}{\sum a} = \frac{1}{2}, \quad \frac{\sum l^2}{\sum a^2} = \frac{1}{3}, \quad \frac{\sum l^3}{\sum a^3} = \frac{1}{4},$$

$$\frac{\sum l^4}{\sum a^4} = \frac{1}{5}, \quad \frac{\sum l^5}{\sum a^5} = \frac{1}{6}.$$

Tabulating results he was thus able to conclude that, for n positive and integral,

$$\frac{\sum l^n}{\sum a^n} = \frac{1}{n+1}.$$

The ingenious approach which Cavalieri devised to arrive at the above conclusions depends in part on the purely geometric concepts of congruence,

† The first published reference to the quadrature of the curves, $y/a = (x/b)^n$ is in the *Cogitata* of Mersenne and refers to Roberval. In a letter to Fermat (11th Oct. 1636) Roberval notes that he has solved the problem (*Oeuvres de Fermat* II, p. 81). Descartes and Fermat possessed the equivalent. (See Mersenne, M., *Cogitata physico-mathematica*, Paris, 1644.)

similarity and equality, but also on the application of an algebraic result regarding binomial coefficients received from Beaugrand† which in modern notation is equivalent to

$$(a+b)^n+(a-b)^n = 2(a^n+nC_2a^{n-2}b^2+nC_4a^{n-4}b^4+ \ldots).$$

For example, in order to determine the sums of the fifth powers,‡ Cavalieri makes use of

$$(a+b)^5+(a-b)^5 = 2a^5+20a^3b^2+10ab^4.$$

Let the parallelogram $ACGQ$ (see Fig. 4.25) be divided by the diagonal CQ into two congruent triangles ACQ and CQG and by the lines BN and HD (intersecting at M) into four congruent parallelograms $ABMD$, $BCHM$,

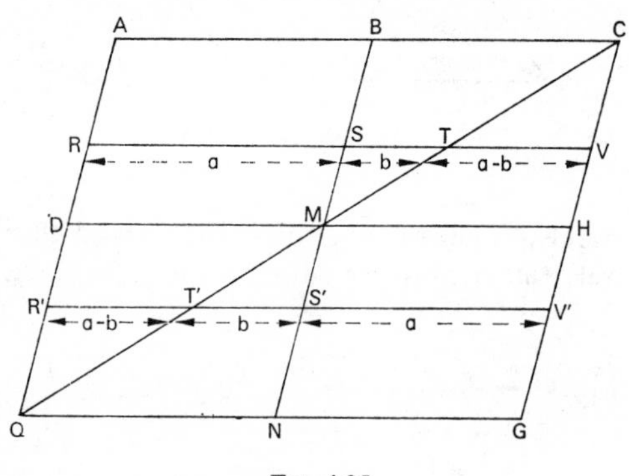

FIG. 4.25

$DMQN$ and $MHGN$. Now let $RSTV$ and $R'T'S'V'$ be drawn parallel to, and equidistant from, DH to cut AQ in R and R', BN in S and S', CQ in T and T' and CG in V and V' respectively. If $RS = S'V' = a$ and $ST = S'T' = b$, then $TV = T'R' = a-b$.

For the parallelogram $ACHD$, we have

$$\sum^{ACMD} (a+b)^5+\sum^{CHM} (a-b)^5 = 2\sum^{ABMD} a^5+20\sum^{BCM} a^3b^2+10\sum^{BCM} ab^4,$$

and for $DHGQ$,

$$\sum^{MHGQ} (a+b)^5+\sum^{DMQ} (a-b)^5 = 2\sum^{MHGN} a^5+20\sum^{QMN} a^3b^2+10\sum^{QMN} ab^4.$$

† *Ex. geom. sex*, p. 283.
‡ *Ibid.*, pp. 286–7.

Taking the summations for $ACHD$ and $DHGQ$ together,

$$\sum^{ACGQ}(RT^5+TV^5) = 2\sum^{ABNQ} a^5+40\sum^{BCM} a^3b^2+20\sum^{BCM} ab^4$$

(since $ABMD \equiv MHGN$ and $\triangle BMC \equiv \triangle QMN$).

But, since $\triangle ACQ \equiv \triangle CQG$,

$$\sum^{ACQ} RT^5 = \sum^{CGQ} TV^5,$$

and
$$\sum^{ACGQ}(RT^5+TV^5) = 2\sum^{ACQ} RT^5.$$

From previous results, we have $\sum^{BCM} b^2 = \frac{1}{3}\sum^{BCHM} a^2 = \frac{1}{6}\sum^{BCGN} a^2$,

and

$$\sum^{BCM} b^4 = \tfrac{1}{5}\sum^{BCHM} a^4 = \tfrac{1}{10}\sum^{BCGN} a^4.$$

Hence,

$$\sum^{ACGQ}(RT^5+TV^5) = 2\sum^{ACQ} RT^5 = 2\sum^{ABNQ} a^5+\tfrac{20}{3}\sum^{BCGN} a^5+2\sum^{BCGN} a^5$$

$$= \tfrac{32}{3}\sum^{ABNQ} a^5 = \tfrac{1}{3}\sum^{ACGQ} AC^5$$

(since $AC = 2a$, $AC^5 = 32a^5$).

Finally,

$$\sum^{ACQ} RT^5 = \tfrac{1}{6}\sum^{ACGQ} AC^5.$$

Cavalieri was thus enabled to draw up a table of integrals† of the form $\int_0^1 x^n\, dx$, n being positive and integral (see Fig. 4.26). The entries in the first column represent the areas $BGCD$, $BHCD$, $BICD$, ..., expressed as ratios of the rectangle $ABCD$: those in the second column represent the volumes of the solids obtained by rotating $BGCD$, $BHCD$, $BICD$,..., about BD. In the third column (and also in subsequent tables) Cavalieri extends the results beyond the field where concrete meaning can be attached to the resulting integral forms: the concepts he is developing relate to n-dimensional space.

The residual areas, $ABGC$, $ABHC$, $ABIC$,..., can be arrived at by subtraction and correspond to the integrals $\int_0^1 (1-x^n)\, dx$: the volumes of solids of revolution formed by rotating the curves about AC give successive integrals as follows: $\int_0^1 (1-x)^2\, dx$, $\int_0^1 (1-x^2)^2\, dx$, $\int_0^1 (1-x^3)^2\, dx$,.... These Cavalieri obtains by combining results from the first set of tables.‡ By the same process

† *Ibid.*, p. 306.
‡ *Ibid.*, pp. 309–10.

he resolves Kepler's problem, to determine the volume of the solid formed by rotating a parabolic segment about an ordinate.†

For the Archimedean spiral‡ ($r = a\theta$) let the line elements of the spiral space S and the circular space C be denoted by l_1 and l_2 respectively. If these

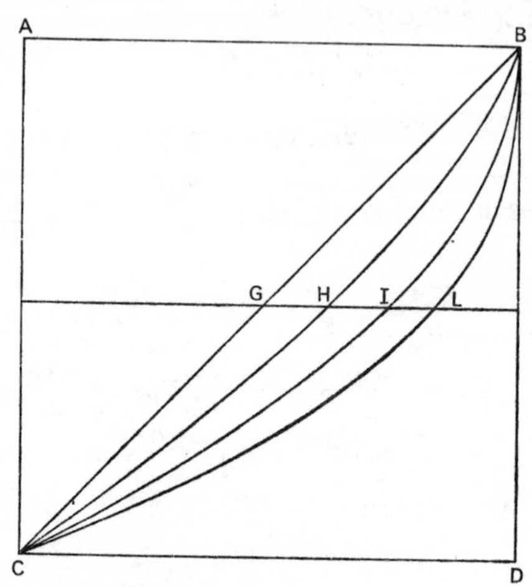

	Sum of the lines	Sum of the squares	Sum of the cubes
1st space	$\frac{1}{2}\left[\int_0^1 x\,dx\right]$	$\frac{1}{3}\left[\int_0^1 x^2\,dx\right]$	$\frac{1}{4}\left[\int_0^1 x^3\,dx\right]$
2nd space	$\frac{1}{3}\left[\int_0^1 x^2\,dx\right]$	$\frac{1}{5}\left[\int_0^1 x^4\,dx\right]$	$\frac{1}{7}\left[\int_0^1 x^6\,dx\right]$
3rd space	$\frac{1}{4}\left[\int_0^1 x^3\,dx\right]$	$\frac{1}{7}\left[\int_0^1 x^6\,dx\right]$	$\frac{1}{10}\left[\int_0^1 x^9\,dx\right]$
4th space	$\frac{1}{5}\left[\int_0^1 x^4\,dx\right]$	$\frac{1}{9}\left[\int_0^1 x^8\,dx\right]$	$\frac{1}{13}\left[\int_0^1 x^{12}\,dx\right]$

FIG. 4.26

elements be *unrolled* in the rectangle $ABCD$ (see Fig. 4.27) in which $BC = R$ and $CD = 2\pi R$, we have,

$$\frac{\text{spiral space } S}{\text{circular space } C} = \frac{\sum l_1}{\sum l_2} = \frac{\sum GX}{\sum NX} = \frac{\text{space } BGDC}{\text{space } BNDC}.$$

† *Ex. geom. sex*, pp. 243–4.
‡ *Geom. ind.*, Book VI.

But, since $R = 2\pi a$, then $l_1 = r\theta = r^2/a = (2\pi/R)r^2$, and the curve *BGD* is a semi-parabola, axis *AB*.

Hence,

$$\frac{\text{space } BGDC}{\text{space } BCD} = \frac{2}{3},$$

spiral space $S = \frac{2}{3}$ circular space C,
and spiral space $A = \frac{1}{3}$ circular space C.

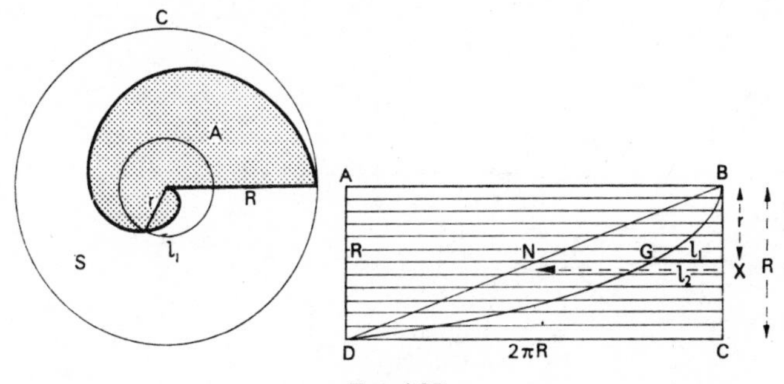

FIG. 4.27

The actual results obtained by Cavalieri were, for the most part, also arrived at independently by a number of his contemporaries and many of the problems which he solved were originally formulated by Fermat. Nevertheless, the indivisible methods which he so stoutly defended were entirely his own and accusations of plagiarism, particularly in relation to Kepler, are clearly unjustified. Guldin, in particular, attempted to bring discredit on Cavalieri by suggesting that the methods were not his own and, furthermore, that the concept of a plane figure as the sum of its lines led to paradoxes. For example,† in the triangle *AHG* (with $AH \neq HG$) let *HD* be drawn perpendicular to the base *AG* (see Fig. 4.28) and let *KM*, *JL*, *NR*, . . ., be a set of lines drawn parallel to *AG* to cut the unequal sides *AH* and *HG* in *K*, *J*, *N*, and *M*, *L*, *R*, respectively. Now, if from these points perpendiculars *KB*, *JC*, *NE*, and *MI*, *LO*, *RF*, be drawn to the base *AG*, we have, $KB = MI$, $JC = LO$ and $NE = RF$, and so on for all such lines. Hence, Guldin argues, by the theory of indivisibles,

$$\sum^{ADH} JC = \sum^{GDH} LG,$$

† *Ex. geom. sex*, pp. 238–9.

and triangle ADH = triangle GDH, which is clearly absurd. Cavalieri points out that this constitutes a misuse of indivisible methods in the sense that the *distribution* of the lines JC, NE,..., in the triangle ADH clearly differs from that of the lines RF, LO,..., in the triangle GDH: the spaces cannot therefore be compared by comparing the sums of the lines. He refers† to the case of two

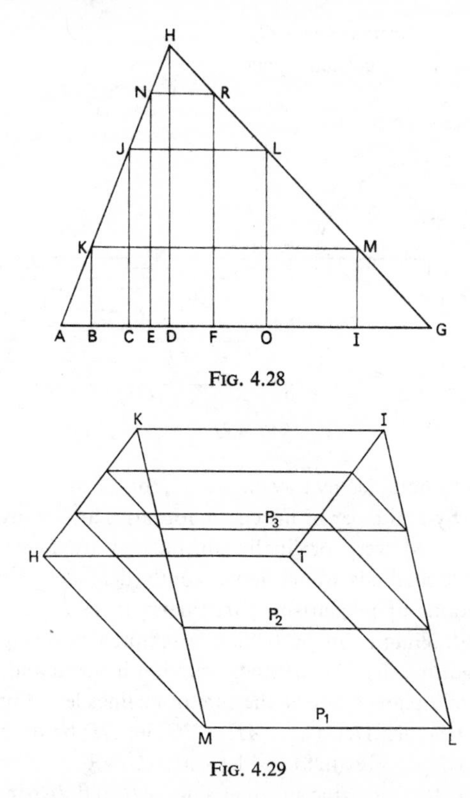

FIG. 4.28

FIG. 4.29

rectangles $KILM$, $KITH$ (see Fig. 4.29), intersecting in a common line KI. Although a set of equidistant planes, P_1, P_2, P_3,..., may make equal intercepts on the planes $KILM$, $KITH$, the rectangles are none the less *unequal*, for their line elements, although equal, are not *equally distributed.* Although this idea of the *distribution* of indivisibles was never very clearly developed by Cavalieri he was careful only to compare figures in which the distribution of

† *Ex. geom. sex.* pp. 15–16.

indivisibles was, in fact, uniform. The summation of lines and planes in the Cavalierian sense is therefore formally equivalent to the addition of rectangular elements of equal thickness,

$$\frac{\sum l_1 b}{\sum l_2 b} = \frac{\sum l_1}{\sum l_2},$$

where $b \to 0$. Many of Cavalieri's contemporaries were quick to see this and to make the required adjustments. Wallis and Roberval, in particular, considered the use of lines and planes entirely justifiable since both derived from infinitely narrow rectangles and sections of uniform thickness. The analogies which Cavalieri drew† between the line indivisibles of a plane surface and the parallel threads of a woven fabric, and the surface indivisibles of a solid and the parallel leaves of a book, did not do much to further the acceptance of Cavalierian indivisibles as valid mathematical concepts. However, the extension of Cavalierian method in which a surface was treated not as a set of parallel ordinates but as a very great number of equally narrow rectangles was rapidly developed by Wallis, Fabri, Lalovera and many others. But the geometric integration methods of the Flemish mathematician, Grégoire de Saint-Vincent, are so different from those of Cavalieri that they certainly merit consideration as an independent contemporary contribution.

4.4. Grégoire de Saint-Vincent (1584–1667)

Grégoire de Saint-Vincent, the most gifted pupil of Clavius (1537–1612), received a sound grounding in Greek mathematics and was certainly acquainted with the works of Stevin and Valerio. The integration methods which he devised, probably during the years 1622–9, constituted an extension of the methods of Archimedes and in no sense a development of the indivisible techniques of Galileo and Cavalieri.‡ Unfortunately the original manuscript was lost for many years and was not prepared for publication until 1647. Even so, the *Opus geometricum*§ attracted a great deal of attention, mainly because it contained yet a further attempt to square the circle, but also because of the systematic approach to volumetric integration developed under the name

† *Ibid.*, p. 3: "Hinc manifestum est figuras planas nobis ad instar telae parallelis filis contextae concipiendas esse: solida vero ad instar Librorum, qui parallelis folÿs coaceruantur."

‡ See Bosmans, H., Grégoire de Saint-Vincent, *Mathésis* 38 (1924), pp. 250–6. Hofmann, J. E., Das Opus geometricum des Gregorius a S. Vincentio und seine Einwirkung auf Leibniz, *Abh. d. Preuss. Ak. d. Wiss.*, 1941: Math.-naturw. Klasse (13), Berlin, 1941.

§ Gregorius a S. Vincentio, *Opus geometricum quadraturae circuli et sectionum coni*, Antwerp, 1647.

ductus plani in planum.† The attempt to square the circle was faulty, the integration methods ingenious and influential but there was, notwithstanding, much other material of value in the book. In seventeenth-century integration methods geometric series played a significant part and we are indebted to Grégoire for the clearest early account of the summation of geometric series‡ (*series geometrica*). A geometric progression is defined as a finite number of terms from a geometric series. However great the number of terms, a progression cannot include the *infinite* part of a geometric series but it can be extended so far that the sum of the terms of the progression differs from the sum of the series by less than any assigned magnitude.

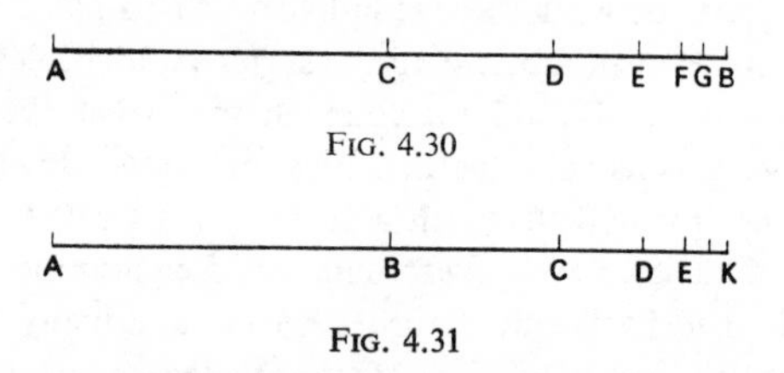

FIG. 4.30

FIG. 4.31

From the relation $AB/CB = CB/DB = DB/EB = \ldots$ (see Fig. 4.30) which determines the geometric series, AB, CB, DB, EB, ..., Grégoire notes that, since

$$\frac{AB-CB}{CB-DB} = \frac{CB-DB}{DB-EB} = \ldots \frac{AB}{BC},$$

then $AC/CD = CD/DE = DE/EF = EF/FG = \ldots$, and the series of differences, AC, CD, DE, ..., also constitutes a geometric series, such that $AC/CD = AB/BC$, where B is a point on the line $ACDEFG\ldots$ beyond all points in the sequence A, C, D, E, F,.... The difference between AB and the sum of any finite number of terms of the series AC, CD, DE,..., can be made less than any assigned magnitude by taking a great enough number of terms. Conversely therefore, given any geometric series AB, BC, CD,... (see Fig. 4.31), it is possible to determine a point K on the line $ABCDE\ldots$ such that, $AB/BC = BC/CD = \ldots = AK/BK$: AK thus represents the sum of the series. Finally the magnitude AK is determined by a geometric construction (see Fig. 4.32).

Although Grégoire's treatment of geometric series lacks the modern analytic approach it was entirely original and shows that he had grasped the idea

† *Opus geom.* VII, pp. 703–864.
‡ *Ibid.* II, pp. 51–166.

of an infinite series as defining a magnitude (the sum of the series) to which the sum of any finite number of terms approaches as the number of such terms becomes progressively greater. He goes on to consider a number of applications of the above including a pleasing exposition† of the paradox of Achilles and the tortoise. Zeno, he notes, in asserting that Achilles could not catch the tortoise, had failed to recognise that the time intervals were in falling geometric progression, and that in this case, although the number of such intervals is infinite their sum is finite.

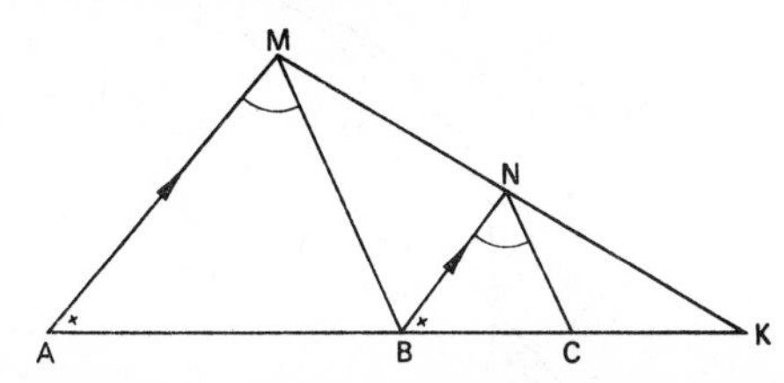

FIG. 4.32. Let $AM/BN = AB/BC$. Since, by similar triangles, $AM/BN = AK/BK$, then $AB/BC = AK/BK$, and AK is the sum of the series.

Grégoire's treatment of the angle of contact‡ (i.e. the angle between curve and tangent at the point of contact) is interesting for it exemplifies the peculiar difficulties attendant upon the formation of any adequate conceptual basis for the development of tangent methods. The unsatisfactory Euclidean concept of curvilinear angle obtruded itself here and prevented any real progress. If two circles be described with centres on the line BH and if BA be their common tangent (see Fig. 4.33) it is intuitively clear that

$$\text{angle of contact } ABC < \text{angle of contact } ABE.$$

Comparing both of these angles with the rectilinear angle ABH, we have

$$\widehat{ABH} > \widehat{ABH}/2 > \widehat{ABH}/2^2 > \widehat{ABH}/2^3 > \ldots \widehat{ABH}/2^n > \widehat{ABC},$$

and similarly,

$$\widehat{ABH} > \widehat{ABH}/2 > \widehat{ABH}/2^2 > \widehat{ABH}/2^3 > \ldots \widehat{ABH}/2^n > \ldots \widehat{ABE}.$$

Clearly, as $n \to \infty$, $\widehat{ABH}/2^n \to 0$, and $\widehat{ABC} = \widehat{ABE} = 0.$

† *Ibid.* II, pp. 101–3.
‡ *Ibid.* VIII, p. 871.

Grégoire, however, found himself unable to accept this formally correct deduction and instead asserted that, in the field of infinitesimals, the fundamental principles of geometry cease to apply. In particular, he challenged the validity of the axiom, "the whole is greater than the part". This passage later aroused the interest of Leibniz and became the starting point for some of his earliest mathematical investigations.†

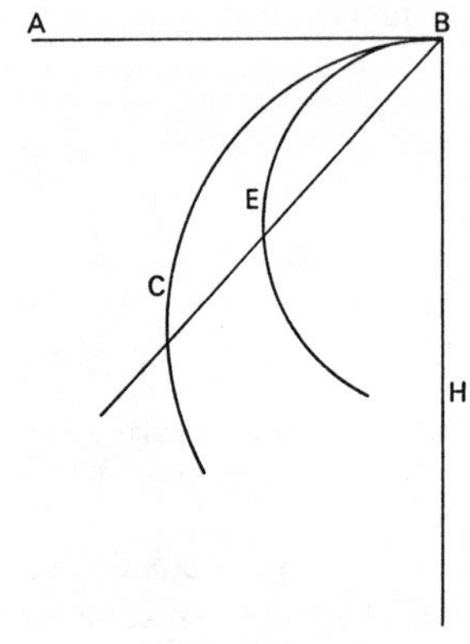

FIG. 4.33

In the general treatment of the conic sections given in Books IV–VI there is not a great deal of interest‡ although the formal approach found favour with those who rejected Cavalierian methods and welcomed a return to the geometric methods of the Greeks. The relation between the Archimedean spiral and the parabola is developed here§ but by 1647 represented nothing new. Much useful and suggestive material does, however, appear in the section on the hyperbola‖ where Grégoire does everything save give explicit recognition to the relation between the area of the hyperbolic segment and the logarithm.†† The method used here is to construct, by means of a sequence

† Cf. Hofmann, J. E., Das Opus geometricum, *Abh. d. Preuss. Ak. d. Wiss.*, 1941, p. 23, n. 60.

‡ See Bopp, K., Die Kegelschnitte des Gregorius a S. Vincentio, *Abh. z. Gesch. d. Math. Wiss.* 20 (1907), pp. 83–314.

§ *Opus geom.* VI, pp. 664–702.

‖ *Ibid.* VI, pp. 583–603.

†† Both Fermat and Torricelli also made use of geometric progressions in the quadrature of the parabolas $(y/a)^m = (x/b)^n$, hyperbolas $(y/a)^m\ (x/b)^n = 1$, and the spirals $(r/a)^m = (\theta/\alpha)^n$.

of ordinates in geometric progression, a set of hyperbolic segments equal in area.

In modern notation, for the rectangular hyperbola, $xy = c^2$, we have

Abscissae: x_1	$x_1 t$	$x_1 t^2$	$x_1 t^3$	...	$x_1 t^r$,
Differences:	$x_1(1-t)$	$x_1 t(1-t)$	$x_1 t^2(1-t)$	...	
Ordinates: y_1	$\frac{y_1}{t}$	$\frac{y_1}{t^2}$	$\frac{y_1}{t^3}$	...	
Areas:	$x_1 y_1(1-t)$	$x_1 y_1(1-t)$	$x_1 y_1(1-t)$	...	

Hence, if

$$\frac{x_{r+m}}{x_r} = t^m, \quad \text{and if} \quad \frac{x_{s+n}}{x_s} = t^n,$$

then,

$$\frac{\int_{x_r}^{x_{r+m}} y\, dx}{\int_{x_s}^{x_{s+n}} y\, dx} = \frac{m}{n} = \frac{\log (x_{r+m}/x_r)}{\log (x_{s+n}/x_s)}.$$

Basically these processes are closely related to the development of the integral calculus and the numerical methods which were subsequently developed for the calculation of logarithms form a fascinating study.†

The *ductus plani in planum*, Grégoire's geometric integration method, is developed in detail in Book VII. Although much of the material is purely systematic the book contains a wealth of interesting examples and some new results. In general, by *ductus in planum* is meant the *construction* of solids by means of two plane surfaces standing on the same *ground line*. At right angles to this ground line a set of equidistant parallels is drawn (see Fig. 4.34).

Let the first figure be *ABCD* and the second *ABG* (ground line *AB*), and let one of the parallels, *MN*, cut *ABCD* in *H*, *I*, and *ABG* in *I*, *K*. Now, taking *ABCD* as the base of a solid, let rectangles be constructed on each of the lines *HI*, in a plane perpendicular to *ABCD* and with altitude *KI*. The resulting solid (see Fig. 4.35) is the ductus of *ABCD* in *ABG*.

Grégoire goes on to consider simple cases in which solids are constructed from rectilinear figures. The cube, for example, is the ductus of a square in a square (*ductus plani in se*); a triangular prism is the *ductus* of a square or

† D. T. Whiteside has recently given an excellent account of the extension of such methods throughout the seventeenth century until their eventual incorporation in the general theory of integration. See Patterns of mathematical thought, *Archive for History of Exact Sciences* I (1961), Chap. III, pp. 214–31.

rectangle in a triangle; a square pyramid is the *ductus* of a right-angled triangle in itself; a triangular pyramid is the *ductus* of a right-angled triangle *subalterne* (see Figs. 4.36–8). In extending the method to the construction of solids bounded by curved surfaces much use is made of the relation between the rectangles formed by segments of intersecting chords of a circle. If, for

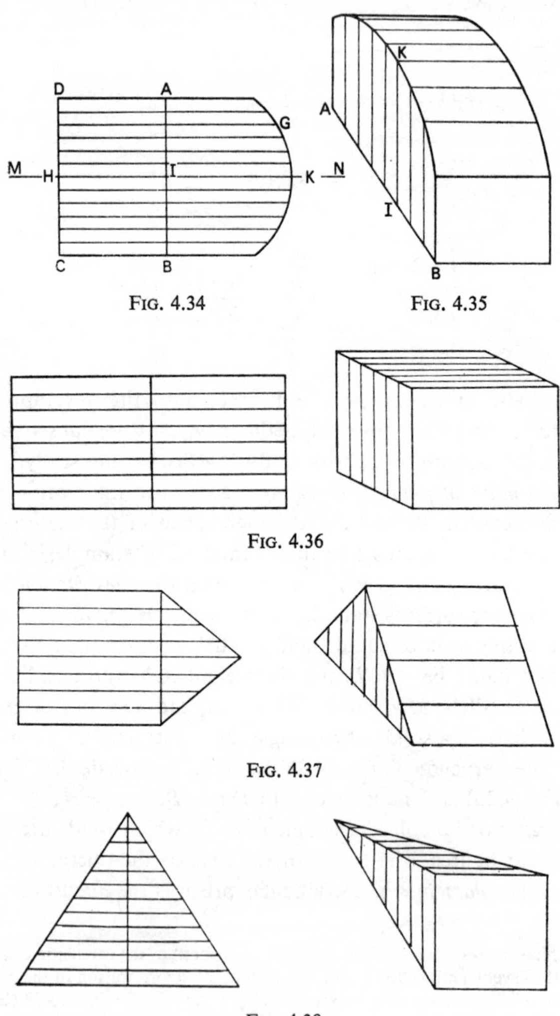

FIG. 4.34

FIG. 4.35

FIG. 4.36

FIG. 4.37

FIG. 4.38

example, two circles intersect in a common chord, AB, then, for any line perpendicular to AB (see Fig. 4.39) cutting AB in P, C_1 in D, C, and C_2 in E, G, we have

$$DP \,.\, PC = AP \,.\, PB = EP \,.\, PG.$$

Hence the *ductus* of AD_1D_2B in ACB equals the *ductus* of AG_1G_2B in AEB (see Fig. 4.39).† If AB be the diameter of the circle C_2 then $EP = PG$, and the *ductus* of AD_1D_2B in ACB equals the *ductus* of the semicircle AEB in itself.‡

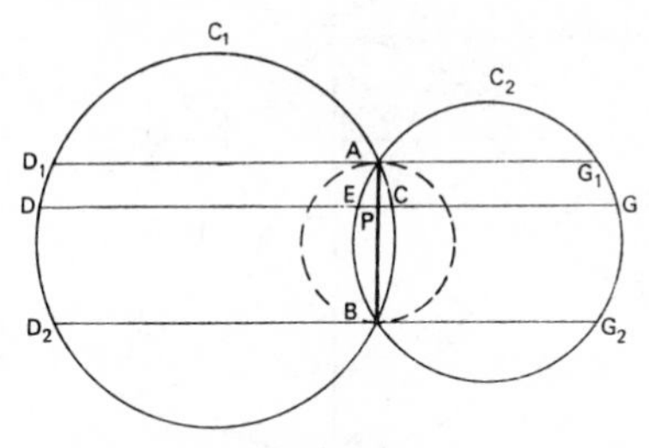

FIG. 4.39

† Let C_1 be $x^2+y^2 = a^2+b^2$, then *ductus* AD_1D_2B in ACB is given by

$$\int_{-a}^{+a} (x+b)\,(x-b)\,dy = \int_{-a}^{+a} (x^2-b^2)\,dy = \int_{-a}^{+a} (a^2-y^2)\,dy = \frac{4}{3}a^3$$

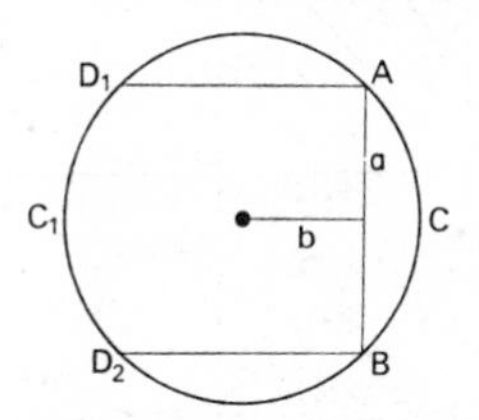

‡ Let C_2 be $x^2+y^2 = a^2$; then the ductus of the semicircle APB in itself is given by

$$\int_{-a}^{+a} x^2\,dy = \int_{-a}^{+a} (a^2-y^2)\,dy = \frac{4}{3}a^3.$$

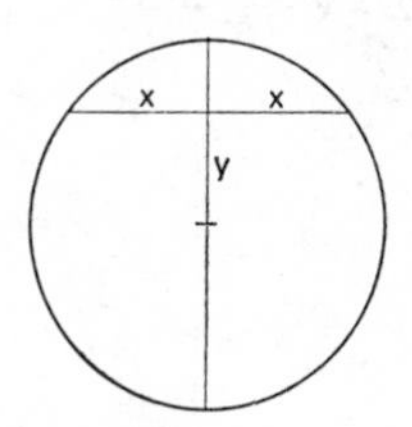

Now, if two right-angled isosceles triangles ABC and ABD be constructed on the diameter AB, then for any lines $UVPWQ$ (see Fig. 4.40) intersecting CB, AD in U, W, the circle in V, Q, and the diameter AB in P, we have

$$UP \,.\, PW = AP \,.\, PB = VP \,.\, PQ = VP^2,$$

and the *ductus* of BAC in BAD is equal to the *ductus* of the semicircle AVB in itself.† Hence, the former being known (a triangular pyramid), the latter —and indeed the entire range of related solids—is determined. Tangent prop-

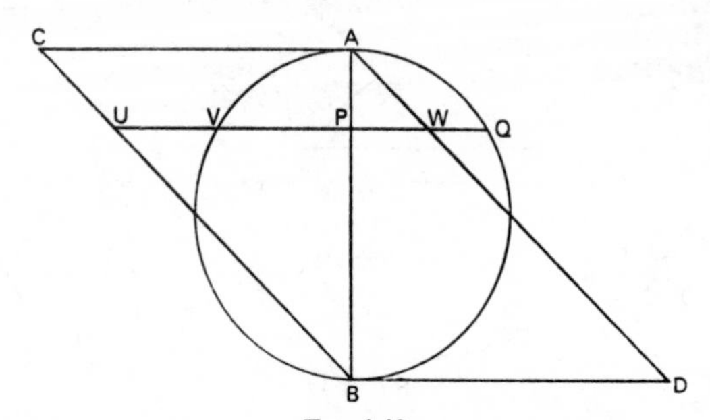

Fig. 4.40

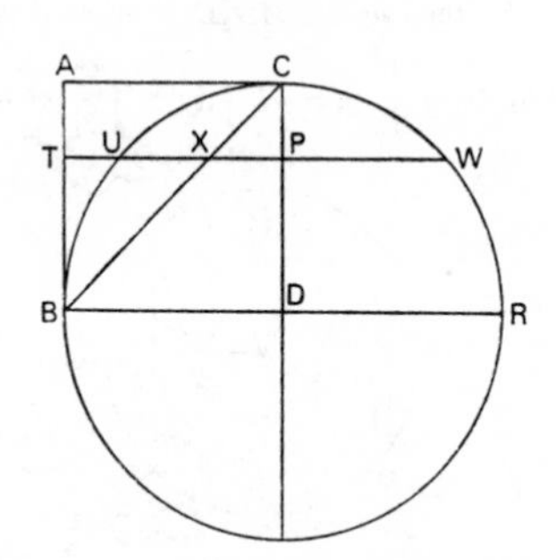

Fig. 4.41

† $\int_0^{2a} x(2a-x)\,\mathrm{d}y = 4a^3/3 \quad (2a-x = y,\ x = 2a-y).$

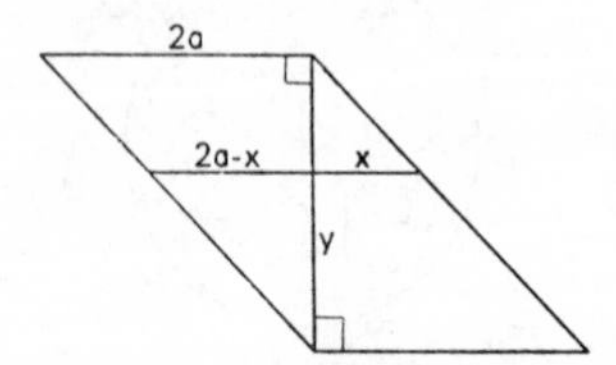

erties also supply further interesting results: if a square *ABDC* be constructed on the radius, *CD*, of a semicircle, *BCR*, then, for any line *TUPW*, cutting the circle in *U*, *W*, the square in *T*, *P*, we have, $TU.TW = TB^2$ (see Fig. 4.41). Now, let the diagonal *BC* cut *TP* in *X*, then $TB = TX$ and $TU.TW = TX^2$. Hence the *ductus* of *BACU* in *BACWR* equals the *ductus* of the triangle *BAC* in itself.†

A single example must suffice to illustrate yet a further extension: to the construction of solids from figures bounded by parabolic segments; let *CAE* (see Fig. 4.42) be a parabolic segment, axis *AB*, inscribed in a double square *DFEC*; let a circle be inscribed on *AB* as diameter and let any line parallel to *CE* cut *AB* in *S*, *CD*, *EF*, in Q_1, Q_2, the parabola in P_1, P_2, the diagonal *AC* in *T* and the circle in R_1, R_2.

$$\begin{aligned} \text{Since } R_1S^2 = AS \,.\, SB &= TS(AB-TS) \qquad (TS = AS) \\ &= TS(Q_1S-TS) \\ &= TS \,.\, Q_1T, \end{aligned}$$

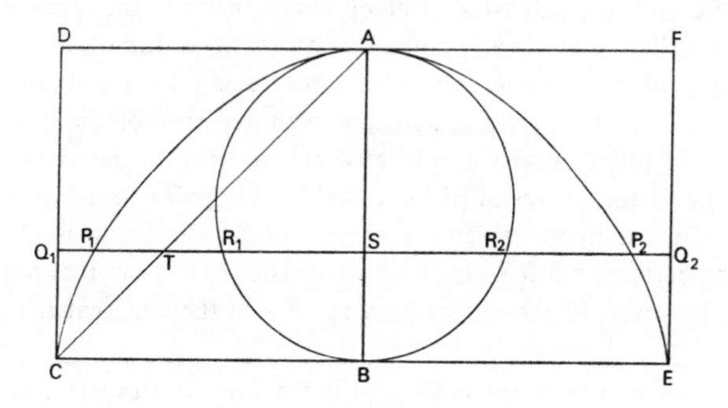

FIG. 4.42

† $\int_0^a (a+x)\,(a-x)\,dy = \int_0^a (a^2-x^2)\,dy = \int_0^a y^2\,dy \quad (x^2+y^2 = a^2)$.

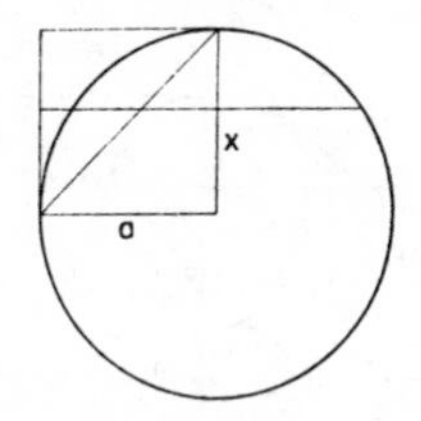

and since, $P_1T \,.\, TP_2 = (P_1S - TS)(P_1S + TS) = P_1S^2 - TS^2$

$$= AS \,.\, Q_1S - AS^2 \qquad (P_1S^2 = AS \,.\, Q_1S)$$
$$= AS(Q_1S - TS)$$
$$= TS \,.\, Q_1T,$$

then $R_1S^2 = P_1T \,.\, TP_2.$

The *ductus* of the parabolic segment, AP_1C in AP_2EC is accordingly equal to the *ductus* of the semicircle, AR_1B in itself.†

In all these examples Grégoire is making extensive use of spatial imagery to create a multitude of solids, the volumes of which reduce to a single construction depending on the *ductus* of a rectilinear figure. Whilst the use of algebraic notation and the integral calculus makes most of his results seem trivial, in the absence of such notation systematic geometric transformation fulfilled an essential role, not only in providing a means by which the volumes of solids could be compared, but also in suggesting a basis for the construction of a further supply of solids whose properties had not previously been studied.

Although, after the extension of algebraic methods to the study of curves, it became possible to create an infinite variety of curves and solids by the mere manipulation of symbols, as long as methods remained purely geometrical the range of known curves and solids was limited to those which could, in some sense, be imaginatively constructed or perceived in the external world. The concept of the *ductus in planum* enabled Grégoire to make use of the geometry of plane figures to construct new solids, the volumes of which he was able to derive from the Euclidean properties of the original figures. The weakness, however, of the whole structure lies in the fact that the power of

---

† If the equation of the parabola be given in the form, $y^2 = ax$, the above integral reduces to

$$\int_0^a (y-x)(y+x)\,dx = \int_0^a (y^2-x^2)\,dx = \int_0^a (ax-x^2)\,dx.$$

For the circle, we have

$$\int_0^a z^2\,dx = \int_0^a x(a-x)\,dx.$$

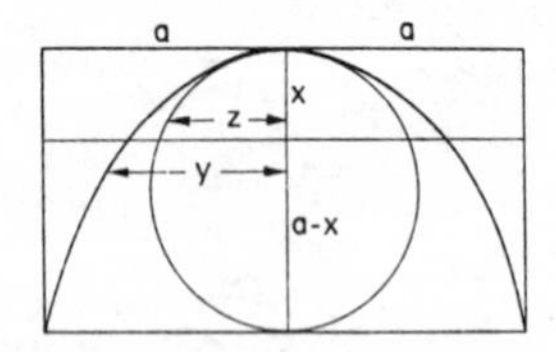

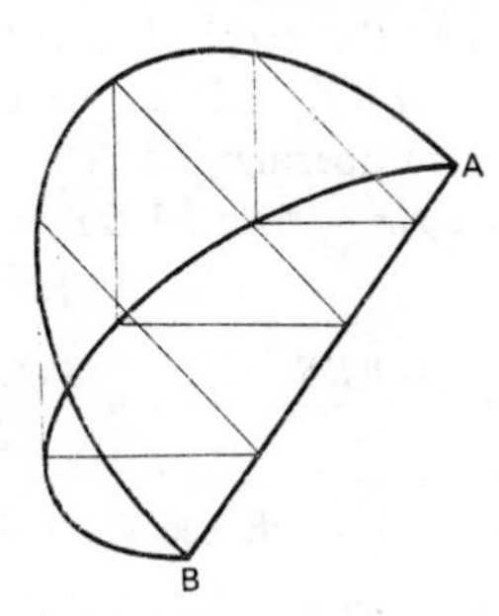

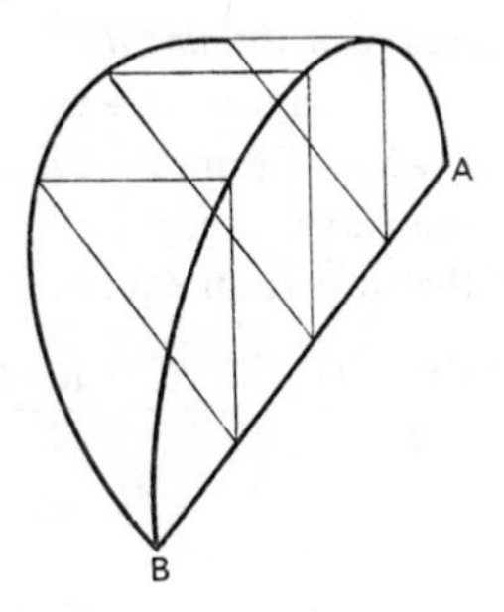

FIG. 4.43

the techniques developed is entirely dependent on the availability, or otherwise, of appropriate Euclidean properties.

In Book IX Grégoire shows how to employ the *ductus in planum* as a practical integration method in determining the volumes of cylindrical sections or *ungulae*. The *ungula* is formed by cutting a right circular cylinder by means of an oblique plane through a diameter of the circular base. In this way a solid is formed which is bounded by, (1) a semicircular base, (2) a semi-ellipse, (3) the portion of the cylindrical surface bounded by (1) and (2) above (see Fig. 4.43). If the height of the *ungula* be equal to the radius of the base, then sections taken perpendicular to the base are right-angled triangles and the sections of a *double ungula*, formed from cylinders with axes at right angles (see Fig. 4.44), are squares. The volume of the double *ungula* is, accordingly, equal to the *ductus* of the semicircle, diameter AB, in itself. To obtain the volume of the single *ungula* over a segment of the semicircle bounded by a chord CD, parallel to the diameter AB (see Fig. 4.45), consider

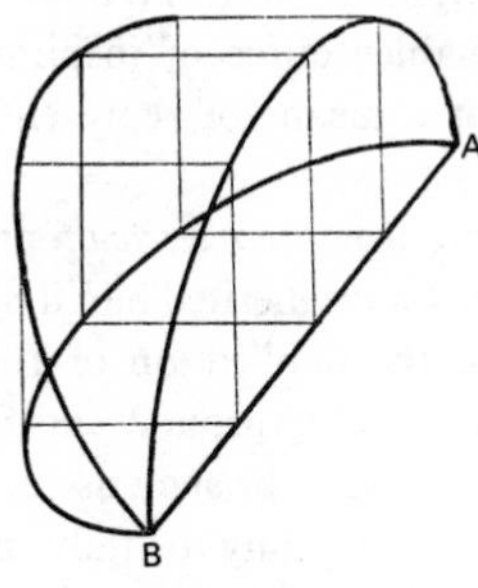

FIG. 4.44

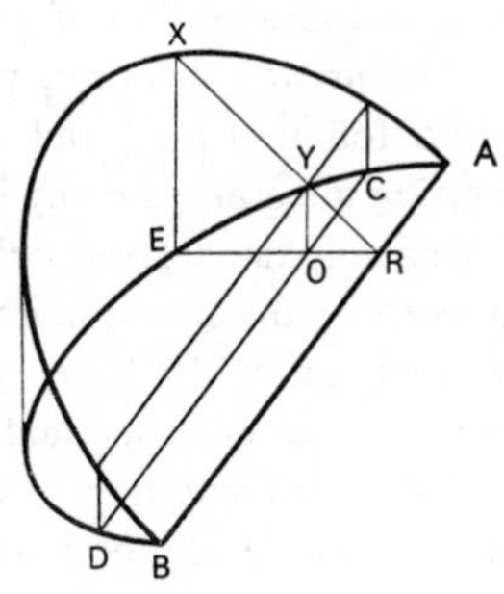

FIG. 4.45

a vertical section of the *ungula* over the line *ER*, perpendicular to *AB*. This section is a right-angled isosceles triangle, *ERX*, and the section over the segment *EO* (where *ER* meets *CD* in *O*) is a trapezium, *EXYO*.

Now, in the trapezium, *EXYO*, we have $EX = ER$ and $OY = OR$, hence the area of the trapezium equals,

$$\frac{EO(EX+OY)}{2} = \frac{EO(ER+OR)}{2} = \frac{EO(RS+OR)}{2} = \frac{EO.OS}{2}$$

($RS = ER$, see Fig. 4.46).

The volume of the segment over the base *CED* is thus equal to one-half of the *ductus* of *DEC* in *DBSAC*, and hence to one-half of the *ductus* of the semicircle, diameter *CD*, in itself.

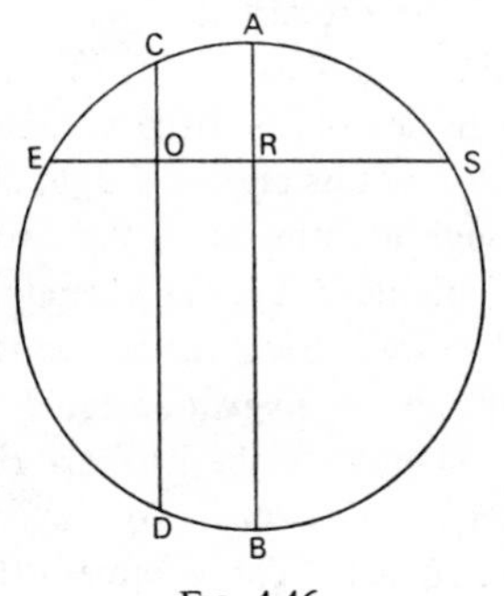

FIG. 4.46

Grégoire himself was primarily concerned to illustrate by reference to the *ungula* that volumetric integration could be reduced, through the *ductus in planum*, to a consideration of geometric relations between the lines of plane figures. The *ungula*, however, proved a valuable source of inspiration for those who followed him, and who saw in it a means of representing and transforming integrals in many ingenious ways.

Unfortunately the delayed publication of the *Opus geometricum* prevented it from receiving the attention it would certainly have merited had it appeared twenty years earlier. In 1647, ten years after the publication of Descartes' *La Géométrie*, algebraic methods were rapidly gaining ground and the form and manner of presentation of Grégoire's work was not such as to make it easy reading. Those who obtained the book did so mainly to study the faulty circle quadrature. Many who read it, however, became fascinated by the geometric integration methods and went on to make a deeper study of the entire work. Amongst those who gained much from the *Opus geometricum* should be counted Blaise Pascal whose *Traité des trilignes rectangles et leurs onglets* is based essentially on the *ungula* of Grégoire. Huygens recommended the section on geometric series to Leibniz who later came to make a thorough study of the entire work.[†] Tschirnhaus, friend and associate of Leibniz during his Paris years, found in the *ductus in planum* a valuable foundation for the development of his own algebraic integration methods.[‡]

Among the pupils and associates of Grégoire, A. A. de Sarasa (1618–67) made specific the relation between the hyperbolic segment and the logarithmic function.[§] Jean-Charles de la Faille (1597–1652) determined, by means of infinitesimals and the principle of moments, the centres of gravity of circular and elliptic segments and sectors.[‖] Paul Guldin (1577–1643), whose attacks on Kepler and Cavalieri have already been discussed, enunciated in general form the well-known theorem which bears his name,[††] i.e. the Pappus–Guldin theorem. The presence of this theorem in the *Collectiones* of Pappus does not appear to have been known until 1660. Antoine de Lalovera[‡‡] (1600–64) determined the centres of gravity of segments of the circle, ellipse, parabola and hyperbola from the relation $\bar{x}\int y\,dx = \int xy\,dx$. Using purely geometrical methods, the second integral, i.e. the turning moment about the axis, was expressed as a pure quadrature by the transformation, $az = xy$: thus, $\int xy\,dx = \int az\,dx$. André Tacquet[§§] (1612–60) returned to the determination of the volumes of cylindrical segments (the *ungulae* of Grégoire) also by purely geometrical methods. Of interest here is the attack he launched on Cavalierian indivisibles: a geometrical magnitude, he says, is made up of homogenea—a solid of solids, a surface of surfaces, a line of lines—not of heterogenea, or parts of lower dimensions as maintained by Cavalieri. Lines cannot be generated from points; solids from surfaces; and indeed no finite quantity can be generated from indivisibles.[‖‖]

† Hofmann, J. E., Das Opus geometricum, *op. cit.*, pp. 8–9.

‡ *Ibid.*, pp. 55–69.

§ Sarasa, A. A. de, *Solutio problematis a Mersenne propositi*, Antwerp, 1649.

‖ Faille, Jean-Charles de la, *De centro gravitatis partium circuli et ellipsis.*

†† Guldin, P., *Centrobaryca.*

‡‡ Lalovera, A. de, *Quadratura circuli et hyperbolae segmentorum*, Toulouse, 1651.

§§ Tacquet, A., *Cylindrica et annularia*, Antwerp, 1651. References given from *Opera mathematica*, Antwerp, 1669.

‖‖ *Ibid.*, p. 13.

In all this work the Jesuits showed a profound knowledge of Greek mathematics and immeasurable skill in the purely geometrical approach to infinitesimal methods. Unfortunately their complete neglect of algebra meant that time and energy was wasted in a search for special geometrical relationships and in the assembly of a collection of particular results. The techniques were all long and unwieldy and no effort was made to avoid repetition or to establish general theorems on which a systematic theory of integration could be built.

CHAPTER 5

Further Advances in France and Italy

5.0. Introduction

In France, during the first half of the seventeenth century, mathematical activity was at a high level. This was, at least in part, due to the enthusiastic interest of Marin Mersenne (1588–1648), like Descartes a pupil of the Jesuit College of La Flèche, who later joined the Minim Friars and organised meetings in his cell in Paris. The group which assembled there pursued a wide range of activities, but, because of Mersenne's own particular enthusiasm for mathematics and natural science, these subjects tended to dominate the discussions. On behalf of the group in Paris, Mersenne corresponded with mathematicians and scientists throughout Europe:† Galileo, Cavalieri and Torricelli in Italy were kept in touch with Roberval in Paris, Fermat in Toulouse and Descartes in Holland. In 1644, Mersenne visited Italy‡ and was able to establish personal contact with the Italian mathematicians.

Besides raising problems for consideration arising out of his own interests and distributing the queries of others, Mersenne made himself responsible for the circulation of manuscripts for comment. Although this may at times have created difficulties, through his good offices conflicting points of view were often resolved and greater clarity achieved.

During this period of intense mathematical activity only a small proportion of the enormous wealth of material produced received the benefit of direct publication: most of it was communicated by verbal means, through correspondence, and by inclusion in the works of others. Under the circumstances priority disputes were inevitable. With these disputes, however, we need not greatly concern ourselves, as the main interest lies rather in the

† Mersenne, M., *Correspondance* (ed. Waard, C. de), Paris, 1932–62.

‡ For a full account of Mersenne's Italian journey see *Oeuvres de Fermat* (ed. Tannery, P., and Henry, C.), Paris, 1891–1912 (4 vols.); *Supplément* (ed. Waard, C. de), Paris, 1922, Introd., pp. xii–xvi.

growth of general methods through which a unified approach to differentiation and integration could occur, than in the success or failure of individual mathematicians in solving particular problems. Even so, the formulation of such problems and the competitive element introduced by attempts to resolve them was of considerable value in focusing attention over periods of time on particular lines of development which subsequently turned out to have far-reaching implications over a wider field. One such problem, relating to the locus of a point attached to the rim of a rolling wheel,[†] was raised by Mersenne, and the study of this locus, the now familiar cycloid curve, led to the development of a great many interesting methods for determining tangent, arc, area and volume of revolution, most of which were subsequently generalised to apply to other curves. The mathematician to whom this problem was, in the first instance, referred by Mersenne, was Roberval.

Gilles Personne de Roberval (1602–75) came to Paris in 1628 and from that year remained in constant touch with Mersenne and his circle.[‡] From 1634 he occupied the Chair of Ramus at the Collège Royal in Paris, an office which was renewable every three years by a contest set by the incumbent. Throughout his lifetime Roberval remained secretive about his methods and this has been attributed to the need to win these triennial contests. Although many of his results were communicated through Mersenne to Fermat[§] and other material seems to have been placed in the hands of his students in the form of lecture notes, no publication took place until 1693, when the Académie des Sciences collected together and published a number of small treatises attributed to him.[‖] According to his own account,[††] Roberval was using infinitesimal methods before 1630 and he therefore claimed priority in the development of such methods over Cavalieri. By 1634, at any rate, Roberval had succeeded in devising a set of ingenious techniques which enabled him to achieve success over a wide range of infinitesimal problems, including that of the cycloid.

The initiator, however, of most of the problems which exercised the skill and ingenuity of French and Italian mathematicians alike during this period was not a member of the intimate group around Mersenne in Paris, but the Toulouse lawyer, Pierre de Fermat (1601–65). A somewhat isolated figure,

† All the references to this problem are given in the *Correspondance* VI, pp. 167–8.

‡ Auger, L., *Un savant méconnu: Gilles Personne de Roberval (1602–75)*, Paris, 1962.

§ *Oeuvres de Fermat* II, pp. 81–4.

‖ *Divers Ouvrages de Mathématique et de Physique, par Messieurs de l'Académie Royale des Sciences*, Paris, 1693.

†† See *Epistola ad Torricellium* in *Divers Ouvrages*, pp. 284–302.

Fermat rarely visited Paris but established communication with the Italians through de Beaugrand and Carcavy and corresponded with Mersenne from 1636.† The correspondence which follows, in which he was able to exchange views with Roberval and subsequently with Descartes, shows that by 1636 he had already made substantial progress in the field of infinitesimal analysis. The well-known tangent process, based on the concept of a small change of variable, was developed by Fermat in 1629. About the same time Fermat conceived the idea of a family of curves, subsequently to become known as the higher parabolas, $(y/a) = (x/b)^p$. These curves, to which Fermat subsequently added the generalised forms, $(y/a)^m = (x/b)^n$ and the hyperbolas, $(y/a)^m(x/b)^n = 1$, were expressly devised to allow scope for the extended use of the techniques of infinitesimal analysis and throughout a great part of the seventeenth century they served as a challenge to mathematicians throughout Europe in the development of new methods. The first approach to the quadrature of these curves was made by Fermat and a little later by Roberval: both were concerned to make use of the power series, $\sum n^p$, in order to establish the areas under the curves, $y = kx^p$, i.e. to determine the integrals $\int_0^1 x^p\,dx = 1/(p+1)$ for p positive and integral.

5.1. The Arithmetisation of Integration Methods

The series $\sum n$ and $\sum n^2$ were, of course, known to the Pythagoreans,‡ and Archimedes made use of them in the quadrature of curves.§ Nicomachus gave an expression for $\sum n^3$ which subsequently appeared in the arithmetic of Boethius. The Arabs‖ (Ibn-al-Haitham) made use of $\sum n^3$ and $\sum n^4$ in finding the volumes of solids of revolution. Stevin was the first to provide an arithmetic method†† based on the power series $\sum n^2$. In the seventeenth century Faulhaber, visited at Ulm by Descartes in 1620, found expressions for the sums of the powers of the natural numbers‡‡ up to 13.

The name of the geometer of Toulouse is, of course, inseparable from the theory of numbers and, having conceived the idea of the higher parabolas $y/a = (x/b)^p$, he had no difficulty in establishing the quadrature of these curves on the basis of the recursive formulae, $2\sum_1^n r = n(n+1)$, $3\sum_1^n r(r+1) =$

† See *Correspondance* VI, pp. 50–4.

‡ See pp. 19, 29, above.

§ See pp. 41–5 above.

‖ See p. 70 above.

†† See pp. 96–101 above.

‡‡ Faulhaber, J., *Miracula arithmetica*, Augsburg, 1622; *Academia algebrae*, Ulm, 1631.

$= n(n+1)(n+2)$, and so on.† Thus, for p positive and integral, we have

$$\lim_{n\to\infty} \sum_1^n \frac{r^p}{n^p} = \frac{1}{p+1}, \quad \int_0^1 x^p\,dx = 1/(p+1).$$

He remained, however, dissatisfied with this approach and soon turned aside to develop a strip method‡ based on division of the abscissa into segments in geometric progression. The only surviving example of an arithmetic quadrature by Fermat is that of the spiral of Galileo,§

i.e.
$$\frac{R-p}{R} = \left(\frac{R\theta}{R\alpha}\right)^2.$$

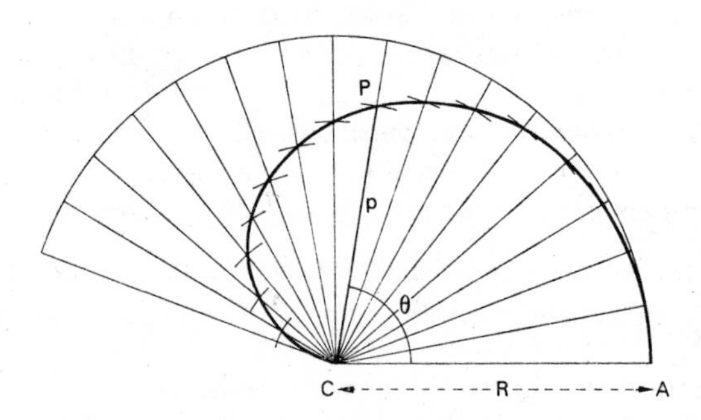

FIG. 5.1. The spiral of Galileo.

If a body, starting from $A(R, 0)$ and moving under the prescribed conditions, reach the point $P(p, \theta)$ in time t and the centre $C(0, \alpha)$ in time T, then $R\theta/R\alpha = t/T$, $(R-p)/R = t^2/T^2$ (see Fig. 5.1). Thus, $(R-p)/R = \theta^2/\alpha^2$, and the path of the body is a spiral curve. To find the area of the spiral space, Fermat divides the arc into sixteen equal parts and inscribes and circumscribes circular sectors in the manner of Archimedes.|| Since the areas of circular

† See *Correspondance* VII, pp. 272–83.

‡ See *Oeuvres de Fermat* I, pp. 255–88.

§ See *Correspondance* VI, pp. 376–82. This problem arose out of Galileo's work on projectiles and concerns the path of a body moving round the centre of the earth with velocity equal to the circular motion of the earth whilst at the same time falling towards the centre with a constant acceleration. Galileo thought the curve was a semicircle. (See Galilei, Galileo, *Dialogue on the Great World Systems*, Salusbury translation ed. Santillana, G. de), Chicago, 1953, pp. 178–81. For an account of the history of the problem see *Oeuvres de Fermat: Supplément*, pp. 1–19. See also Koyré, A., A documentary history of the problem of fall from Kepler to Newton, *Trans. Amer. Phil. Soc.* 45, 4 (1955), pp. 329–95.

|| See Archimedes, *On Spirals*, *op. cit.*, Prop. 24.

sectors are as the squares of their radii, we have

$$\frac{\text{inscribed figure}}{\text{circular sector}} = \sum_{0}^{15} \frac{p^2}{R^2} = \frac{\sum_{0}^{15} (16^2 - p^2)^2}{16 \times 16^4},^{\dagger}$$

and
$$\frac{\text{circumscribed figure}}{\text{circular sector}} = \sum_{1}^{16} \frac{p^2}{R^2} = \frac{\sum_{1}^{16} (16^2 - p^2)^2}{16 \times 16^4}.$$

Calculating the ratios for the inscribed and circumscribed figures respectively, Fermat shows that the first is less than 8/15 and the second greater: by increasing the number of divisions indefinitely the difference between the two can be made less than any arbitrarily chosen small quantity. Thus the area of the spiral space is 8/15 of the circular sector.‡ Fermat is clearly using the arithmetic method only to confirm a value already arrived at by the use of power series.

Although Fermat himself did not regard power series as an adequate foundation for integration methods§ his contemporary Roberval seems to have remained satisfied throughout his entire life with methods based on geometric intuition and incomplete induction. His published treatises|| give no indication that he had at any time made constructive efforts to develop a formal basis for his infinitesimal techniques. His style is obscure, verbose and difficult for, although he abandons any attempt to adhere to the rigorous geometric methods of his predecessors, he makes no move to order his ideas through the use of algebraic symbols. None the less, his techniques are easy enough to follow for he contents himself, for the most part, with a rough sketch of his own *discovery* method, making no attempt to wrap it up in any kind of formal proof structure.

Although Roberval defended Cavalieri's line and surface indivisibles against the attacks of others,†† his own approach was founded rather on the division of surfaces (or solids) into strips (or slices) with unit bases. In this way, by

† From $(R-p)/R = \theta^2/\alpha^2$, we have, $p/R = 1-\theta^2/\alpha^2 = (\alpha^2-\theta^2)/\alpha^2$, and $p^2/R^2 = (\alpha^2-\theta^2)^2/\alpha^4$. Since the values of θ increase arithmetically, i.e. $\theta = \alpha/16, 2\alpha/16, 3\alpha/16, \ldots, \alpha$, then

$$\sum_{0}^{15} p^2/R^2 = \sum_{0}^{15} (16^2-p^2)^2/16^4.$$

‡ $$\lim_{n\to\infty} \sum_{1}^{n} \frac{(n^2-r^2)^2}{n^4} = \lim_{n\to\infty} \sum_{1}^{n} \frac{(n^4-2n^2r^2+r^4)}{n^4} = 1-2/3+1/5 = 8/15.$$

§ See *Correspondance* VI, pp. 145–6.

|| *Divers Ouvrages: Traité des indivisibles*, pp. 190–245.

†† *Ibid., Epistola ad Torricellium*, pp. 285–6.

increasing the number of divisions indefinitely the measure of the ordinate in linear units became the measure of the surface in square units.

A surface, he says,† is divided into an infinite number of small surfaces which are equal, or have equal differences, or maintain some regular progression such as from squares to squares, from cubes to cubes, and so on. And, since the surfaces are enclosed by lines, instead of comparing the surfaces one compares the lines. The infinity of lines represents the infinity of little surfaces making up the total surface: the infinity of surfaces represents the infinity of little solids making up the total solid.

Thus, having established the sum of an infinite number of lines as a measure of the surface area, Roberval is in a position to avail himself of the power series $\sum_1^n r^p$, to find the areas under the curves, $y = kx^p$. We have, for example, $\sum_1^n r = n^2/2+n/2$, and $\sum_1^n r - n^2/2 = n/2$. Thus, if n be increased indefinitely, $\sum_1^n r = n^2/2$, for a line $(n/2)$ has no ratio to a square $(n^2/2)$. Now, if $\sum_1^n r$ represents the sum of the *lines* of a triangle erected at equal intervals along the base b (see Fig. 5.2), then, in some sense, $n \times 1 = b$: at the same time $n \to \infty$, so that,

$$\text{area of triangle} = \sum_1^n r = n^2/2 = b^2/2.$$

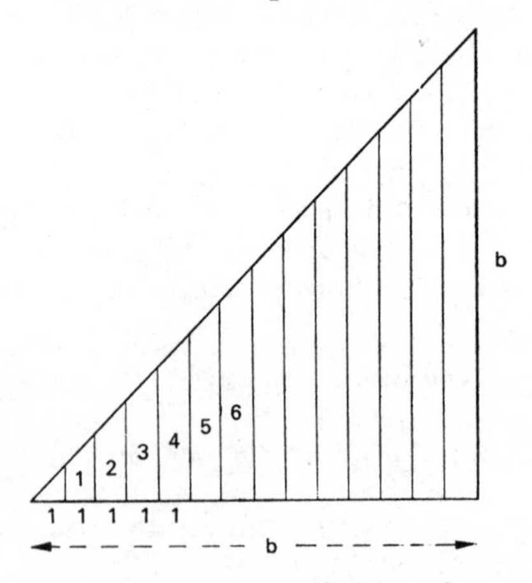

FIG. 5.2. 1 = one unit, $1 \times n = b$.

† *Traité des indivisibles*, pp. 190–2.

The discarded quantity, $n/2$, can be conceived as equal in magnitude to one half the altitude of the triangle, or one half of an infinitely narrow rectangle (breadth $= 1 = b/n$). Part of the difficulty here is that it is never entirely clear whether Roberval is thinking in terms of lines or of rectangles and in any case he considers the matter unimportant since all rectangles have unit breadth; a further difficulty arises from defective or non-existent symbolism. In following through any of Roberval's quadratures it is always necessary to bear in mind that, whilst n increases indefinitely, $n \times w = b$, where $w = 1$ unit, and b is some finite quantity. Thus n is used in two different and distinct ways in the same problem.

For the quadrature of the parabola, $y/a = (x/b)^2$, we have $\sum_1^n r^2 = n^3/3 + n^2/2 + n/6$, and these latter quantities can be neglected since neither a square nor a line have any ratio to a cube. Thus, as $n \to \infty$,

$$\sum_1^n r^2 \to n^3/3.$$

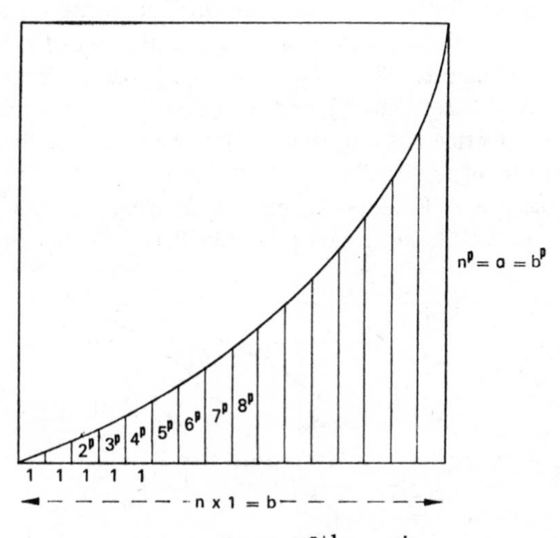

FIG. 5.3. $\sum_1^n r^p = \dfrac{n^{p+1}}{n+1} = \dfrac{ab}{n+1}$.

Roberval concludes that, in general, the difference,

$$\sum_1^n r^p - \frac{n^{p+1}}{p+1},$$

can be neglected, provided n be increased indefinitely. In this case we have

$$\lim_{n \to \infty} \sum_1^n \frac{r^p}{n^p} = \frac{1}{p+1},$$

and if $\sum_1^n r^p$ be taken to represent the lines of the space under the curve $y/a = (x/b)^p$, drawn parallel to the y-axis, we have $\int_0^b y\,dx = ab/(p+1)$ (see Fig. 5.3). Roberval discusses in detail only the cases, $p = 1$, $p = 2$, and the extension which he makes to other values of p is given without any kind of justification: no general expression is given for $\sum_1^n r^p$, although in correspondence addressed to Torricelli, Roberval mentions that he had made use of the inequality,

$$\sum_1^{n-1} r^p < \frac{n^{p+1}}{p+1} < \sum_1^n r^p.$$

5.2. First Investigations of the Cycloid

When Mersenne referred the cycloid problem to Roberval for his consideration he suggested that the path of a point on the rim of a rolling circle might be a semi-ellipse.† Roberval established the form of the curve $[x = a(\theta - \sin\theta),\ y = a(1 - \cos\theta)]$, and went on to determine the area under the curve using a Cavalerian approach in which he compared the *lines* of one figure with the *lines* of another.‡

Let the semicircle APB, radius a, centre O (see Fig. 5.4) move through a distance OO' ($= AA'$) in a direction perpendicular to the bounding diameter

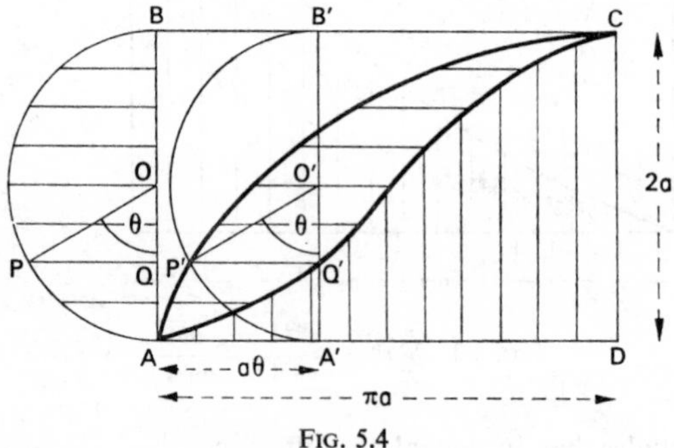

Fig. 5.4

† See p. 150 above.

‡ *Traité des indivisibles*, pp. 191–2.

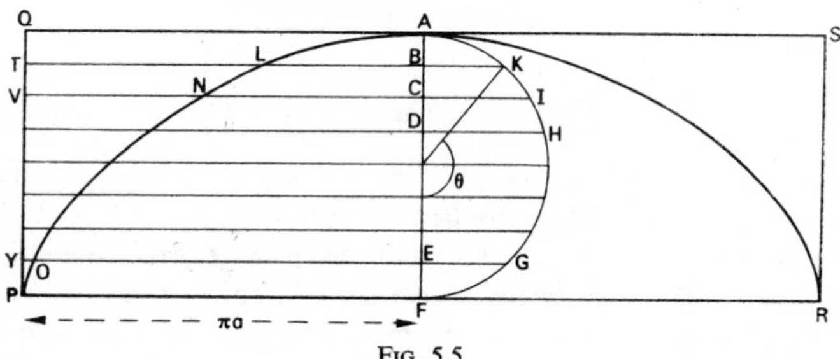

FIG. 5.5

AB. The semi-chord *PQ* (parallel to *OO'*) moves into position *P'Q'*, where $PP' = QQ' = AA'$ $(= a\theta)$. *AP'C* is a cycloidal arc $(x = a\theta - a\sin\theta,\ y = a - a\cos\theta)$, and *Q'* lies on another curve $(x = a\theta,\ y = a - a\cos\theta)$ which Roberval named the *Companion* to the cycloid. Since $P'Q' = PQ$, and this is true for all such lines, the space *AP'CQ'A* between the two curves is equal to the semicircle *APB*.† But, by symmetry, the lines of the space *AQ'CD* (drawn parallel to *CD*) are equal to the corresponding lines of the space *ABCQ'*, and, since the area of the rectangle *ABCD* equals twice the space of the rolling circle, the space *AQ'CD* is equal to the space of the rolling circle, and the area of the cycloidal segment *AP'CDA* equals $\frac{3}{2}\pi a^2$.

In 1638, provoked by Mersenne's constant references to Roberval's success with the cycloid, both Fermat and Descartes made contributions. The method given by Fermat runs as follows:‡ let *PLA* be a cycloidal arc inscribed in a rectangle *PQAF* and let *AKIF* be one-half of the rolling circle (see Fig. 5.5). The spaces *PLAF* and *PQAF* may be compared by comparing *all the lines*§ of the cycloid, *LB*, *CN*, ..., *OE*, *PF*, with *all the lines* of the rectangle, *TB*, *VC*, ..., *YE*, *PF*. Since, however, for any line *TB* which cuts the rectangle in *T*, *B*, the cycloid curve in *L*, *B*, and the rolling circle in *B*, *K*, we have,

$$LB = BK + \text{arc } AK,\|$$

then

$$\frac{\text{cycloid segment } PLAF}{\text{rectangle } PQAF} = \frac{\sum LB}{\sum TB} = \frac{\sum BK + \sum \text{arc } AK}{\sum TB}.$$

† $AP'CQ' = a^2 \int_0^{\pi} \sin\theta \, d(\cos\theta) = \pi a^2/2, \quad AQ'CD = a^2 \int_0^{\pi} (1 - \cos\theta) \, d\theta = \pi a^2.$

‡ See *Correspondance* VII, pp. 376–80.

§ *Ibid.* VII, p. 378, "il faut comparer toutes les lignes...".

‖ $TL = a\theta - a\sin\theta,\ LB = a(\pi - \theta) + a\sin(\pi - \theta) = AK + BK.$

But,

$$\frac{\sum BK}{\sum TB} = \frac{\text{semicircle}}{\text{rectangle}} = \frac{1}{4}.$$

It remains, therefore, to determine the ratio, $\sum \text{arc } AK/\sum TB$. But here Fermat encountered a difficulty, for he notes that, if the arcs AK, AI, $AH, \ldots$, increase *arithmetically*, i.e. if the arc be divided equally at the points K, I, H, $\ldots$, G, then $\sum AK = \pi a^2$ (the result obtained by Roberval) (see Fig. 5.6). The divisions, however, being *unequal*, he fears that Roberval may have erred.

Returning to the problem a week later,† "pour justifier Mr Roberval contre la censure trop précipitée", Fermat finds the required sum by considering the small portions of the arc in pairs (assuming an *even* number of divisions).

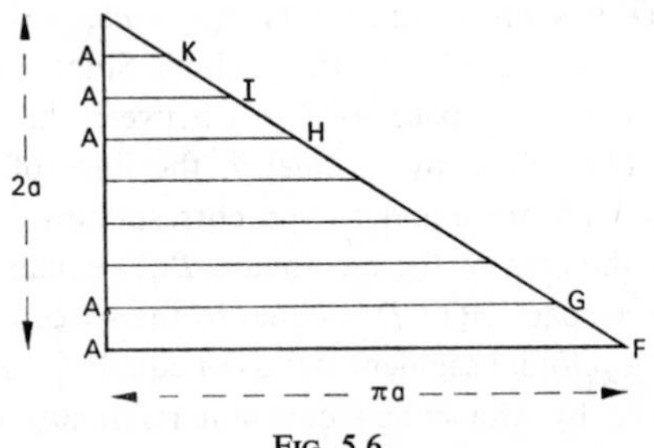

FIG. 5.6

Since $AK+KF = \pi a$, $AI+IF = \pi a$, and so on for all such pairs (see Fig. 5.5), then, $AK+AI+AH+\ldots AF = n\pi a/2$, where n is the number of lines. Hence,

$$\frac{\sum AK}{\sum TB} = \frac{n\pi a}{2n\pi a} = \frac{1}{2},$$

and $\sum AK = \frac{1}{2}\sum TB$. Thus, the space of the cycloid is three times the space of the rolling circle.

Descartes also felt impelled to submit a solution,‡ "de faire voire à ceux qui en font grand bruit, qu'elle est tres facile". Clumsily expressed, Descartes' method rests primarily on a correspondence he established between the *lines* of a cycloidal segment and those of a semicircle, diameter AB (see Fig. 5.7). If AFC be the cycloidal arc, inscribed in the rectangle $ABCD$, then $\triangle ABC = \pi a^2$, the area of the rolling circle. It remains to establish the area of the

† *Correspondance* VII, pp. 397–9.
‡ *Ibid.* VII, pp. 407–12.

cycloidal segment, $AGCA$. Descartes now considers two lines, GI and KM, parallel to AB and such that $CI = MB$. Thus, in modern notation, if $BM = a(1-\cos\theta)$, $BI = a(1+\cos\theta)$, $KM = a(\pi-\theta)+a\sin\theta$, $GI = a\theta+a\sin\theta$, $KM+GI = 2a\sin\theta+\pi a$. Since, however, $HI+LM = \pi a$, we have, $GH+$ $+KL = 2a\sin\theta$. If, now, the cycloidal sector, $AFGCEA$, be replaced by two sectors, $FGCE$, and FAE, the latter being inverted and placed alongside the former along the line FE (see Fig. 5.8), then *all the lines GHK* of this figure,

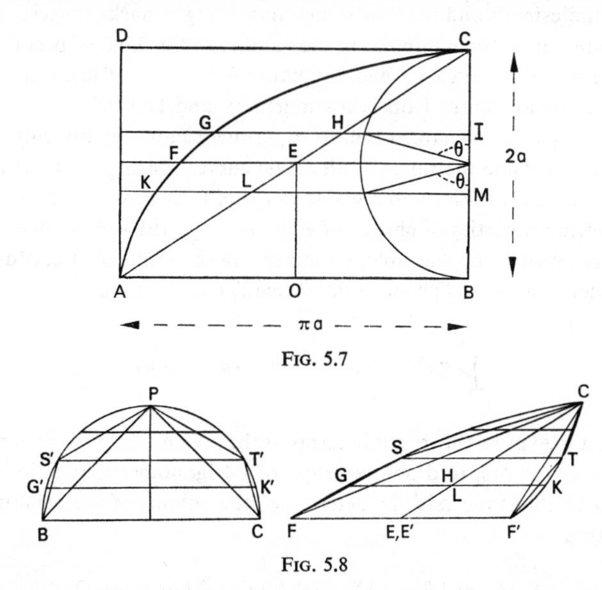

FIG. 5.7

FIG. 5.8

$FGCKF'$, correspond with, and are equal to, *all the lines* of the semicircle, diameter BC ($= 2FE$).† But, says Descartes, because this theorem is not perhaps accepted by everyone, I will develop it otherwise.‡ He then develops an alternative proof by repeated bisection of the ordinates, showing that triangles inscribed in the figure $FGCF'$ are equal in area to the corresponding figures inscribed in the semicircle $BG'PK'C$: thus, $\triangle BPC = \triangle FCF'$, $\triangle BS'P + \triangle PT'C = \triangle FSC + \triangle CTF'$ (see Fig. 5.8), and so on for all such triangles.

† *Ibid.* VII, p. 410: "... pour ceux qui sçavent que generalement, lorsque deux figures ont mesme baze et mesme hauteur, et que toutes les lignes droites, paralleles à leurs bazes, qui s'inscrivent en l'une, sont egales à celles qui s'inscrivent en l'autre à pareilles distances, elles contiennent autant d'espace l'une que l'autre."

‡ *Ibid.* VII, p. 411: "Mais pource que c'est un theoresme qui ne seroit peut-estre pas avoué de tous, je poursuis en cete sorte."

In consequence, he says, the space of the cycloidal segment is equal to the space of the semicircle, for, if all the parts of a quantity are equal to all the parts of another quantity, then the whole is necessarily equal to the whole.† He concludes: "et c'est une notion si evidente, que je croy qu'il n'y a que ceux qui sont en possession de nommer toutes choses par des noms contraires aux vrais, qui soient capables de la nier, et de dire que cela ne conclud qu'à peu près." The above example represents one of Descartes' few ventures into the field of infinitesimals and the tone of his concluding remarks suggests a degree of uncertainty in his own mind as to the validity of this kind of proof. Indeed, neither Fermat nor Descartes had any liking for the use of line infinitesimals in the Cavalierian sense. Roberval's ingenuity and lack of concern for the rigour of his methods enabled him to experiment more readily with integration methods outside the range of algebraic curves. Basing his studies on the cycloid he was able to develop a whole range of trigonometric integrals from the geometric properties of chords of a circle. From the well-known proposition of Archimedes‡ he was able, by increasing the number of chords indefinitely, to derive a general proposition§ formally equivalent to

$$\int_{\alpha}^{\beta} a \sin \theta \, d(a\theta) = a^2(\cos \alpha - \cos \beta).$$

He went on to express this result in terms of the area under the sine curve|| and was probably the first to draw the graph of a trigonometric function. Rotating the curve about the axis†† he determined the volume of the resulting solid of revolution:

$$\sum_{1}^{n} a^2 \sin^2 k\theta = a^2 \sum_{1}^{n/2} (\sin^2 k\theta + \cos^2 k\theta) = a^2 n/2,$$

where n is the number of divisions of the base. Since, therefore, $n \times 1 = \pi a/2$, we have

$$\sum_{1}^{n} a^2 \sin^2 k\theta = \pi a^3/4 \qquad \left(\int_{0}^{\pi/2} \sin^2 \theta \, d\theta = \pi/4\right).$$

† *Correspondance* VII, p. 411: "Car toutes les parties d'une quantité estant egale à toutes celles d'une autre, le tout est necessairement egal au tout."

‡ Archimedes, *On the Sphere and Cylinder* I, Prop. XXI *et seq.*

§ *Traité des indivisibles*, pp. 193–4.

|| *Ibid.*, p. 194.

†† *Ibid.*, pp. 216–17.

To find the volume of the solid formed by rotating the *Companion to the Cycloid* (i.e. the curve of versines) about the axis† (see Fig. 5.9), let $AS = a(1+\cos\theta)$, $SD = a(1-\cos\theta)$, $SR = a\sin\theta$. In the circle, diameter AD, we have,

$$\sum AD^2 = \sum (AS+SD)^2 = \sum AS^2+\sum SD^2+2\sum AS.SD$$
$$= \sum AS^2+\sum SD^2+2\sum SR^2.$$

But, by symmetry, $\sum AS^2 = \sum SD^2$, and also $2\sum SR^2 = \pi a^3$,‡

therefore, $4\pi a^3 = 2\sum AS^2+\pi a^3 \quad (n\times 1 = \pi a)$

and $3\pi a^3 = 2\sum AS^2$.

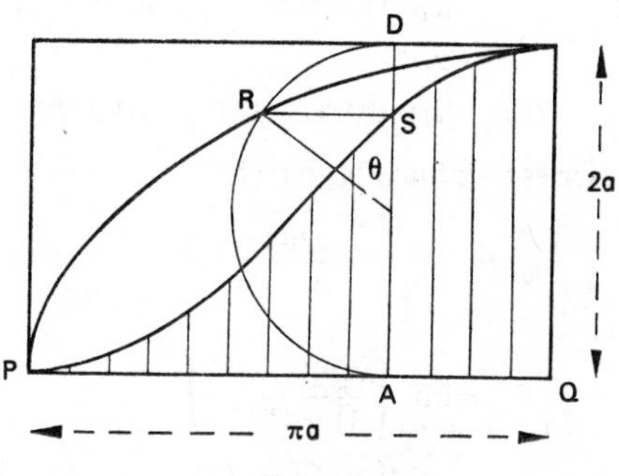

FIG. 5.9

The volume of the solid of revolution is thus equal to $\frac{3}{2}\pi^2a^3$. By combining this with the volume of the solid formed by rotating the space between the curves about the axis (equal to the volume of the solid formed by rotating the semicircle ARD about the line PQ) Roberval was able to determine the volume of the solid formed by rotating the area under the cycloidal arc.§

Besides being of interest in themselves, the above methods illustrate particularly well how Roberval, having made a study of a special curve and its related problems, was enabled to develop a whole range of trigonometric integrals.

† $a^2\int_0^\pi (1+\cos\theta)^2\, d(a\theta) = \frac{3\pi}{2}a^3$.

‡ $2a^2\int_0^\pi \sin^2\theta\, d(a\theta) = \pi a^3$.

§ *Ibid.*, p. 221.

5.3. Another Integration Method

The method devised by Fermat† for the quadrature of the parabolas, $(y/a)^m = (x/b)^n$, and the hyperbolas, $(y/a)^m (x/b)^n = 1$, depends on the construction of a set of inscribed rectangles with areas in geometric progression. Fermat was thus able to obtain an expression for the sum of the geometric series and, by considering the limit of this sum as the number of divisions became infinite, to obtain the quadrature of the curve.

Using algebraic notation the method runs as follows:‡ for the curve, $(y/a)^m = (x/b)^n$, we have

Abscissae:	x_1	$x_1 t^m$	$x_1 t^{2m}$	$x_1 t^{3m}$	...
Differences:	$x_1(1-t^m)$	$x_1 t^m(1-t^m)$	$x_1 t^{2m}(1-t^m)$		...
Ordinates:	y_1	$y_1 t^n$	$y_1 t^{2n}$	$y_1 t^{3n}$	...
Areas:	$x_1 y_1(1-t^m)$	$x_1 y_1 t^{m+n}(1-t^m)$	$x_1 y_1 t^{2(m+n)}(1-t^m)$		...

The sum of this series to infinity is given by,

$$S_t = \frac{x_1 y_1(1-t^m)}{1-t^{m+n}}.$$

Hence,

$$\lim_{t\to 1} S_t = \lim_{t\to 1}\left\{\frac{x_1 y_1(1-t^m)}{(1-t^{m+n})}\right\}$$

$$= \lim_{t\to 1}\left\{x_1 y_1 \frac{(1-t^m)}{(1-t)}\,\frac{(1-t)}{(1-t^{m+n})}\right\}$$

$$= x_1 y_1 \frac{m}{m+n}.\text{§}$$

Fermat was able to extend these results to the spirals, $(r/a)^m = (r\theta/a\alpha)^n$ and $[(R-r)/R]^m = (r\theta/R\alpha)^n$. He knew that, in the case of the rectangular hyperbola, $(x/a)(y/b) = 1$, the division of the abscissa at points in geometric progression resulted in equal divisions of the area under the curve.

† *Oeuvres de Fermat* I, pp. 255–85: "De aequationum localium transmutatione et emendatione ad multimodam curvilineorum inter se vel cum rectilineis comparationem, cui annectitur proportionis geometricae in quadrandis infinitis parabolis et hyperbolis usus."

‡ Cf. p. 139 above.

§ Although both Torricelli and Fermat use this result neither of them indicates how it was actually arrived at. We have, however,

$$1+t+t^2+t^3+\ldots+t^{n-1} = \frac{1-t^n}{1-t},$$

and it is easy to see that, as $t \to 1$, the left-hand side, and consequently the expression on the right, approaches n.

5.4. The Concept of Tangent

René Descartes (1596–1650) was introduced to the works of Stevin by Isaac Beeckman (1588–1637)† and in his youth used indivisibles to determine the laws of falling bodies.‡ He later solved problems sent to him by Fermat through Mersenne, but rarely displayed his methods.§ Although there can be no doubt that he was fully conversant with the use of indivisibles, or line infinitesimals,‖ and he was prepared on occasion to use mechanical methods bringing in ideas of motion, he was careful always to distinguish between *precision* methods and *approximate* methods in mathematics.†† For him the whole range of work with indivisibles came into this latter category. Finite algebraic processes constituted the precision tool through which he hoped finally to resolve all mathematical problems capable of resolution. The classification of curves which he adopted in *La Géométrie* (1637)‡‡ enabled him to accept the use of solutions involving motion in the case of the so-called mechanical curves whilst rejecting any other than finite algebraic processes in determining the tangents to algebraic curves. The problem of rectification he dealt with by reiterating Aristotle's assertion that no curve is capable of rectification, i.e. the arc cannot be expressed as a finite ratio of a given straight line. At the same time he seems to have accepted that mechanical curves, such as the spirals could, by mechanical means, be expressed in such terms.§§

To determine the tangent to the cycloid, Descartes developed the idea of an instantaneous centre of rotation:‖‖ if a polygon rolls on a given straight

† Descartes, René, *Oeuvres* (ed. Adam, C. and Tannery, P.), 12 vols. and supplement, Paris, 1897–1913, X, pp. 75–8. See also Milhaud, G., *Descartes savant*, Paris, 1921, pp. 163–4.

‡ *Oeuvres* X, p. 76.

§ *Ibid.* II, pp. 120–1, 246–76. Tannery lists the problems solved (p. 252). See also Milhaud, G., *op. cit.*, pp. 164–8.

‖ The ingenious solution provided for De Beaune's inverse tangent problem (*Oeuvres* II, pp. 513–17) shows clearly the deployment of infinitesimal techniques followed by a mechanical construction (see *Oeuvres* II, pp. 520–3). See also Scriba, C. J., Zur Lösung des 2 Debeauneschen Problems durch Descartes, *Archive for History of Exact Sciences*, I (4) (1961), pp. 406–19.

†† Cf. Hofmann, J. E., Descartes und die Mathematik (in *Descartes:* Scholz, H., Kratzer, A. and Hofmann, J. E., Munster, 1951, pp. 55–6).

‡‡ *Oeuvres* VI, pp. 369–485. English translation: *The Geometry of René Descartes*, Smith, D. E. and Latham, M. L., Dover Pubs., New York, 1925.

§§ Cf. Hofmann, J. E., Descartes und die Mathematik, *op. cit.*, pp. 67–8. See *Oeuvres* VI, p. 412. In effect Descartes says: "On the other hand, geometry should not include lines that are like strings, in that they are sometimes straight and sometimes curved, since the ratios between straight lines and curved lines are not known, and I believe cannot be discovered by human minds, and therefore no conclusion based upon such ratios can be accepted as rigorous and exact."

‖‖ *Oeuvres* II, pp. 307–13.

line the curve described by each vertex will be made up of a succession of circular arcs and the tangent to the curve at each point on it will be at right angles to the radius of the arc (see Fig. 5.10). If a circle rolls on a straight line then the circle can be considered as a polygon made up of "cent mil milions" of sides and the tangent at each point will be perpendicular to the line joining that point to the point of contact of the generating line with the base line.†

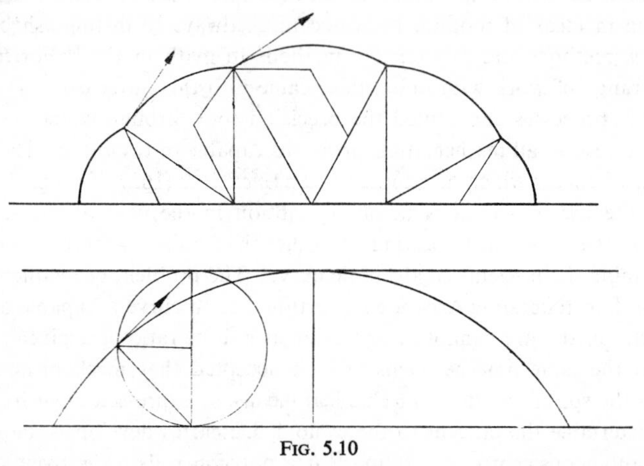

FIG. 5.10

† If $x = a(\theta - \sin\theta)$, $y = a(1-\cos\theta)$, $dx/d\theta = a(1-\cos\theta)$, $dy/d\theta = a\sin\theta$, $dy/dx = \cot\theta/2$, $d^2y/dx^2 = -(1/4a)\operatorname{cosec}^4(\theta/2)$,

$$\varrho = \frac{(1+(dy/dx)^2)^{3/2}}{d^2y/dx^2} = 4a\sin(\theta/2) = PI = 2PC.$$

Hence, the instantaneous centre of rotation is at I and not at C. The direction, however, is correct.

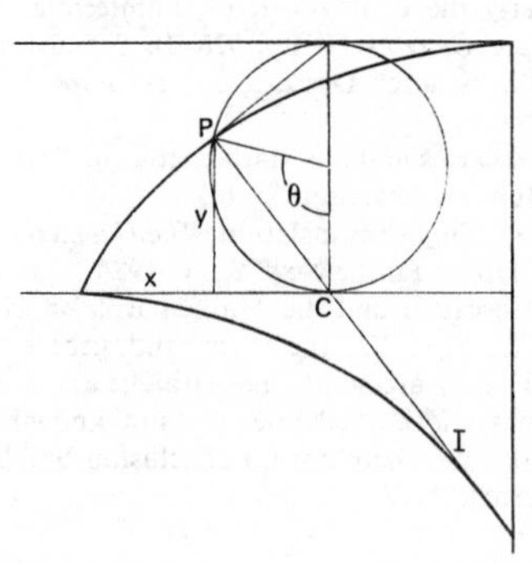

For algebraic curves, Descartes developed a technique for determining the normal to a curve at a given point† which avoided the use of infinitely small quantities. This method, which he published in *La Géométrie*, was so successful and at the same time so general, that it sustained still further his belief that the use of infinitely small quantities in geometry was quite unnecessary.

If, says Descartes,‡ a circle cut a curve in two distinct points, then it is possible to form an equation (in x or y) with two roots corresponding to the points. As these points approach each other the difference between the roots of the equation diminishes; and when the points become coincident the roots become equal and in this case the circle will touch the curve.

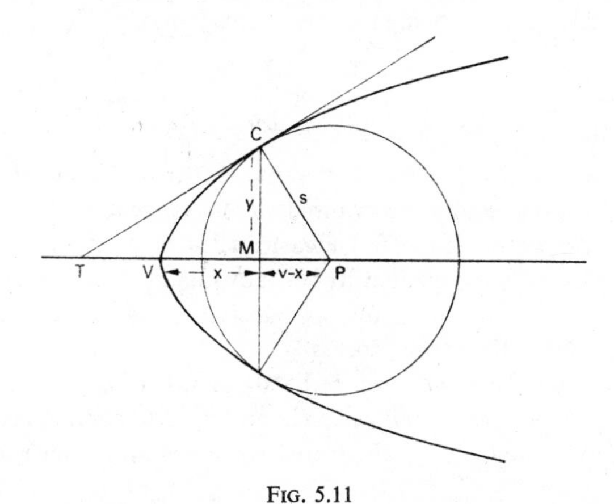

Fig. 5.11

Thus, to determine the normal to the parabola, $y^2 = kx$ (vertex V), let a circle be drawn with centre at a point P on the axis of the parabola, to touch the curve at C (see Fig. 5.11). Let $PV = v$, $CP = s$, $VM = x$, $MC = y$.

From the right-angled triangle, CMP, we have

$$(v-x)^2+y^2 = s^2, \quad \text{and since} \quad y^2 = kx,$$

then
$$(v-x)^2+kx = s^2,$$

and
$$x^2+x(k-2v)+(v^2-s^2) = 0.$$

Since, however, this is an equation with equal roots, we can compare it

† *Oeuvres* VI, pp. 413–24.
‡ *Ibid.* VI, pp. 417–18.

with the equation, $(x-e)^2 = 0$,†

i.e. $$x^2-2ex+e^2 = 0.$$

Thus, equating coefficients, $k-2v = -2e = -2x$,

and $$v-x = k/2.$$

The gradient of the normal (and consequently of the tangent) is accordingly obtained.

In the notation of the calculus, if an equation, $f(x) = 0$, has a pair of equal roots ($x = \alpha$), then the equation, $f'(x) = 0$, has also a root, $x = \alpha$.‡ Thus, for $y = f(x)$, we have $y^2+(v-x)^2 = s^2 = (f(x))^2+(v-x)^2$. Differentiating with respect to x (s and v being treated as constants),

$$0 = f(x)\,f'(x)-(v-x),$$

and $$(v-x) = f(x)f'(x) = yf'(x).§$$

Descartes attached considerable importance to this process‖ and, from 1637 onwards, seventeenth-century discussion of tangent methods centred on it. In practice, however, although Descartes selected the worked examples in *La Géométrie* to show the method to best advantage,†† the solution of equations formed in this way was often extremely tedious and this soon led to attempts to simplify the entire process.‡‡

In 1637, through the good offices of Beaugrand, Fermat was enabled to see the proofs of *La Géométrie*, and was thus provoked to communicate his own tangent method which he had discovered some years previously.§§ Fermat's

† *Oeuvres* VI, p. 423: "I desire rather to tell you in passing that this method, of which you have here an example, of supposing two equations to be of the same form in order to compare them term by term and so to obtain several equations from one, will apply to an infinity of other problems and is not the least important feature of my general method." (This is the method of detached coefficients.)

‡ Let $f(x) = (x-\alpha)^2\varphi(x)$, then $f'(x) = 2(x-\alpha)\,\varphi(x)+(x-\alpha)^2\,\varphi'(x)$
$= (x-\alpha)\,(2\varphi(x)+(x-\alpha)\,\varphi'(x))$.

§ If $y^2 = kx$, $dy/dx = k/2y = (v-x)/y$, and $v-x = k/2$.

‖ *Ibid.* VI, p. 413: "Et i'ose dire que c'est cecy le problesme le plus utile & le plus general, non seulement que ie sçache, mais mesme que i'aye iamais desiré de sçavoir en Geometrie".

†† *Ibid.* VI, pp. 413–24.

‡‡ *Oeuvres de Fermat: Supplément*, p. 105: Beaugrand refers to the tangent method of Descartes as "un labyrinte dont l'issue seroit extraordinairement difficile".

§§ For Fermat's own description of his achievements up to about 1644 see *Oeuvres de Fermat* I, pp. 195–8; III, pp. 169–71. See also *Supplément*, pp. 99–100, for an account of his communications with the Italian mathematicians.

method occupies a significant place in the history of the calculus for in it occurs, apparently for the first time, the idea of a slight change of variable.[†] When first introducing the method Fermat associated it with his method of finding the extreme value of certain variables.[‡] If, for example, it is required to maximise the expression $x(b-x)$, represented by the rectangle contained by the segments x and $b-x$, of a given straight line, then we allow x to become $x+e$ and the product becomes $(x+e)(b-x-e)$. Putting the two expressions approximately equal,[§] we have,

$$x(b-x) \simeq (x+e)(b-x-e).$$

Hence, removing common terms,

$$0 \simeq be-2xe-e^2,$$

and, dividing by e, we have,

$$0 \simeq b-2x-e.$$

To obtain the extreme value we now delete the term e,[‖] and finally,

$$b = 2x.$$

In response to queries raised during the tangent controversy regarding the validity of this approach Fermat clarified the process and developed it further.[††] He notes that, in general, if the product $x(b-x)$ is denoted by a^2, there are two distinct values of x, for each value of a. If these values be denoted by x and y, we have

$$x(b-x) = y(b-y) = a^2.$$

Now, let $y = x+e$, then $x(b-x) = (x+e)(b-x-e)$. Hence,

$$0 = be-2xe-e^2,$$
$$0 = b-2x-e,$$
$$2x = b-e.$$

For an extreme value, $y = x+e$, $e = 0$ and $x = b/2$. In all these considerations, although the notion of limit is not in any formal sense present the idea

† It is primarily on the strength of this method and the applications which he developed for it that Fermat has been declared the inventor of the calculus by Lagrange, Laplace and Tannery among others. See Cajori, F., Who was the first inventor of the calculus?, *Amer. Math. Monthly* 26 (1919), pp. 15–20.

‡ *Oeuvres de Fermat* I, pp. 133–4.

§ *Diophantus* (V, 14 and 17) employs a term for an approximate equality which Xylander and Bachet translate as *adaequalitas, adaequale*. Fermat uses the term *adaequantur*, referring directly to Diophantus (*Oeuvres* I, p. 133).

‖ *Elidatur.*

†† *Oeuvres de Fermat: Supplément*, pp. 120–5 (probably 1643–4).

is emerging of two converging values so that, as $y \to x$, $e \to 0$, and the extreme value is reached. In another example† Fermat shows how to distinguish between a maximum and a minimum.

Let $f(x) = x^2(b-x)$,

then $$f(x\pm e) = x^2(b-x)\pm e(2bx-3x^2)+e^2(b-3x)\mp e^3.$$

Now, at an extreme value, the coefficient of e in the above expression $(f'(x))$ vanishes (by the previous result). Hence, $f(x\pm e) \gtreqless f(x)$ according as the coefficient of e^2 (i.e. $b-3x$) $\gtrless 0$. Since, however, $b > 0$, $x = 2b/3$ makes $x^2(b-x)$ a maximum ($b-3x = b-2b = -b$). The term in e^3 (e being small) does not influence the result.

To determine the tangent to a curve we follow broadly the same method.‡ Let BDN be a parabola, vertex D, and let the tangent at B meet the axis, CD produced, in E (see Fig. 5.12). Now, let OI be drawn parallel to the ordinate

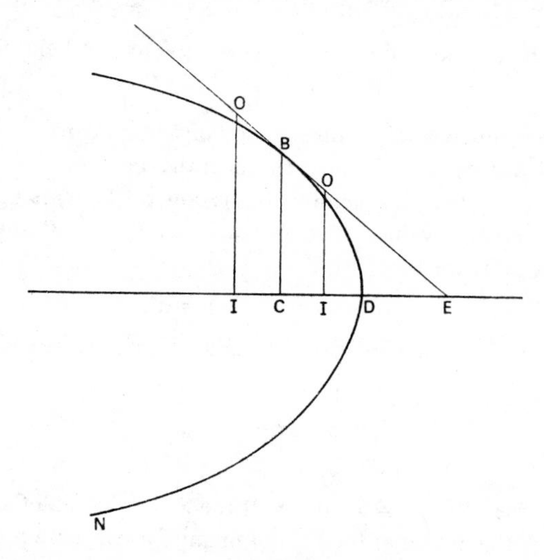

FIG. 5.12

† *Oeuvres de Fermat: Suppl.*, pp. 123–5. This material is discussed in Wieleitner, H., Bemerkungen zu Fermats Methode der Aufsuchung von Extremwerten und der Bestimmung von Kurventangenten, *Jahresbericht der Deutschen Mathematiker-Vereinigung* 38 (1929), pp. 24–35.

‡ *Oeuvres de Fermat* I, pp. 134–6. See also: Duhamel, J. M. C., Mémoire sur la méthode des maxima et minima de Fermat et sur les méthodes des tangentes de Fermat et Descartes, *Mémoires de l'Académie des Sciences de l'Institut Imperial de France* 32 (1864), pp. 269–330; Aubry, A., Sur l'histoire du calcul infinitésimal entre les années 1620 et 1660, *Annaes Scientificos da Academia Polytechnica do Porto* 6 (1911), pp. 82–9.

CB (at B), to meet the tangent BE at O.† Since OI is greater than the ordinate to the curve through I, then $CD/DI > BC^2/OI^2$. But, from similar triangles, $BC^2/OI^2 = CE^2/IE^2$; hence

$$CD/DI > CE^2/IE^2.$$

Thus, if $CD = d$, $CE = a$, $CI = e$, then

$$\frac{d}{d \pm e} > \frac{a^2}{a^2+e^2 \pm 2ae},$$

and

$$d(a^2+e^2 \pm 2ae) > a^2d \pm a^2e.$$

Now, putting these expressions approximately equal (as in the method of extreme values) and removing the common terms, we have

$$de^2 \mp a^2e \simeq \mp 2dae.$$

Dividing by e,

$$de \mp a^2 \simeq \mp 2da.$$

Finally, removing terms containing e, there remains $a = 2d$. The reasoning in this method is not clear in the sense that it is not given. Fermat does not say that e approaches zero or becomes zero or is put equal to zero: he simply instructs that the terms containing e be removed.

The process can be generalised as follows: let the curve be $y = f(x)$, with $DC = x$, $DI = x+h$, $CE = t$, $EI = t+h$, $CB = f(x)$, $IO > f(x+h)$ (see Fig. 5.12).

Now,

$$\frac{f(x+h)}{f(x)} \simeq \frac{t+h}{t},$$

and

$$tf(x+h) \simeq (t+h)f(x),$$

hence

$$\frac{f(x+h)-f(x)}{h} \simeq \frac{f(x)}{t} \quad (= dy/dx).$$

The formation of an approximate equation, discarding of terms, and so on, is at this stage Fermat's only conception of a limit and it is scarcely surprising that Descartes found much to object to in the method so presented.‡ He suspected that the success of Fermat's method depended on the existence of an explicit relation between y and x of the form, $y = f(x)$, and immediately

† Fermat conceived the tangent to a curve as a line meeting the curve in one point only. Every other point on the tangent lay outside the curve.

‡ See, *Correspondance* VII: Descartes à Mersenne, Jan. 1638, pp. 13–21; Descartes à Mydorge, 1st Mar. 1638, pp. 58–73; Descartes à Mersenne, 3rd May 1638, pp. 189–202; Descartes à Hardy, 29th June 1638, pp. 288–92.

challenged him[†] to find the tangent to the curve, $x^3+y^3 = n\,xy$ (the *folium of Descartes*). Fermat had little difficulty in satisfying this demand.[‡] Thus, in general, if $f(x, y) \simeq f\left(x+e, \frac{t+e}{t}y\right)$,[§]

then
$$f(x, y) \simeq f(x, y)+e\frac{\partial f}{\partial x}+\frac{ey}{t}\frac{\partial f}{\partial y}+\ \dots$$

Removing common terms, dividing by *e*, and deleting terms containing *e*, we have,

$$0 = \frac{\partial f}{\partial x}+\frac{y}{t}\frac{\partial f}{\partial y},$$

and
$$\frac{y}{t} = -\frac{\partial f/\partial x}{\partial f/\partial y} \qquad (= dy/dx).$$

Descartes also went to some lengths in investigating the foundations of the method and was not satisfied that the solution was, in fact, a case of extreme value. He worked the problem in a variety of different ways and was unable to discover the quantity maximised by Fermat's process. Fermat does not appear to have found it easy to reply to these objections although he could well have interpreted the method in terms of extreme value by maximising the angle *BEC* (see Fig. 5.13). Thus, given an external point *E*, situated on the axis *CE*, to draw a line *BE* meeting the curve in *B* and such that the angle

[†] See *Correspondance* VII, p. 18.
[‡] *Ibid.* VII, p. 327. End of June 1638.
[§] For a point on the tangent,

$$\frac{y+\delta y}{t+e} = \frac{y}{t}$$

and $\delta y = ey/t$.

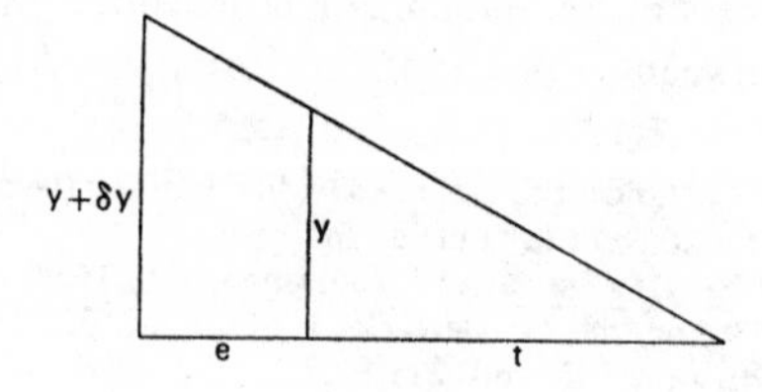

BEC is a maximum.† He was, however, hampered in the development of this kind of solution by the fact that he adopted as standard procedure the formation of an equation relating two points on the tangent rather than on the curve‡ and he only subsequently applied, through the formation of an approximate equation, the property of the curve to points on the tangent. Had Fermat been able to conceive the tangent as the limiting position of the

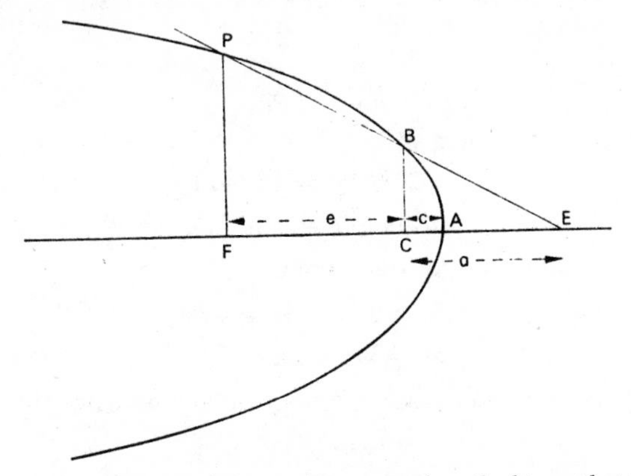

FIG. 5.13. *Note:* Descartes is presenting a general method to apply to a whole class of curves and is therefore content to sketch a curve which is parabolic in form. In any case at this stage there was little general interest in the special forms of curves.

† $y_1/a > y_2/(a-e)$, $y_1^2/a^2 > y_2^2/(a-e)^2$, but $y_1^2/y_2^2 = d/(d-e)$, hence, $d/(d-e) > a^2/(a-e)$. The solution follows as before.

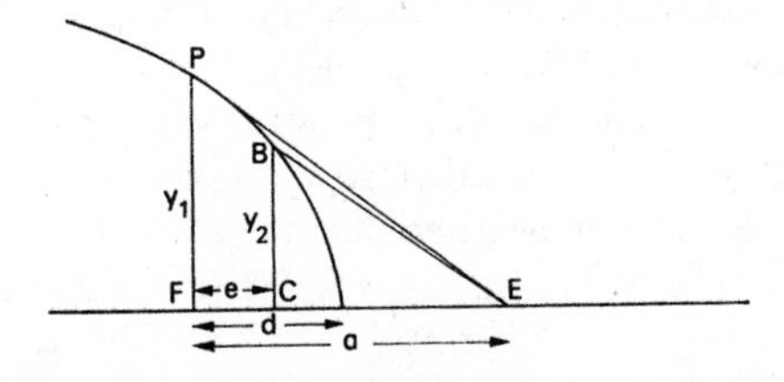

‡ *Oeuvres de Fermat* I, p. 159: "Consideramus nempe in plano cujuslibet curvae rectas duas positione datas, quarum altera diameter, si libeat, altera applicata nuncupetur. Deinde, jam inventam tangentem supponentes ad datum in curva punctum, proprietatem specificam curvae, non in curva amplius, sed in invenienda tangente, per adaequalitatem consideramus et, elisis (quae monet doctrina de maxima et minima) homogeneis, fit demum aequalitas quae punctum concursus tangentis cum diametro determinat, ideoque ipsam tangentem."

secant, then he would have been able to form an exact equation with roots approaching equality as $e \to 0$ and in this way would have established the method on a more secure foundation. Descartes was able from the outset to approach the problem in this way and consequently to effect an improvement in Fermat's method. Let B, P be two points on the curve, $y^3 = kx$, with ordinates BC and PF respectively (see Fig. 5.13) and let the chord PB meet the axis AF in E. If $AC = c$, $AF = c+e$, $EC = a$, we have, from similar triangles,

$$\frac{a}{a+e} = \frac{BC}{PF}.$$

From the curve equation,

$$BC^3/PF^3 = c/(c+e),$$

hence

$$a^3/(a+e)^3 = c/(c+e).$$

Expanding and removing common terms,

$$a^3e = 3a^2ce + 3cae^2 + ce^3,$$

$$a^3 = 3a^2c + 3ace + ce^2 \quad \text{(i)}$$

Now, in general, PF and BC are in a ratio h/g and we have

$$h/g = (a+e)/a,$$

$$ha = ga + ge. \quad \text{(ii)}$$

From equations (i) and (ii) a and e can be determined. If BP is a tangent to the curve, then $h = g$, $e = 0$ and consequently $a = 3c$.†

A curious feature of the tangent controversy is the role played by Beaugrand whose friendship with Fermat dated from before 1635 and who received in Paris most of Fermat's mathematical communications.‡ Beaugrand was a vigorous critic of Descartes and he made himself responsible for interpreting and distributing an alternative tangent method which is clearly derived from that of Fermat but which contains significant variations.§ In 1638 he made some interesting changes in notation, replacing the 'e' of Fermat by a small 'o'.|| In an example in which he determines the tangent to the ellipse, we have the following (see Fig. 5.14) : let $BD = b$, $DA = d$, $DE = a$, $DG = o$.

† See *Correspondance* VII: Descartes to Hardy, 29th June 1638, pp. 288–92.

‡ See *Oeuvres de Fermat: Supplément*, pp. 98–114.

§ London, British Museum, Harleian MSS. 6796, art. 18, fol. 155. Copy in the hand of Thomas Hobbes. Published in the *Bulletin des Sciences Mathématiques* 42 (1918), pp. 169–77. See also *Oeuvres de Fermat: Supplément*, pp. 98–114.

|| *Ibid.*, p. 103. "Or si la ligne *HCE* touche l'ellipse au poinct *C*, il est nécessaire que la ligne *GD* soit *o*, c'est a dire nulle, auquel cas il est très evident que toutes les quantitez qu'elle aura multipliées sont nulles, . . . "

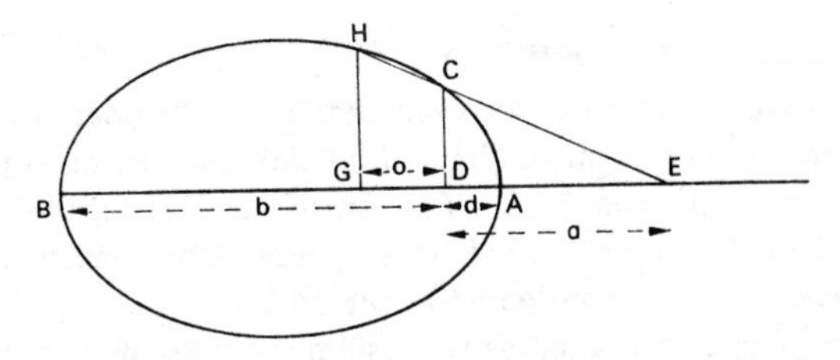

FIG. 5.14

From similar triangles,

$$GH/GE = DC/DE,$$

and from the property of the ellipse,

$$\frac{BD \cdot DA}{BG \cdot GA} = \frac{DC^2}{GH^2}.\dagger$$

Hence,

$$\frac{bd}{(b-o)(d+o)} = \frac{a^2}{(a+o)^2},$$

$$bd(a^2+2ao+o^2) = a^2(bd-od+ob-o^2)$$

$$2bdao+bdo^2 = oba^2-oda^2-o^2a^2$$

$$2bda+bdo = ba^2-da^2-oa^2.$$

But if the line touches the curve, then $GD = 0$, and

$$2bda = a^2b-a^2d,$$

$$\frac{2bd}{b-d} = a.$$

The introduction of this unfortunate symbol to denote a quantity which was ultimately to be put equal to zero, i.e. $o = 0$, might have been of little significance had the matter ended here. Beaugrand, however, travelled extensively in Italy where he had many contacts and it seems likely that James Gregory, who subsequently made use of the symbol, saw a copy of this method distributed in Italy by Beaugrand. Isaac Newton appears to have hit on the same symbol by chance but his use of it gave rise to much controversy.

† $\frac{x^2}{a^2}+\frac{y^2}{b^2} = 1, \quad \frac{y^2}{b^2} = 1-\frac{x^2}{a^2}, \quad \frac{y^2}{b^2} = \frac{a^2-x^2}{a^2} = \frac{(a-x)(a+x)}{a^2},$

hence $$\frac{y_1^2}{y_2^2} = \frac{(a-x_1)(a+x_1)}{(a-x_2)(a+x_2)}.$$

5.5. The Composition of Motions

Roberval participated in a correspondence with Fermat from 1636 onwards concerning tangent methods† and, at this time, he may have been in possession of an analytic method. Unfortunately no trace of this remains and the method he was using in 1638‡ is what is now termed mechanical—in the sense that every curve is considered as the path of a moving point and the tangent to the curve at a point, as the resultant of two or more motions impressed upon it simultaneously. This idea did not, of course, originate with Roberval but with the Greeks; it was developed further in the Middle Ages and exploited by Galileo and later by Torricelli.

Actually the validity of the approach to the tangent by the composition of motions depends very much on the physical independence of these motions. If, for example, a point move at a certain rate along a given straight line whilst in the same time the straight line moves parallel to itself, then the resultant path of the moving point is correctly determined by compounding the two motions. If, however, the two motions are not physically (or mathematically) independent, then only in certain special cases can the correct path be arrived at in this way. Roberval had no general method for determining the true motion of a point on a curve and the fact that he arrived at

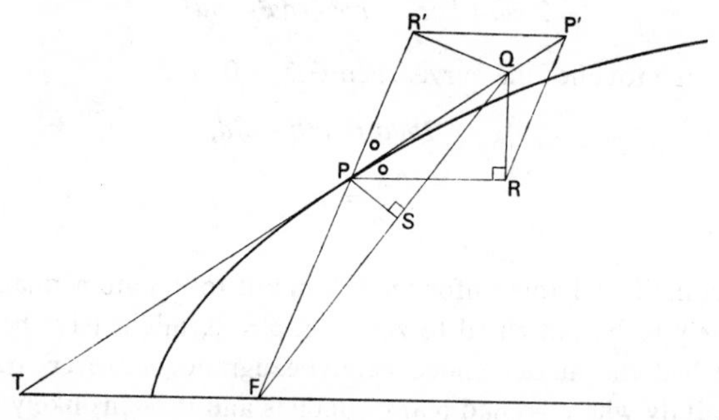

Fig. 5.15

† *Oeuvres de Fermat* II, p. 81.

‡ *Divers Ouvrages*, pp. 69–111: ***Observations sur la composition des mouvemens, et sur le moyen de trouver les touchantes des lignes courbes***. Although the above work was identified and published as the work of Roberval it was apparently found in the hands of his pupil, Du Verdus. There can be no doubt that the ideas presented are those of Roberval himself but the wording and the introductory propositions were supplied by Du Verdus. It is not possible, therefore, to say with any degree of certainty the exact point of view Roberval took on the compounding of motions. See Mersenne, M., *Correspondance* VI, p. 173, fn. 1.

correct results was sometimes due rather to a happy instinct than to any real understanding of the nature of the motion. In the case of the parabola, for example,† because of the focal directrix property, the resultant motion of a point on the curve is said to be compounded of two equal motions and it is therefore only necessary to bisect the angle between these two directions in order to obtain the tangent line. A consideration of the actual displacement of the point P (see Fig. 5.15) shows that, although in this case the tangent line is correctly drawn, the resultant motion is not correctly represented by compounding equal displacements along the lines PR and PR', but by compounding two displacements at right angles, e.g. along and perpendicular to the radius vector, or along and perpendicular to the axis of the parabola.‡

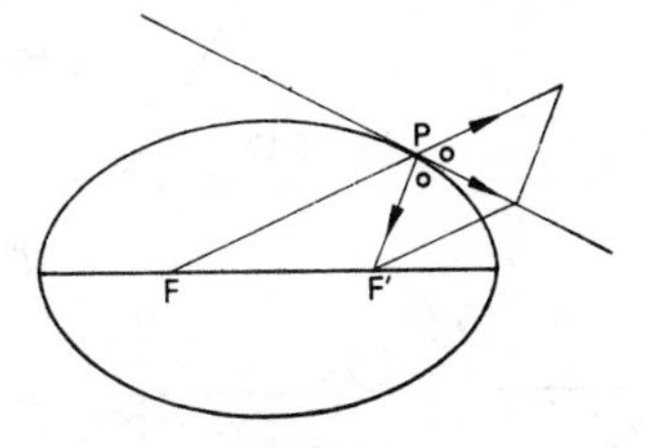

FIG. 5.16

For the ellipse and the hyperbola the tangent constructions are similarly derived from the focal properties (see Figs. 5.16 and 5.17). For the cycloid, Roberval gives to the moving point P a velocity equal to the velocity of the centre C and an equal velocity perpendicular to the radius CP (see Fig. 5.18). The resultant motion is thus determined by bisecting the angle between the two. This again gives the correct tangent line but is not strictly an accurate description of the resultant motion.§

For the conchoid of Nicomedes ($(r-b)\cos\theta = a$, see Fig. 5.19) the point P on the curve is considered to have two motions, (i) perpendicular to PC and proportional to PC, (ii) along PC. Since $GP = b$ (= constant) this motion

† *Divers Ouvrages*, pp. 80–1.

‡ See Fig. 5.15. $\overline{PQ} = \overline{PR}+\overline{RQ} = \overline{PS}+\overline{SQ}$,

$$\overline{PP'} = \overline{PR}+\overline{RP'} = \overline{PR}+\overline{PR'} \neq \overline{PQ}.$$

Since, however, in this case, $|\overline{PR'}| = |\overline{PR}|$, $\widehat{RPQ} = \widehat{R'PQ}$, and $\overline{PP'} = \lambda\overline{PQ}$, then $\overline{PP'}$ is the correct tangent line.

§ See Fig. 5.18. $\overline{PQ} = \overline{PR}+\overline{RQ} = \overline{PR'}+\overline{R'Q} \neq \overline{PP'}$.

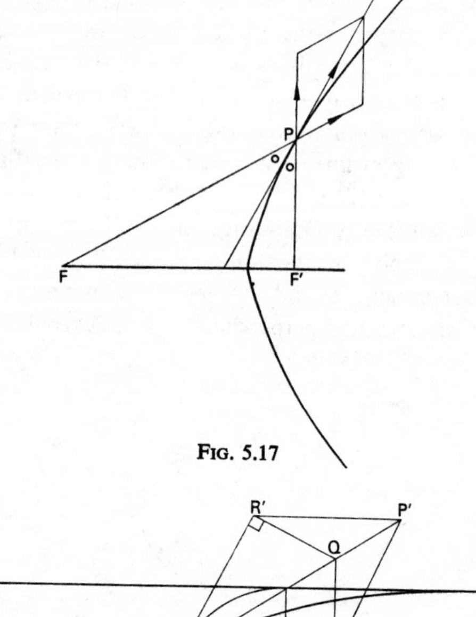

Fig. 5.17

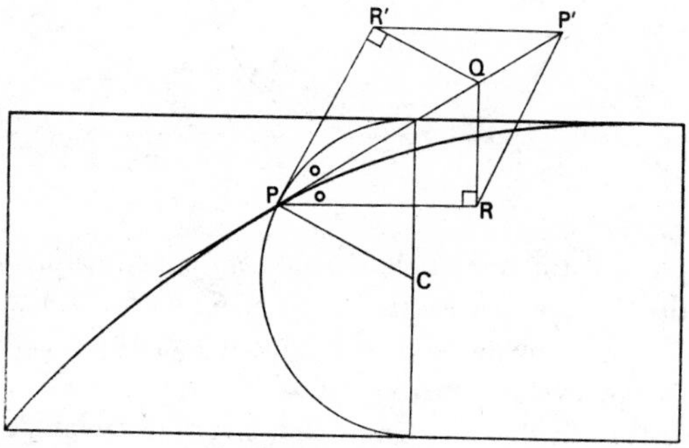

Fig. 5.18

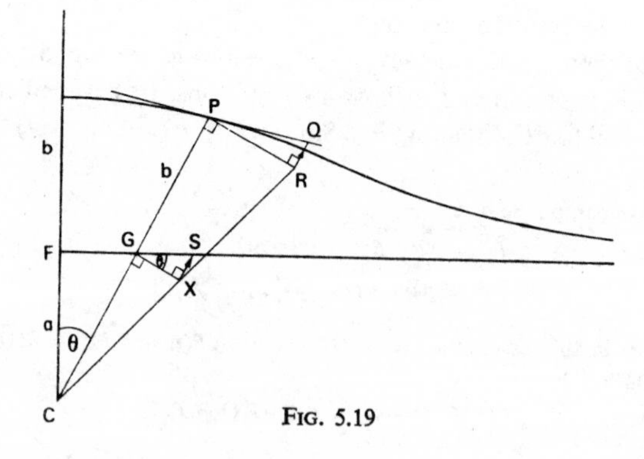

Fig. 5.19

is equal to the motion of the point G along CG. The point G, however, being constrained to move along the line FG, the resultant motion is along this line. Hence, to construct the tangent at the point P on the conchoid, draw PR and GX perpendicular to CP, such that $PR/GX = PC/GC$. Now, draw XS parallel to CG to meet the line FS in S. XS represents the motion of the point G along CG and hence is also the motion of the point P along CP. Thus, to complete the construction, if RQ be drawn equal and parallel to XS, then PQ represents the tangent line at P.†

5.6. The Link Between Differential and Integral Processes

One of the earliest links between differential and integral processes was provided by Fermat who extended the method of small variations, not only to the construction of tangents to curved lines, but also to the determination of the centre of gravity of a conoid.‡ This problem, from the time of Archimedes, had always been tackled by some kind of summation process.§

If a paraboloid of revolution, formed by rotating the curve $y^2 = kx$ about the x-axis, be cut by a plane parallel to the base (see Fig. 5.20), we have, in general,

$$V = \tfrac{1}{2}\pi y^2 x.$$

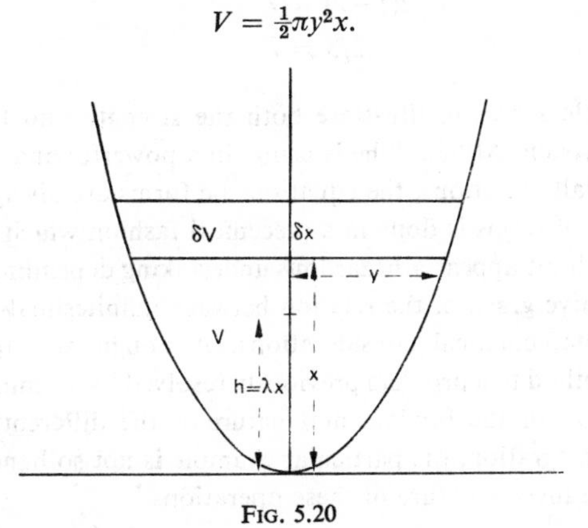

Fig. 5.20

† See Fig. 5.19. $\overline{PQ} = \overline{PR}+\overline{RQ} = \overline{PR}+\overline{XS}$.

$$\overline{PR} = \frac{r}{r-b}\overline{GX}, \qquad \overline{GX} = \overline{GS}+\overline{SX} = \overline{GS}-\overline{XS}, \qquad \overline{XS} = \overline{GS}-\overline{GX} = \overline{RQ}.$$

Also, $QR/PR = \dfrac{GX \tan\theta}{r/(r-b)GX} = \dfrac{r-b}{r}\tan\theta \left(= \dfrac{1}{r}\dfrac{dr}{d\theta}\right)$.

‡ See *Oeuvres de Fermat* I, pp. 136–9; III, pp. 124–6.

§ See pp. 90–6, above.

Now, if h be the distance of the centre of gravity from the vertex, then $h = \lambda x$, where λ remains to be determined. For any small increase in x, we have

$$\delta V \simeq \pi y^2 \delta x,$$
$$\delta h = \lambda \delta x.$$

But, from the principle of moments,

$$Vh + \delta V\left(x + \frac{\delta x}{2}\right) \simeq (V + \delta V)(h + \delta h).$$

Hence, disregarding second-order infinitesimals,

$$x\delta V \simeq h\delta V + V\delta h,$$
$$\frac{\delta V}{V} \simeq \frac{\delta h}{x-h},$$
$$\frac{\pi y^2\, \delta x}{\frac{1}{2}\pi y^2 x} \simeq \frac{\lambda\, \delta x}{x(1-\lambda)},$$
$$2(1-\lambda) = \lambda$$
$$2/3 = \lambda.$$

This example serves to illustrate both the strength and the weakness of Fermat's approach. Although he is using, in a powerful and general way, the concept of small variations, the equations he forms are always approximate. The dropping of terms is done in a piecemeal fashion which makes the approach to the limit appear a hazardous undertaking depending for its success upon an intuitive grasp of the relation between infinitesimals rather than on any formal mathematical considerations. Although the application of a differential method to a problem previously resolved by a summation indicates an appreciation of the fundamental nature of the differential process and its link with integration, the particular example is not so handled as to bring out clearly the inverse nature of these operations.†

† In general, if $y^2 = f(x)$, then $h = \dfrac{\int_0^x y^2 x\, dx}{\int_0^x y^2\, dx}$,

$$h\int_0^x f(x)\, dx = \int_0^x x f(x)\, dx.$$

[Footnote continued on next page

Roberval, on the other hand, approaching the determination of centres of gravity in the traditional way as a summation process, was led to explore the use of visual imagery in geometric transformations and in this way was able to relate the tangent and area properties of curves.

By considering the volume of a solid of revolution (see Figs. 5.21 and 5.22) Roberval established the relation, $\int_0^b (x^2/2)\, dy = \int_0^a xy\, dx$, and utilised this result in the determination of centres of gravity,† i.e.

$$\int_0^b (x^2/2)\, dy = \int_0^a xy\, dx = \bar{x} \int_0^a y\, dx,$$

and

$$\bar{x} = \frac{\int_0^b (x^2/2)\, dy}{\int_0^a y\, dx}.$$

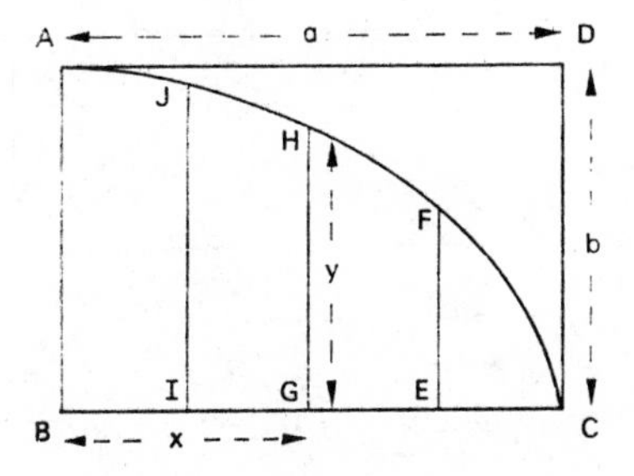

FIG. 5.21

Hence,

$$\frac{dh}{dx}\int_0^x f(x)\, dx + hf(x) = xf(x).$$

$$\frac{dh}{dx} = \lambda = \frac{f(x)\,(x-h)}{\int_0^x f(x)\, dx} = \frac{xf(x)\,(1-\lambda)}{\int_0^x f(x)\, dx},$$

or

$$\frac{\int_0^x f(x)\, dx}{xf(x)} = \frac{1-\lambda}{\lambda} \qquad [= \tfrac{1}{2}, \text{ if } y^2 = kx].$$

† *Divers Ouvrages: Traité des indivisibles*, pp. 233–4.

This, of course, is very little more than Guldin's theorem and was perfectly understood by Saint-Vincent and by Torricelli who made use of it in a more sophisticated way. Fermat had no need for this kind of representation for he was rapidly able to develop a whole range of analytic transformations which required neither geometric representation nor visual imagery.† An in-

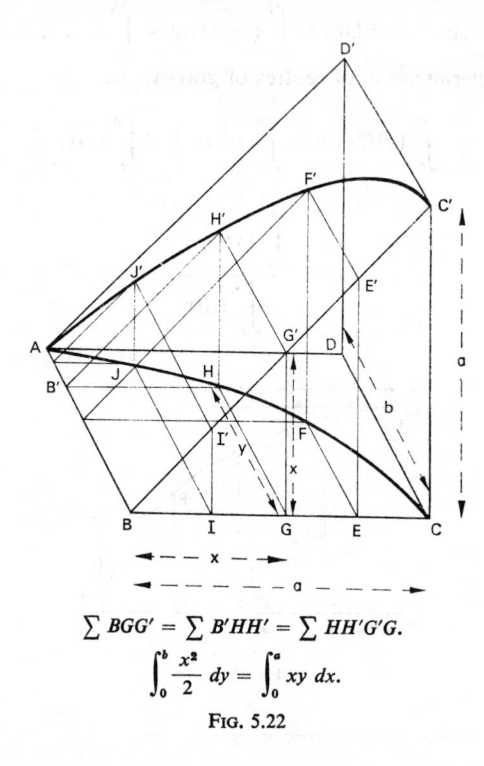

$$\sum BGG' = \sum B'HH' = \sum HH'G'G.$$

$$\int_0^b \frac{x^2}{2}\, dy = \int_0^a xy\, dx.$$

FIG. 5.22

teresting application of the above result arose in connection with the volume of the hyperboloid of revolution formed by rotating the rectangular hyperbola ($xy = ab$), about the y-axis.

We have (see Fig. 5.23)

$$\int_b^\infty x^2\, dy = 2\int_0^a x(y-b)\, dx = a^2b.$$

† See *Oeuvres de Fermat* I, pp. 255–88.

The volume of an infinitely long solid is thus shown to be equal to a finite quantity.

More important, however, is the geometrical transformation through which, by means of the relation, $t/x = dy/dx$, Roberval transforms the integral $\int_0^a x\,dy$ into $\int_0^b t\,dx$.[†]

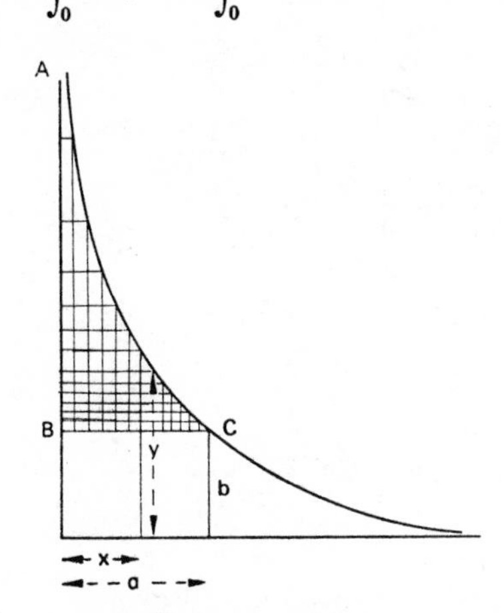

$$\int_b^\infty x^2\,dy = 2\int_0^a x(y-b)\,dx = a^2b \qquad (xy = ab).$$

FIG. 5.23

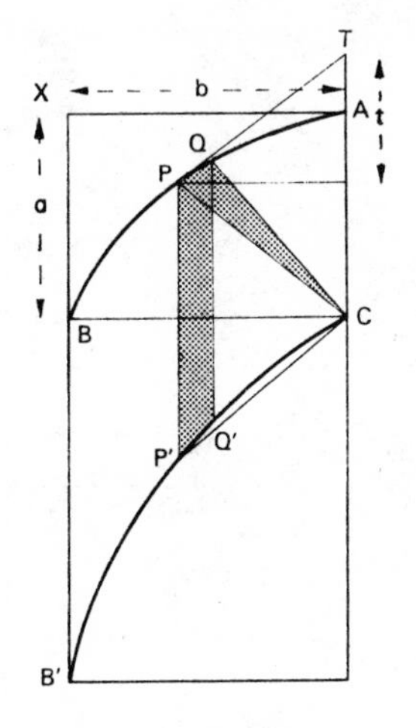

FIG. 5.24

If the arc APB of a curve be divided into an infinite number of parts, of which PQ is one, then the arc PQ corresponds with the tangent line at P (see Fig. 5.24). Let PQ be produced to meet the axis CA at T and let a second curve, $CQ'P'B'$, be constructed such that, for each point P' on it, $PP' = CT$. Since $PTCP'$ is a parallelogram, $\Delta PQC = \frac{1}{2}$ space, $PQQ'P'$, and, summing for all such triangles, we have

$$\text{area } AQPBC = \tfrac{1}{2} \text{ area } BPACP'B',$$

and area $AQPBC$ = area $BCP'B'$ $\left(\int_0^b t\,dx = \int_0^a x\,dy, \text{ where } t/x = dy/dx\right)$.

Interesting applications follow in which the tangent and area properties of the parabolas, $(y/a) = (x/b)^n$, are clearly linked. Let the curve APB be of the form, $y/a = (x/b)^n$ (see Fig. 5.25). From the tangent properties of these

† *Traité des indivisibles*, pp. 240–5.

curves, we have $t = ny$.† Hence, if area $AXBP = A_1$, area $CBB'P' = A_2$, area $APBC = A_3$, then $A_2 = nA_1$, $A_3 = nA_1 = A_2$, $A_1+A_3 = ab = (n+1)A_1$, $A_1 = ab/(n+1)$ and $A_3 = nab/(n+1)$. Thus, from the tangent properties the quadratures can be derived‡ and conversely. By applying the Pappus–Guldin

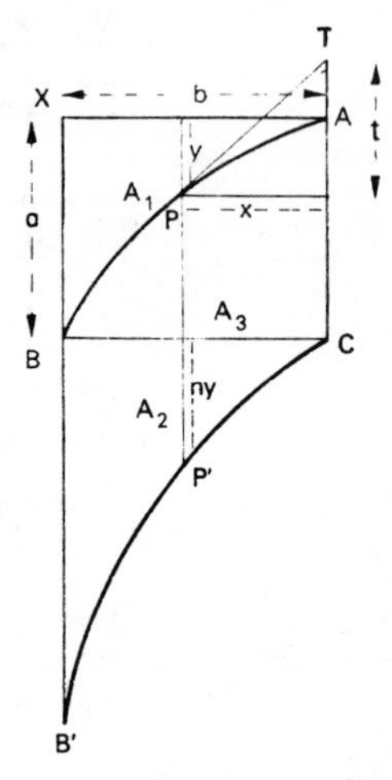

Fig. 5.25

theorem to the innumerable solids formed by rotating the above figures about the axis§ Roberval was able to determine the volumes of the corresponding solids of revolution.

5.7. Evangelista Torricelli: Tangent and Quadrature

Evangelista Torricelli, Italian mathematician and physicist, pupil of Castelli and friend of Cavalieri, worked with Galileo for three months in Rome and succeeded him as mathematician to the Duke of Tuscany. He is particularly remembered for his work on the barometer carried out with Viviani in association with the *Accademia del Cimento*. He died in 1647, leaving be-

† $dy/dx = t/x = ny/x$, $t = ny$.

‡ $\int_0^1 x^n\,dx = 1/(n+1)$.

§ *Traité des indivisibles*, p. 244: "Nous voyons qu'il se fait plusieurs cylindres, rouleaux de cylindres, cônes, ou rouleaux de cônes."

hind him many unpublished manuscripts.[†] Torricelli fully recognised the advantages and disadvantages of indivisible methods and far outdid Cavalieri in flexibility and perspicuity. He was somewhat catholic in applying the ideas and methods of his contemporaries and hence became the centre of many disputes concerning priorities. Although he acknowledged the help he received from his friends in Italy, he was perhaps not so forthcoming as to the assistance he received through Mersenne and Carcavy from sources beyond the Alps. In the *Opera geometrica*, which he published in 1644,[‡] he gave a full account of indivisible methods which was helpful and illuminating to many who found the works of Cavalieri obscure and difficult. Along with the Cavalierian methods, which he greatly admired, he supplied full demonstrations according to the methods of the ancients, and speculated, as did Isaac Barrow later, on the likelihood that Archimedes himself had used similar methods in order to discover the theorems for which he supplied such elegant demonstrations.[§]

In discussing the demonstrations of Valerio[||] he showed himself, in some respects, closer to the concept of a limit than Valerio. In *De dimensione parabolae*[††] he supplied twenty-one demonstrations of the quadrature of a parabolic segment, ten according to the methods of the Greeks and eleven making use of indivisibles. Of the latter, one is almost identical with that given by Archimedes in *The Method*.[‡‡]

Torricelli was a skilful geometer and in all his work he recognises the supreme importance of continued proportions, i.e. geometric series, and includes an interesting geometric treatment together with a construction for the sum of the series.[§§]

Given two concurrent lines, DG and CG (see Fig. 5.26), let a *flexilineum* be constructed, so that the points $C, E, I, K, M, \ldots$ lie on CG and the points $D, F, H, \ldots$ on DG and the segments $DC, CF, FE, EH, HI, \ldots$ are alternately parallel to each of two given directions. The lines $DC, FE, HI, JK, \ldots$ are in continued proportion, i.e. $DC/FE = FE/HI = HI/JK = \ldots.$ By a *reductio* proof Torricelli shows that all the terms in the series, $DC, FE, HI, JK, \ldots$, continued to infinity, are represented in the *flexilineum* and that,

Torricelli, Evangelista, *Opere* (ed. Loria, G. and Vassura, G.), Faenza, 1919 (4 vols.); *De infinitis spiralibus* (ed. Carruccio, E.), Pisa, 1955.
‡ Torricelli, Evangelista, *Opera geometrica*, Florence, 1644.
§ *Opere* I, p. 140.
|| *Ibid.* I, p. 95.
†† *Ibid.* I, pp. 102–62.
‡‡ *Ibid.* I, pp. 160–1.
§§ *Ibid.* I, pp. 147–8.

conversely, every segment of the flexilineum parallel to DC represents a term in the progression. Thus, in order to find the sum of the series we proceed as follows: through G, draw GL, parallel to CF, to meet DC produced in L: DL represents the sum of the series, for (see Fig. 5.26)

$$DL = DC+CE'+E'I'+I'K'+\ldots$$
$$= DC+FE+HI+JK+\ldots.$$

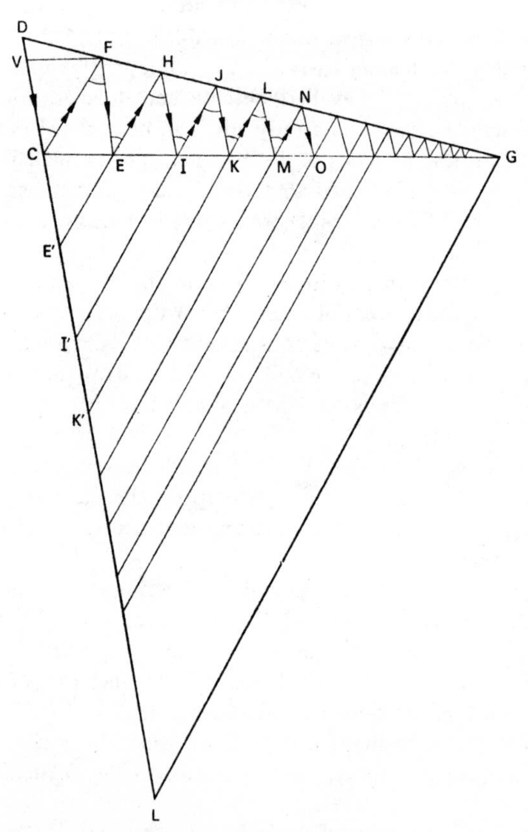

Fig. 5.26. From similar triangles,

$$DC/CF = FE/EH = HI/IJ = \ldots$$
$$CF/FE = EH/HI = IJ/JK = \ldots$$

Hence

$$DC/FE = FE/HI \quad (= CF/EH).$$

Similarly

$$FE/HI = HI/JK\ldots, \text{ and so on.}$$

Hence, if FV be drawn parallel to EC to meet DC in V, we have

$$DL/DC = DG/DF = DC/DV = DC/(DC-FE),$$

i.e. the first term of the progression is a mean proportional between the first difference and the sum to infinity.†

Torricelli makes use of this result in determining the area of the segment of a parabola by inscribing a sequence of triangles in the manner of Archimedes.‡ Thus, instead of giving the usual *reductio* proof§ he proceeds directly to the limit, i.e.

Space of parabolic segment $= \Delta\left(1+\frac{1}{4}+(\frac{1}{4})^2+\ldots\right) = \Delta(4/3)$.

Of much greater significance, however, is the skilful way in which Torricelli manipulates continued proportions in the quadrature of the hyperbolas, $(y/a)^m(x/b)^n = 1$. Although there can be no question that Fermat was the first to propose this problem, or indeed that he was the first to find a general solution, in the hands of Torricelli the work developed along different lines.‖ In the history of the calculus these investigations of Torricelli are supremely important, for the manner in which he expressed his results enabled him to perceive, and to formulate, the tangent construction from the integral. Amongst all those who contributed to the development of infinitesimal processes before Newton and Leibniz, Torricelli exhibited most clearly the link between the two operations now known as differentiation and integration. Unfortunately much of this work remained unpublished until 1918, and in consequence the importance of his contribution has scarcely, even now, been fully recognised.

To appreciate fully the role played by continued proportions in Torricelli's quadratures†† it is necessary to bear in mind that his approach was wholly geometrical. Until Wallis extended the index notation to include negative and fractional exponents it was necessary to make use of continued proportions to construct and represent expressions of the form, $a^{m/n}$.

† If $S = a+ar+ar^2+ar^3+ar^4+\ldots$, then Torricelli has $S/a = a/(a-ar)$, i.e. $S = a/(1-r)$.

‡ *Opere* I, pp. 150–1.

§ See p. 36 above.

‖ *Opere* I, Pt. 2; *De infinitis hyperbolis*, pp. 227–74; *De infinitis parabolis*, 275–328; *De infinitis spiralibus*, pp. 349–99. See also Bortolotti, E., La memoria "De infinitis hyperbolis" di Torricelli: *Archeion*, 6 (1925), pp. 49–58, 139–52. Bortolotti, E. La scoperta e le successive generalizzazioni di un teorema fondamentale di calcolo integrale, *Archeion* 5 (1924), pp. 204–27. Jacoli. F. Evangelista Torricelli ed il metodo delle tangenti detto metodo del Roberval, *Bulletino di Bibliografia e di Storia delle Scienze Matematiche e Fisiche* 8 (1875), pp. 265–304.

†† *Opere* I, Pt. 2, pp. 231–42.

Thus, if $x_0, x_1, x_2, x_3, \ldots, x_n$, denote a sequence of numbers in continued proportion, then,

$$x_n/x_0 = (x_1/x_0)^n, \quad x_1/x_0 = \sqrt[n]{(x_n/x_0)},$$

and

$$x_m/x_0 = \left(\sqrt[n]{(x_n/x_0)}\right)^m = (x_n/x_0)^{m/n}.$$

For Torricelli, the curves now expressed analytically in the form, $y = kx^n$ (n being positive, negative, integral or fractional), had to be considered separately as, $(y/a)^m = (x/b)^n$, and $(y/a)^m(x/b)^n = 1$. In each case construction of the curves depended upon purely geometric relations between line segments and it was not only natural, but also necessary, to use continued proportions to represent the relations, $(y/a) = (x/b)^{n/m}$, $(y/a) = (b/x)^{n/m}$.

Let P and Q be the points (x_1, y_1), (x_2, y_2) on the hyperbola $(y/a)^m(x/b)^n = 1$; then, $(y_1/y_2)^m = (x_2/x_1)^n$, and $y_1/y_2 = (x_2/x_1)^{n/m}$. Now, by the use of continued proportions, it is possible to find a point B in the line OX (see Fig. 5.27) such that, $OB = x_1(x_2/x_1)^{n/m}$, $(n < m)$. Thus, $OB = x_1(y_1/y_2)$, and $x_1y_1 = OB \, . \, y_2$.

Hence rectangle $AOEP$ = rectangle $OBRC$,

and, subtracting rectangle $OEVC$,

rectangle $APVC$ = rectangle $VRBE$.

Now,

$$DE/BE = \frac{DO-OE}{BO-OE} = \frac{DO/OE-1}{BO/OE-1} = \frac{x_2/x_1-1}{(x_2/x_1)^{n/m}-1}.$$

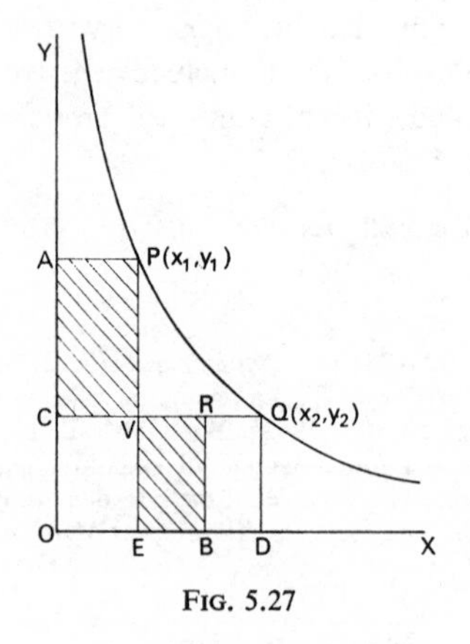

FIG. 5.27

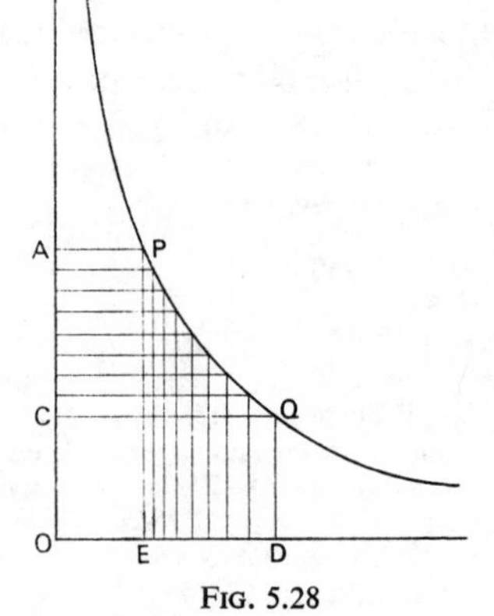

FIG. 5.28

Thus, as $P \to Q$, we have, $DE/BE \to \lim_{x_1 \to x_2} \left\{ \frac{x_2/x_1 - 1}{(x_2/x_1)^{n/m} - 1} \right\} = m/n$,†

and consequently,

$$\text{rectangle } DEVQ \to m/n \text{ rectangle } VRBE$$
$$\to m/n \text{ rectangle } APVC.‡$$

Thus, in the notation of the calculus, by inscribing successive figures within the curve, we have, in general,

$$\frac{\int_P^Q y\,dx}{\int_P^Q x\,dy} = \frac{m}{n} \quad \text{(see Fig. 5.28).}$$

Torricelli works in terms of inequalities and establishes the above relation in the form,

$$\frac{\text{space } QDEP}{\text{space } APQC} = \frac{m}{n}, \quad \text{by } \textit{reductio ad absurdum.}$$

The extension is made to the infinite case where we have,

$$\frac{\int_0^P y\,dx}{\int_P^\infty x\,dy} = \frac{m}{n}, \quad \text{and} \quad \frac{\text{area } OEPCH}{\text{area } APCH} = \frac{m}{n} \quad \text{(see Fig. 5.29)}$$

$$\frac{\text{area } OEPA}{\text{area } APCH} = \frac{m-n}{n}.§$$

Now, if the areas A_1 and A_2 (see Fig. 5.30) be rotated about the axis, OH, we have (by the Pappus–Guldin theorem),

$$\frac{V_2}{V_1} = \frac{A_2(x_1+x_2)/2}{A_1 x_1/2},$$

† If $x_2/x_1 = r$, then $\frac{x_2/x_1 - 1}{(x_2/x_1)^{n/m} - 1} = \frac{r-1}{r^{n/m} - 1}$,

and $\lim_{r \to 1} \left\{ \frac{r-1}{r^{n/m} - 1} \right\} = \frac{m}{n}$ $(n < m)$. See p. 162 above.

‡ Equivalent to $dy/dx = -ny/mx$. If $y_1 = y$, $y_2 = y + \delta y$, $x_1 = x$, $x_2 = x + \delta x$, we have, $\delta x(y+\delta y) = (m/n)(-\delta y)x$, $dy/dx = \lim_{\delta x \to 0} (\delta y/\delta x) = -ny/mx$.

§ $\int_y^\infty x\,dy = \frac{n}{m-n} xy$ (provided $(y/a)^m (x/b)^n = 1$, $n < m$).

and, since
$$\frac{A_2}{A_1} \to \frac{m}{n} \quad \text{as} \quad P \to Q,$$

then
$$\frac{V_2}{V_1} \to \frac{2m}{n} \quad \text{as} \quad P \to Q \quad (\text{provided } n < 2m).$$

Hence, by inscribing figures within the curve,† we have

$$\frac{\text{solid formed by rotating } PEOH \text{ about } OH}{\text{solid formed by rotating } PAH \text{ about } OH} = \frac{2m}{n},$$

the proof again being established on a *reductio* basis.

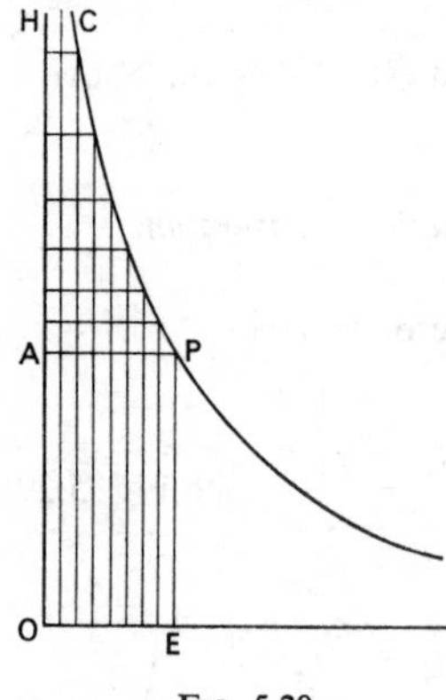

FIG. 5.29

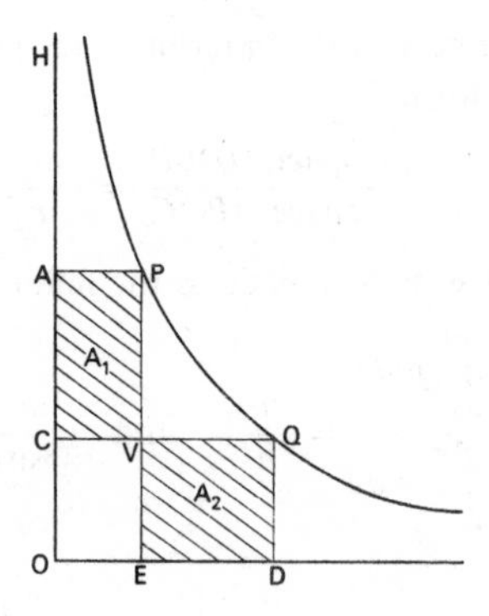

FIG. 5.30

Thus, not only did Torricelli establish an effective means of integration but was also able to lay down conditions for the existence of such integrals in cases where the ordinate became infinite. The processes outlined above were, however, not only important as a means of integration, but also as a significant transition from an integral to a differential operation. In the notation

† If (for simplicity),

$$y = kx^p = kx^{-n/m}, \quad V_2 = 2\pi \int_0^x xy\,dx = 2\pi k \int_0^x x^{p+1}\,dx = \frac{2\pi k}{p+2}\,x^{p+2} = \frac{2\pi y x^2}{p+2},$$

$$V_1 = V_2 - \pi x^2 y = \pi x^2 y\left(-\frac{p}{p+2}\right), \quad V_2/V_1 = (2/-p) = 2m/n;$$

V_1 exists provided $p+2 > 0$, i.e. $n < 2m$.

of the calculus, the relation

$$\frac{A_2}{A_1} \to \frac{m}{n}, \quad \text{as} \quad P \to Q,$$

becomes

$$\lim_{\delta y \to 0} \left| \frac{y\,\delta x}{x\,\delta y} \right| = \frac{m}{n}, \quad \text{or} \quad \lim_{\delta x \to 0} \left| \frac{\delta y}{\delta x} \right| = \frac{ny}{mx}.$$

Thus, if the tangent at P be regarded as the limiting position of the chord PQ as $P \to Q$, to draw the tangent at the point P it is only necessary to determine a point T on OY (see Fig. 5.31), such that $TA/AP = ny/mx$, i.e. $TA/n = y/m$. Torricelli supplies this construction together with a standard *reductio*

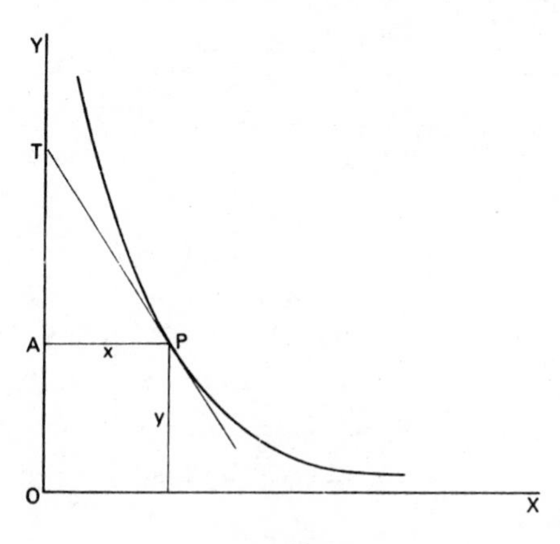

FIG. 5.31

proof. Although he does not refer explicitly to the basis for this construction it appears to arise so readily and in such close association with the previous work on integration that it is difficult not to conclude, with Bortolotti,† that Torricelli understood clearly the inverse relation between integration and differentiation in the sense that he used the relation,

$$\lim_{\delta x \to 0} \frac{\delta y}{\delta x} = \frac{ny}{mx},$$

† Bortolotti, E., La scoperta, *op. cit.* Unfortunately final judgement on the merits of such claims made on behalf of Torricelli depends very much on the order in whieh the unpublished manuscripts were arranged for publication by the editors.

both to draw the tangent and to find the area under the curve, i.e.

$$m\int_P^Q x\,dy = n\int_P^Q y\,dx.$$

Elsewhere Torricelli develops an interesting and completely different tangent method† in which he links the indivisible methods used by Galileo for rectilinear motion and projectiles‡ with the quadrature of curves of the form

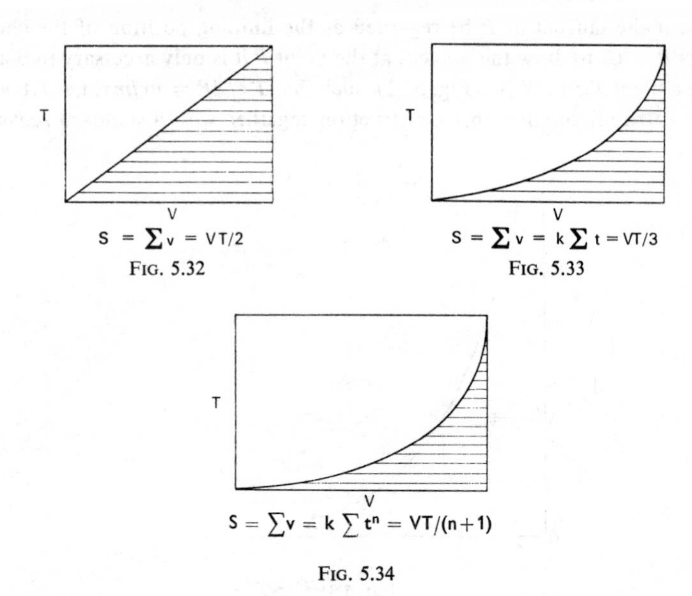

FIG. 5.32

FIG. 5.33

FIG. 5.34

$y = kx^p$ (p integral). The basic concept here is the mediaeval idea of the velocity–time graph§ in which the total distance covered (S) is represented by the area under the curve. Thus, if the area be conceived as the sum of the ordinates (v), $S = \sum v$. For motion under constant acceleration the graph is a straight line ($v \propto t$), and $S = VT/2$, where V is the velocity attained after a time interval T (see Fig. 5.32); if $v \propto t^2$, the curve is a parabola and $S = \sum v = VT/3$. In general, if $v \propto t^n$, then $S = \sum v = k\sum t^n = VT/(n+1)$ (see Figs. 5.33 and 5.34).

† *Opere* I, Pt. 2: *De infinitis parabolis*, pp. 275–328.
‡ See pp. 120–2 above.
§ See pp. 81–9 above.

Now, consider a point $P(x, y)$ moving along a curve from B to A with two velocities, one parallel to CP and the other parallel to BC (see Fig. 5.35). Let the velocity parallel to CP be uniform and such that $CP = t = x$: let the velocity parallel to BC be denoted by v and such that, $v = kt^n = kx^n$. Now, at the point P, we have,

$$y = \sum v = k \sum t^n = \frac{kx^{n+1}}{n+1}.$$

Conversely, if a point P describe the curve, $y = kx^n$, with a uniform velocity parallel to CP, such that $x = t$, and a velocity, $v = f(t) = f(x)$, parallel to BC, then, since $\sum v = \sum f(t) = y = kx^n = kt^n$, and since $k \sum t^{n-1} = kt^n/n$, then $v = nkt^{n-1} = ny/x$.

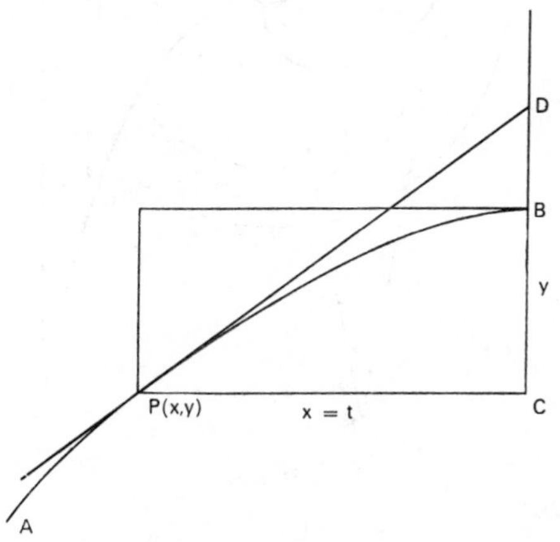

Fig. 5.35

To determine the tangent to the curve, $y = kx^n$, it is therefore only necessary to mark off a point D on the line CB produced, such that $CD/CP = ny/x$.

Torricelli was able to extend these methods to curves of the form $(y/a)^m = (x/b)^n$ (the higher parabolas) and thus to draw tangents to curves of the general form, $y = kx^\alpha$ (α integral and fractional).

Perhaps the most fascinating of all the examples worked out by Torricelli is the relation he develops, through a geometrical transformation, between the spiral curves of the general form $(r/a)^m = (r\theta/a\alpha)^n$ and the higher parabolas $(y/a)^m = (x/b)^n$.† This work is complicated, entirely geometrical and

† *Opere* I, Pt. 2; *De infinitis spiralibus*, pp. 349–99.

difficult to follow in the original form. In modern notation the method runs briefly as follows: for simplicity let the equation of the spiral be given as $r^m = k\theta^n$ and let a point P move along the curve from A to C (see Fig. 5.36) with two velocities, the first a progressive motion (v) along the radius vector, and the second a circular motion (u) in which the radius vector rotates uni-

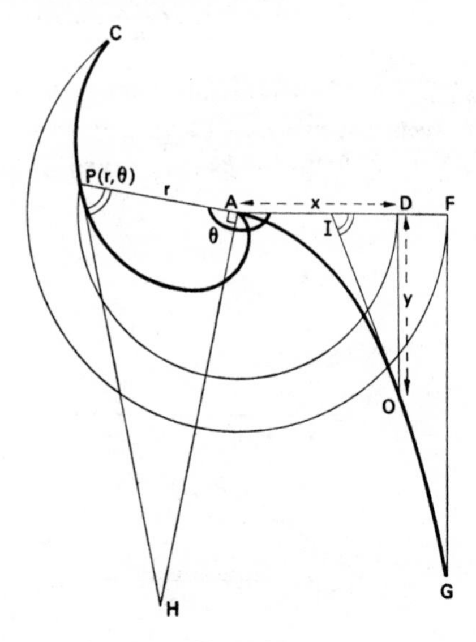

FIG. 5.36

formly about the centre A. By analogy with the previous methods for the parabolas, we have $\theta \propto t$, $r^m \propto t^n$, $u[= r\, d\theta/dt] = rc$ (where $\theta = ct$), $v[= dr/dt] = nr/mt$. Hence,

$$v/u = \frac{nr/mt}{rc} = \frac{n}{m\theta}\left[= \frac{1}{r}\frac{dr}{d\theta}\right].$$

Thus, to construct a tangent to the curve at the point P, we draw AH perpendicular to AP, such that, $AH/AP = m\theta/n$, i.e. $AH = mr\theta/n = (m/n)PD$, where PD is the circular arc, radius r (see Fig. 5.36).

Writing the equation of the spiral curve in the form $r^{m+n} = k(r\theta)^n$, and substituting

$$x = r, \quad y = \frac{m}{m+n}(r\theta),$$

the spiral is transformed into the parabola,

$$x^{m+n} = k\left(\frac{m+n}{m}y\right)^n.$$

The tangent to this curve, constructed as before, makes an angle (OID) with AD given by,

$$\tan OID = DO/DI = \frac{(m+n)\,y}{nx} = \frac{m\theta}{n},$$

and consequently angle OID = angle HPA. From the correspondence between the points on the parabola and on the spiral and from the similarity of the triangles AHP and DOI Torricelli deduces the equality of the spiral arc AC and the parabolic arc AG† (see Fig. 5.36).

The quadrature of the spiral space is achieved by making use of the parabolic space, or *trilineum*, ADO. Thus, if angle $CAH = \alpha$, we have

$$\frac{\sum r\theta}{\sum r\alpha} = \frac{\text{sector } AFMCPA}{\text{circular sector } AFMCA} = \frac{\sum DO}{\sum DX} = \frac{\text{space } AFG}{\text{triangle } AFG} = \frac{2n}{m+2n}\ ‡$$

(see Fig. 5.37).

The controversies regarding the use Torricelli made of suggestions received through Mersenne and other intermediaries from Roberval and Fermat in France resounded through France and Italy for centuries after his death. The unpublished papers which he left behind him, however, indicate that in his thinking he had progressed far beyond Cavalieri in the sophisticated use he was making of rectangular elements and in the scope of the tangent methods he was able to develop by relating them to the integration process. Although unquestionably Fermat, Roberval and Torricelli covered a great deal of

† *Opere* I, Pt. 2, p. 388: "ergo omnia et singula puncta parabolae *AOG* aequalia sunt et numero et magnitudine, sive specie omnibus punctis spiralis *DEB*, propterea parabolica linea *AOG* equalis erit lineae spirali *DEB*."

$$s = \int_0^r \sqrt{(1+r^2(d\theta/dr)^2)}\,dr = \int_0^x \sqrt{(1+(dy/dx)^2)}\,dx.$$

‡ $y^n = A^n x^{m+n}$, $\displaystyle\int_0^x y\,dx = A\int_0^x x^{(m+n)/n}\,dx = \frac{nA}{m+2n}\,x^{(m+2n)/n} = \frac{nxy}{m+2n}$

$$= \frac{2n}{m+2n}\,(xy/2).$$

common ground, the work of each has its own distinctive character and in consequence provides new insight into the discovery process. The ideas of Torricelli were transmitted through his pupil Angeli to Gregory and to Barrow during their sojourn in Italy and in this way directly influenced the development of infinitesimal methods in England.†

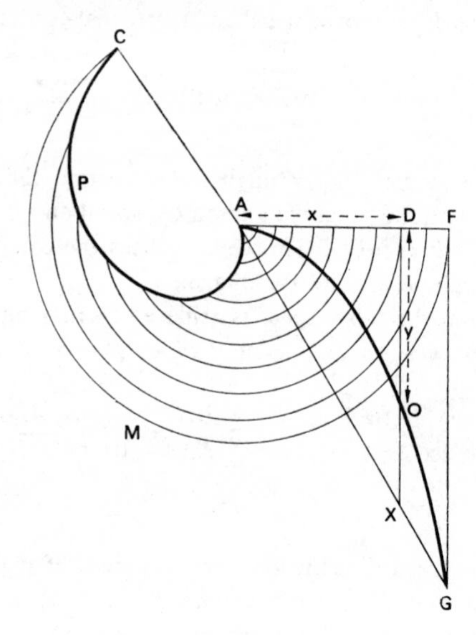

FIG. 5.37

† Torricelli's friends and pupils included Viviani (1622–1703), Angeli (1623–97) and Ricci (1619–82). Angeli wrote many small treatises on infinitesimal methods and probably had access to the unpublished manuscripts of Torricelli. James Gregory (1638–75) spent four years closely associated with Angeli in Rome.

CHAPTER 6

Consolidation of Gains: France, England and the Low Countries

6.0. Introduction

From 1648 onwards mathematical work in the field of infinitesimals intensified and in reviewing this work problems of selection become more difficult.

The spread of Cartesian notation, through the publication of Latin translations of *La Géométrie*, meant that an increasing volume of geometrical work received the benefits of algebraic symbolism. The conic sections were studied algebraically as plane curves and other curves were invented whose properties could be expressed neatly in the now familiar (x, y) notation. The symbols lent themselves to the development of general methods and analytic transformations became commonplace, at least in the case of algebraic curves. Certain perceivable patterns became apparent in the forms of the results and this led to an intensified search for simplified methods including the formulation of general rules through which tangents, quadratures and centres of gravity could be written down directly from the curve equations.

The benefits of algebraic notation were so apparent to all who used it that the introduction of new symbols to cover operations, exponents, concepts of infinity and the like, became fairly widespread. None the less, the strain of finding a path through the maze of new symbols devised to represent obscurely defined terms could be considerable and led to a certain amount of exasperation. Thomas Hobbes, for instance, says: "the Symboles serve only to make men go faster about, as greater Winde to a Winde-mill".†

More serious, however, was the widening of the gap between the Archimedean demonstration which most mathematicians of repute still regarded as the ultimate fount of authority in the field of infinitesimal methods and the discovery methods founded upon intuitive concepts on the one hand and

† Hobbes, T., *Six Lessons*, London, 1656, Preface.

the standardised rules and techniques on the other. Intensified competition linked with considerations of national prestige and supported by the growth of the influential scientific societies meant that ultimately every mathematician must face the problem of the presentation of his work to the public. Bitter controversies arose regarding the style and manner of presentation of new results. Fermat, for example, kept his own counsel and published little but had a great deal to say about Wallis's *Arithmetica infinitorum*. Wallis stoutly defended himself on the grounds that, by publishing his own discovery methods rather than by exhibiting proofs in the manner of the ancients, he hoped to further the researches of others.

It follows that during the period under consideration the problem of methods and techniques remained a central and a common interest. Apart from certain notable exceptions, published work consisted of sketches of effective ways of attacking and resolving new problems and simplifying old ones. In some cases the problem of presentation was resolved by the publication of rules and results whilst the methods themselves were withheld. This served to establish priority whilst preserving secrecy and deferring indefinitely the issue of formal proof structure.

From the fascinating variety of work in progress during this period it has not been easy to make a selection. I have resisted the temptation to make this chapter into a catalogue of achievement or even an account of outstanding individual successes but have tried instead to include work which is representative of an important trend or direction and which, as such, exerted a powerful influence on further developments and more especially on the work of of Newton and of Leibniz. Even so, limitations of time and space prohibit a comprehensive treatment and the ultimate choice must remain a personal one.

6.1. Blaise Pascal (1623–1662)

Blaise Pascal was, in his early youth, introduced by his father, Étienne Pascal, himself a distinguished mathematician, to the "salon scientifique" of Mersenne, Mydorge, Gassendi, Desargues and Roberval. Tradition has it that, without any formal schooling in mathematics,† Pascal discovered geometry for himself and within a short space of time was able to make original contributions.‡

† See *Oeuvres de Blaise Pascal* (ed. Brunschvicg, L. and Boutroux, P.), Paris, 1908–25 (references for Vols. I–III are given from the 2nd ed., 1923), I, pp. 53–7.

‡ *Essay pour les coniques*, Paris, 1640: *Oeuvres* I, pp. 252–60.

Itn association with Fermat and Roberval, Pascal developed a strong interest in number theory and on this he founded his integration methods. Although the techniques he developed in this field were in no sense original he was, unlike his contemporaries, Fermat and Roberval, interested in exposition and his clear and lucid mind enabled him to throw some light on the conceptual basis for such methods.[†] His published work was an important influence in the early studies of Leibniz.[‡]

In an early work *Potestatum numericarum summa*[§] (probably 1654) Pascal outlined a general method for finding the sums of the integral powers of any sequence of numbers in arithmetic progression.[||] Let the terms of the progression be a, $a+d$, $a+2d$, $a+3d$, $\ldots$, $a+(r-1)d, \ldots$, $a+(n-1)d$, then, since in general (for p positive and integral),

$$(\overline{a+rd}+d)^{p+1} = (a+rd)^{p+1} + {}_{p+1}C_1(a+rd)^p\, d + \ldots \quad d^{p+1},$$

$$\sum_{0}^{n-1} (a+rd+d)^{p+1} = \sum_{0}^{n-1} (a+rd)^{p+1} + {}_{p+1}C_1 \sum_{0}^{n-1} (a+rd)^p\, d + \ldots \quad n\, d^{p+1},$$

we have (by subtraction)

$$(a+nd)^{p+1} - a^{p+1} = {}_{p+1}C_1 \sum_{0}^{n-1} (a+rd)^p\, d + \ldots \quad n\, d^{p+1}.$$

Accordingly, for the sums of the powers of the natural numbers ($a = 1$, $d = 1$),

$$(n+1)^{p+1} - 1 = {}_{p+1}C_1 \sum_{1}^{n} r^p + {}_{p+1}C_2 \sum_{1}^{n} r^{p-1} + \ldots \quad n.$$

The general rule, equivalent to ${}_nC_r = n(n-1)\ldots(n-r+1)/(1.2.3\ldots r)$, for deriving the binomial coefficients (n positive and integral) was established by Pascal on the basis of a proof by "complete" induction,[††] i.e. since it is

† See Bosmans, H., La notion des "indivisibles" chez Blaise Pascal, *Archivio di storia della Scienza* 4 (1923), pp. 369–79; Sur l'oeuvre mathématique de Blaise Pascal, *Mathésis* 38 (1924), *Supplément*, pp. 1–59; Pascal et son traité du triangle arithmétique, *Mathésis* 37 (1923), pp. 455–64.

‡ See Gerhardt, C. I., Leibniz und Pascal, *Sitz. der König. Preuss. Akad. der Wiss.*, Berlin, 1891, pp. 1053–68. Given in Child, J. M., *The Early Mathematical Manuscripts of Leibniz*, Chicago and London, 1920.

§ *Oeuvres* III, pp. 343–67.

|| The demonstration is given entirely by numbers. The rules are enunciated verbally (see *Oeuvres* III, p. 361).

†† Pascal, B., *Traité du triangle arithmétique*, Paris, 1665 (probably written in 1654): *Oeuvres* III, pp. 445–503.

evident that the proposition holds for $n = 2$, and since it can be established that, if the proposition holds for some value of n it necessarily holds for $n+1$, then, in fact, it is valid for all values of n (for, since it holds for $n = 2$ it holds for $n = 3$ and for $n = 4$ and so on for all values).†

Pascal had no difficulty in applying these results in the quadrature of curves for, he says,‡

> Ceux qui sont tant peu au courant de la doctrine des *indivisibles* ne manqueront pas de voir quel parti on peut tirer des resultats qui précèdent pour la détermination des aires curvilignes. Ces résultats permettront de quarrer immédiatement tous les genres de paraboles et une infinité d'autres courbes.

The approach to the limit,

i.e.
$$\lim_{n\to\infty} \frac{\sum_1^n r^p}{n^{p+1}} = \frac{1}{p+1},$$

is achieved by discarding lower order infinitesimals which Pascal pictures in geometric terms.§ The sum of a certain number of lines is to the square of the greatest as 1 : 2; the sum of the squares is to the cube of the greatest as 1 : 3; the sum of the cubes is to the fourth power of the greatest as 1 : 4; for, in truth, a point adds nothing to a line, a line to a surface, a surface to a solid, or, to speak of numbers, the roots do not contribute to the squares, the squares

† *Oeuvres* III, pp. 456–7: "que si cette proportion se trouve dans une base quelconque, elle ce trouvera nécessairement dans la base suivante." The binomial coefficients (for positive integral exponents) were well known in the West in the sixteenth century and references and tables occur in the works of Stifel (*Arithmetica integra*, Nuremberg, 1544), Rudolff (*Die Coss*, Königsberg, 1553–4), Tartaglia (*General trattato di numeri et missure*, Venice, 1556–60), Stevin (*L'Arithmétique*, Leyden, 1585), Girard (in an edition of the works of Stevin, Leyden, 1625); Oughtred (*Clavis mathematicae*, London, 1631), Hérigone (*Cursus mathematicus*, Paris, 1634–7). See *Oeuvres* III, pp. 438–43. What is significant with Pascal is the provision of a proof by *complete* induction. Bosmans (Sur l'oeuvre mathématique, *op. cit.*, p. 31) claims that this is the earliest use of *complete* induction in the history of mathematics. Vacca, G. (Maurolycus, the first discoverer of the principle of mathematical induction, *Bull. Amer. Math. Soc.* 16 (1909), pp. 70–3) claims that Maurolyco made systematic use of induction (in the *Arithmeticorum libri duo*, Venice, 1575). Cantor gave the credit for this method to James Bernoulli (*Acta Eruditorum*, 1686) in that he made rigorous the *incomplete* induction of John Wallis. On this see Scott, J. F., *The Mathematical Work of John Wallis*, London, 1938.

‡ *Oeuvres* III, p. 365: "Those who know something of the doctrine of indivisibles will be able to see at once that the above results enable them to square a parabola of any degree and an infinity of other curves."

§ *Oeuvres* III, pp. 365–7.

to the cubes, and so on. In brief, it is possible to neglect as nothing the quanties of inferior degree.†

In the summer of 1658 Pascal, motivated by his own enthusiasm and increasing success with infinitesimal processes, sponsored a competition for the quadrature, cubature and centres of gravity of figures bounded by cycloidal arcs.‡ This contest aroused great interest, for the existence of many unresolved problems in this field rendered it eminently suitable as a testing ground for new approaches to infinitesimal methods. Huygens and Sluse held back after some initial success; Wallis and Lalovere entered for the prize but the solutions submitted were not regarded by Pascal as sufficiently general;§ Wren provided the first rectification of the arc.|| In the *Histoire de la roulette* (1658/9) Pascal reviewed the history of the cycloid problem and in the *Lettre de A. Dettonville* (1659) he published his own researches.†† These publications engendered a great deal of bitterness amongst the Italians for the attacks on Torricelli and in England for the failure to appreciate the merit of Wallis' contribution.

Although Pascal was undoubtedly attracted by the power of indivisible methods he was impressed by the careful geometrical approach of Grégoire de Saint-Vincent‡‡ and swayed by the vigorous criticism of Cavalierian indivisibles launched by André Tacquet in his work *Cylindrica et annularia* (Antwerp, 1651).§§ Pascal was accordingly impelled to examine carefully the basis for the use of indivisibles in geometry. Although he defended the use of

† *Oeuvres* III, p. 367: "En sorte qu'on doit négliger, comme nulles, les quantités d'ordre inférieur."

We have, for the sums of the powers (see p. 197 above)

$$(n+1)^{p+1}-1 = {}_{p+1}C_1 \sum_1^n r^p + {}_{p+1}C_2 \sum_1^n r^{p-1} + \ldots n.$$

Since $\sum_1^n r^{p-1}, \sum_1^n r^{p-2}, \ldots,$ contribute nothing to $\sum_1^n r^p$, we have $\lim_{n\to\infty} \sum_1^n \frac{r^p}{n^p} = \frac{1}{p+1}$ $\left(\text{equivalent to } \int_0^1 x^p\,dx = 1/(p+1)\right)$.

‡ *Oeuvres* VII, pp. 343–7: *Première lettre circulaire relative à la cycloïde* (June 1658). *Oeuvres* VIII, pp. 17–19: *Seconde lettre circulaire relative à la cycloide* (July 1658).

§ *Oeuvres* VIII, pp. 241–6: *Récit de l'examen et du jugement des écrits envoyés pour les prix.*

|| Wren submitted only the result: the method was published by Wallis in *Tractatus, duo*, Oxford, 1659.

†† *Histoire de la roulette*, Paris, 1658: *Oeuvres* VIII, pp. 195–223. *Lettre de A. Dettonville:* Paris, 1958/9: *Oeuvres* VIII, pp. 327–384; IX, pp. 3–133.

‡‡ See pp. 135–46 above.

§§ See p. 147 above.

such expressions as "the sum of the lines", "the sum of the planes", and so on, he was at some pains to explain that in making use of such terms he intended to imply the sum of an infinite number of rectangles or the sum of an infinite number of parallelepipeds. Whenever, he says, one speaks of the sum of an infinite multitude of lines one must always have regard for the line from which the equal portions are taken by which the ordinates are multiplied. In the case of the semicircle, for example, by the "sum of the ordinates" is meant the sum of an indefinite number of rectangles formed by each ordinate with the small equal portions into which the diameter is divided. This sum is a plane which differs from the space of the semicircle by less than any given quantity. This, he says, is the natural division; if it is intended to multiply by equal portions of any other line it is necessary to specify the line.

In developing the concept of double and triple integrals Pascal has recourse to the simple, triangular and pyramidal numbers of Pythagorean number theory. A set of rectangles inscribed in a given curve is considered as a set of weights balanced at equal distances along a lever arm.† If the curve be of the form $y = f(x)$, then the *weights* can be considered as proportional to $f(1)$, $f(2), f(3), \ldots, f(n)$. The simple sum, $\sum_1^n f(r)$, is the area under the curve. The triangular sum is given by the following table:

$$\begin{array}{ccccccc}
f(1) & f(2) & f(3) & f(4) & f(5) & \ldots & f(n) \\
 & f(2) & f(3) & f(4) & f(5) & \ldots & f(n) \\
 & & f(3) & f(4) & f(5) & \ldots & f(n) \\
 & & & f(4) & f(5) & \ldots & f(n) \\
 & & & & & \ldots & f(n) \\
 & & & & & \ldots & f(n) \\
 & & & & & & f(n) \\
\hline
\multicolumn{7}{c}{f(1)+2f(2)+3f(3)+4f(4)+\ldots\ nf(n)}
\end{array}$$

Hence, the triangular sum,

$$\sum_1^n \sum_1^r f(r) = \sum_1^n rf(r).$$

Since the *simple sum* is a plane then the *triangular sum*, being the sum of planes, is a solid (see Fig. 6.1): the pyramidal sum,

$$\sum_1^n \sum_1^r \sum_1^r f(r),$$

† *Oeuvres* VIII, pp. 334–50.

is the sum of solids. Pascal makes it clear that he is not thinking purely in terms of geometrical problems, i.e. quadratures and cubatures, but is utilising geometrical imagery as a means to an end; for, he argues, in the absence of a fourth dimension it is feasible enough to give this a representation in two dimensions.† Thus, for Pascal the concept of *simple*, *triangular* and *pyramidal*

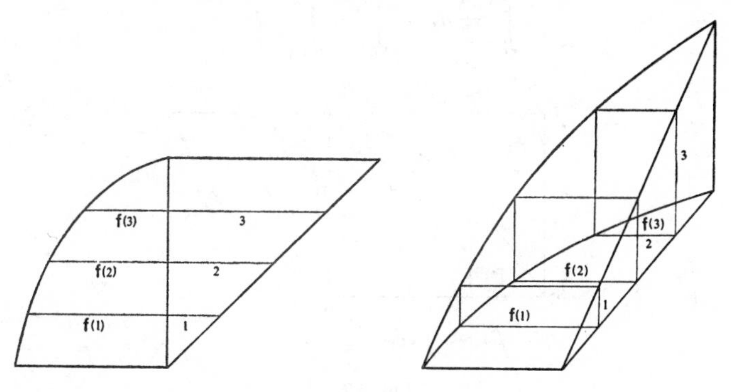

FIG. 6.1

sums serves a valuable purpose in that it enables him to relate, through a geometric representation in the manner of the *ductus in planum* of Grégoire de Saint-Vincent, the concept of moment (and correspondingly the determination of centres of gravity) with volumetric integration.

In the *Traité des trilignes rectangles et de leurs onglets*‡ Pascal carries through constructions in the manner of Saint-Vincent and systematically builds up a whole range of general integral forms expressed in geometric terms. In the *triligne ABC* (see Fig. 6.2) the sum of the ordinates *to the base* equals the sum of the ordinates *to the axis*.§ Now, on the *triligne ABC*, construct the *adjointe KBA*, and let a solid be erected on the base *ABFC* (see Fig. 6.3) such that any section *FDOH*, perpendicular to the base *ABFC* and parallel to *AC*, is a rectangle with altitude $FH = DO$. The solid is bounded by the plane *ABFC*,

† *Oeuvres* VIII, p. 367: "Et l'on ne doit pas estre blessé de cette quatrième dimension, puis que, comme je l'ay dit ailleurs, en prenant des plans au lieu des solides, ou mesme de simples droittes, qui soient entr'elles comme les sommes triangulaires particulières qui font toutes ensemble la somme pyramidale, la somme de ces droittes fera un plan qui tiendra lieu de ce plan-plan."

‡ *Oeuvres* IX, pp. 3–45.

§ $\int_0^a y\,dx = \int_0^b x\,dy.$

the plane *ABKO*, the cylindrical surface *CBKH* and the cylindrical surface *ACHKO*. The section *HFP* (see Fig. 6.3) by a plane parallel to *ABK* is congruent to *ODA*. Hence, the sum of the rectangles *FDOH* equals the sum of the sections *FPH*, i.e. the *triangular* sum of the spaces *ODA*. In the notation of the calculus, if $OD = z$, $AD = x$, $HO = FD = y$, we have

$$\int_0^a yz\, dx = \int_0^b dy \int_0^x z\, dx.$$

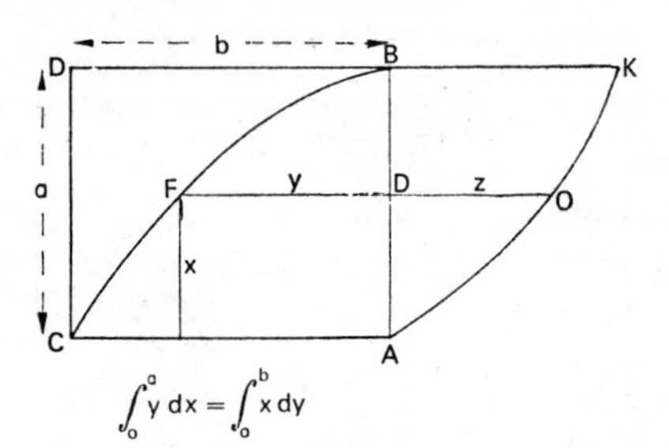

FIG. 6.2

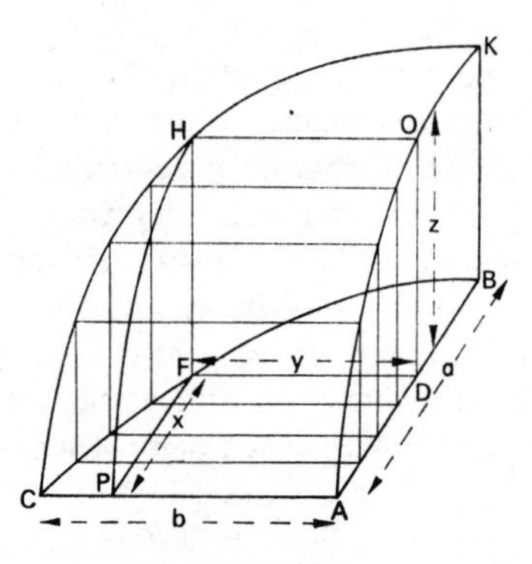

FIG. 6.3

Now, if the *adjointe* be an isosceles right-angled triangle, i.e. $z = x$, we have

$$\int_0^a yz\, dx = \int_0^a xy\, dx = \int_0^b dy \int_0^x z\, dx = \int_0^b dy \int_0^x x\, dx = \int_0^b (x^2/2)\, dy,$$

and

$$2\int_0^a xy\, dx = \int_0^b x^2\, dy.$$

If $z = x^2$, i.e. the *adjointe* is a parabola, we have

$$\int_0^a yz\,dx = \int_0^a x^2y\,dx = \int_0^b dy \int_0^x z\,dx = \int_0^b dy \int_0^b x^2\,dx = \int_0^b (x^3/3)\,dy,$$

and
$$3\int_0^a x^2y\,dx = \int_0^b x^3\,dy.$$

In general, if $z = x^m$,

$$\int_0^a yz\,dx = \int_0^a x^my\,dx = \int_0^b dy \int_0^x z\,dx = \int_0^b dy \int_0^x x^m\,dx$$
$$= \int_0^b \frac{x^{m+1}}{m+1}\,dy,$$

and
$$(m+1)\int_0^a x^my\,dx = \int_0^b x^{m+1}\,dy.$$

The *onglet* of Pascal is simply a variant of the *ungula* of Saint-Vincent. A *triligne ABC* is placed in a vertical plane with base AC horizontal and axis AB vertical (see Fig. 6.4). Through any point E on the curve BEC (ordinate

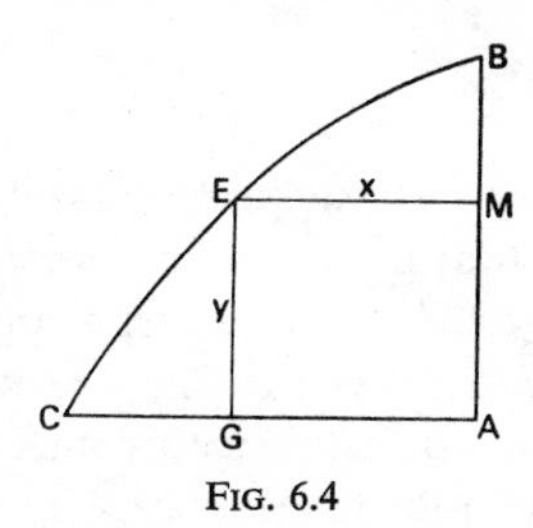

FIG. 6.4

$EG = y$, abscissa $EM = GA = x$), a line FEF' is drawn perpendicular to the plane ABC and such that $FE = EF' = EG = y$: let the line through B be DBD' with $DB = BD' = BA$. A solid is thus formed bounded by planes DAD', CAD', CAD and the cylindrical surface $DD'C$.

The volume of the *onglet* is thus given as

$$V = \sum \Delta FGF' = \sum \text{rectangle } FHH'F' \text{ (see Fig. 6.5).}$$

Thus, if the equation of the curve be available in the form $y = f(x)$, we have

$$V = \int_0^a y^2\,dx = 2\int_0^b xy\,dy.$$

By considering moments, or triangular sums, we have,

$$\int_0^a y^2x\,dx = \int_0^b x^2y\,dy.$$

The diagram which particularly attracted the attention of Leibniz when he was introduced to the work of Pascal by Huygens† was contained in the small

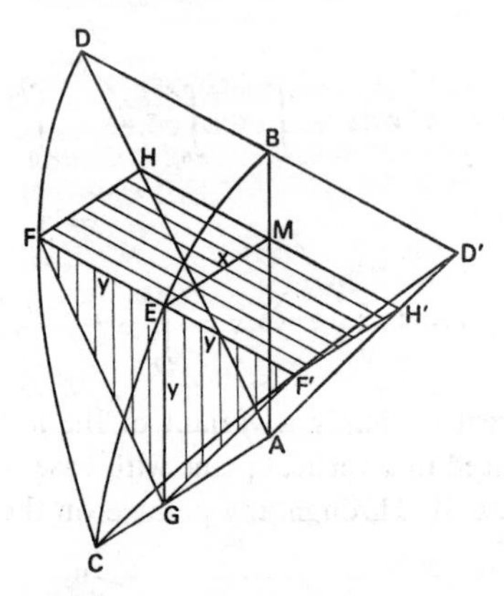

Fig. 6.5

treatise entitled *Traité des sinus du quart de cercle*‡ in which Pascal uses a species of characteristic triangle to obtain a series of results equivalent to trigonometric integrals by dividing the arc of a circle into equal parts. From the similar triangles, *EKE* and *DIA* (see Fig. 6.6), we have $DI/DA = RR/EE$, and $DI.EE = DA.RR = AB.RR$. Hence, summing for all such rectangles and identifying the arc with the tangent (since by taking n great enough the arc differs from the tangent by less than any given quantity, however small),

$$\sum DI.EE = AB\sum RR, \quad \text{or} \quad \int_\alpha^\beta r\sin\theta\,d(r\theta) = r^2(\cos\alpha - \cos\beta).$$

Multiplying by DI, DI^2, DI^3, and so on, successive sums are obtained, thus,

$$\sum DI^2EE = AB\sum DI.RR, \quad \text{or} \quad \int_\alpha^\beta r^2\sin^2\theta\,d(r\theta) = r\int_\alpha^\beta r\sin\theta\,d(r\cos\theta),$$

$$\sum DI^3EE = AB\sum DI^2\,RR, \quad \text{or} \quad \int_\alpha^\beta r^3\sin^3\theta\,d(r\theta) = r\int_\alpha^\beta r^2\sin^2\theta\,d(r\cos\theta).$$

† See Gerhardt, C. I., Leibniz und Pascal, *Sitz. der König. Preuss. Akad. der Wiss.*, Berlin, 1891, pp. 1053–68.

‡ *Oeuvres* IX, pp. 60–76.

Although the use of a characteristic triangle in this way might well have suggested the inverse relation between tangents and quadratures Pascal showed no interest in tangent methods and Leibniz wrote later that he seemed to have had a bandage over his eyes.† His mathematical work was unfortunately only pursued intermittently and he never seriously studied the algebra of

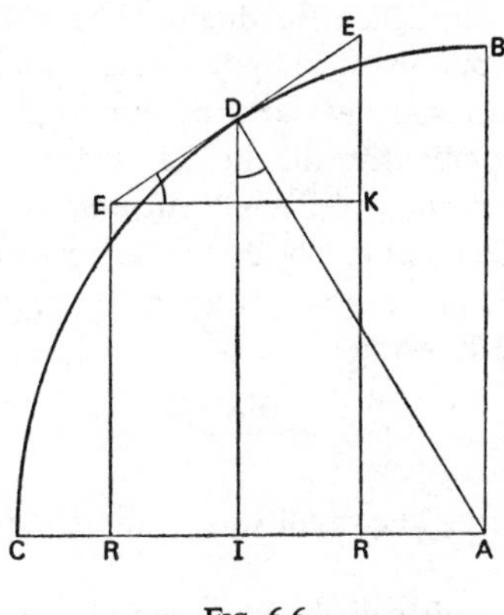

FIG. 6.6

Viète or Descartes so that he missed the benefits of any kind of suggestive symbolism and in this field did not eclipse his contemporaries.

6.2. Infinitesimal Methods in England

John Wallis (1616–1703), one of the original Fellows of the Royal Society of London,‡ became interested in mathematics through a chance perusal of the *Clavis mathematicae* (1631) of William Oughtred and after his installation as Savilian Professor of Geometry at Oxford in the year 1649 was able to devote himself uninterruptedly to the study of mathematics for nearly half a century.

† Leibniz repeatedly states that he was led to the invention of the calculus by a study of the works of Pascal. The evidence in support of this was first assembled by Gerhardt in *Leibniz in London* and *Leibniz und Pascal*. Both of these articles are given by Child, J. M., in *The Early Mathematical Manuscripts of Leibniz*: Chicago and London, 1920. See also Hofmann, J. E., Die Entwicklungsgeschichte der Leibnizschen Mathematik, *op. cit.*, pp. 28–36.

‡ See Scott, J. F., The Reverend John Wallis, F.R.S., *Notes and Records of the Royal Society of London* 15 (1960), pp. 57–67. Also, Scott, J. F., *The Mathematical Work of John Wallis*, London, 1938.

Wallis always lamented the poor instruction he received during his student years in mathematics and in the early stages of his own mathematical researches his reading was not extensive. He retained a great respect for his own countryman, Thomas Harriott, and studied the algebra of Viète and of Oughtred. Initially he only knew of the Cavalierian method of indivisibles through the works of Torricelli.† His knowledge of developments in France remained slight and, throughout his lifetime, he was unable to appreciate the value of the contribution made by Descartes to mathematics.‡ From the outset, however, he seems to have grasped intuitively the importance of the use of appropriate symbolism and in the *Mathesis universalis* (1657) he attempted the arithmetisation of Euclid's *Elements* without a very clear idea of the extent of the difficulties involved.§ In this work he included an algebraic treatment of geometric progessions giving the terms in the familiar form, A, AR, AR^2, AR^3, ..., AR^{n-1} and the sum as

$$S = \frac{VR-A}{R-1},$$

where V is the last term. The result was established by long division (up to $R = 4$).||

In *De sectionibus conicis*†† Wallis made considerable advances in the study of the conic sections in that he treated the parabola, ellipse and hyperbola as plane curves without the "embranglings of the cone". In this work he made use of Cavalierian indivisibles introducing the symbol ∞ for the infinitely many lines (or parallelograms) making up a plane surface. Unlike Tacquet Wallis did not consider the distinction between lines and parallelograms of any great importance in the sense that parallelograms whose altitudes are supposedly infinitely small, that is, having no altitude (since a quantity infinitely small is no quantity), scarcely differ from a line. He does, however, make the proviso that, the line is to be regarded as having so much thickness that, by infinite multiplication, it becomes capable of acquiring an altitude equal to that of the figure in which it is inscribed.‡‡ Hence, for the area of the tri-

† See *Arithmetica infinitorum*, Oxford, 1656; *Opera mathematica*, Oxford, 1693–9, I, p. 357.

‡ See *A Treatise of Algebra*, London, 1685, Preface.

§ See *Opera mathematica* I, pp. 13–228.

|| *Opera* I, pp. 169–83. Schooten, F. van (1615–60), also published an algebraic treatment of geometric progressions in the *Exercitationes mathematicae*, Leyden, 1657 (see pp. 447–50).

†† *Opera* I, pp. 291–354.

‡‡ *Opera* I, p. 297: "Nam Parallelogrammum cujus altitudo supponitur infinite parva, hoc est, nulla, (nam quantitas infinite parva perinde est atque non-quanta,) vix aliud est

angle[†] we have (see Fig. 6.7): let B be the base and A be the altitude and let the number of lines in the surface be ∞, then the total length of the lines $= \infty B/2$;[‡] but the altitude of each parallelogram inscribed in the figure $= A/\infty$, hence the area of the triangle $= (\infty B/2)\times(A/\infty) = AB/2$. The symbol ∞ introduced by Wallis is clearly playing a dual role since, in the first instance it represents

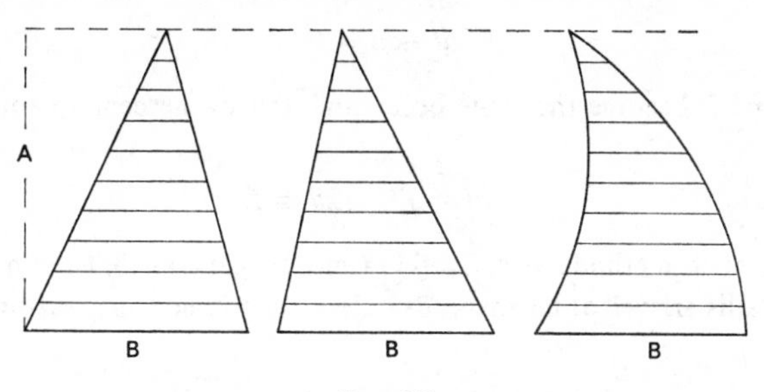

FIG. 6.7

the number (n) of lines drawn in the surface and also blankets a limit operation in which $n \to \infty$ and through which the difference between the area of the triangle and the sum of the areas of the inscribed parallelograms becomes less than any finite quantity.

Wallis had some interest in tangent methods but offered nothing in advance of Fermat.[§] To determine the tangent to a convex curve he made use of the concept of the tangent as a line meeting the curve in a single point and remaining outside it at every other point. Thus, for the parabola $A\alpha$ (see Fig. 6.8), if αF be the tangent at the point α and if $P\alpha = p$, $PA = d$, $DA = d \pm a$, $PF = f$, $DF = f \pm a$, we have $PA/DA = P\alpha^2/DO^2$, $DO^2 = p^2(d\pm a)/d$.

From similar triangles, $PF^2/DF^2 = P\alpha^2/DT^2$, hence,

$$DT^2 = \frac{f^2 \pm 2fa + a^2}{f^2} p^2,$$

and since $$DT > DO, \quad DT^2 > DO^2:$$

quam linea. (In hoc saltem differunt, quod linea haec supponitur dilatabilis esse, sive tantillam saltem spissitudinem habere ut infinita multiplicatione certam tandem altitudinem sive latitudinem possit acquirere, tantam nempe quanta est figurae altitudo.)" Prop. 1.

† *Opera* I, pp. 297–9.

‡ The sum of an arithmetic progression, or the aggregate of all its terms equals the sum of the extremes multiplied by one-half the number of terms. No. of terms $= \infty$, extremes equal 0 and B, hence, sum $= ((0+B)/2)\times\infty = (\infty B)/2$.

§ *Opera* I, pp. 322, 331–2, 340–1.

then,

$$\frac{f^2 \pm 2fa + a^2}{f^2} p^2 \gtrless \frac{d \pm a}{d} p^2,$$

$$df^2 \pm 2dfa + da^2 \gtrless df^2 \pm f^2 a.$$

Removing df^2 from both sides and dividing by a we have,

$$\pm 2df + da \gtrless f^2.$$

But if D and P become the same point and the evanescent quantity a disappears,

then $$2df = f^2, \quad 2d = f.$$

The use of such methods was, by this time, so general that the question of whether Wallis arrived at his methods independently has no great significance.

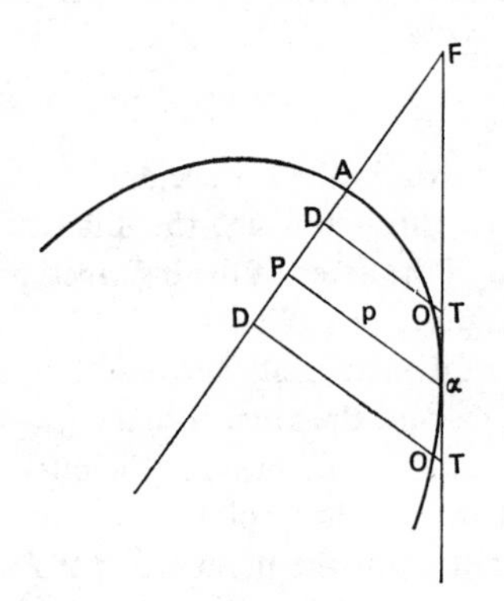

FIG. 6.8

In the *Arithmetica infinitorum*† (1656) Wallis uses the limit sum of the integral powers of the natural numbers to arithmetise the methods of Cavalieri and so determine the areas under curves and the volumes of solids without resort to geometric imagery. Thus, we have‡

$$\frac{0+1}{1+1} = \frac{1}{2}, \quad \frac{0+1+2}{2+2+2} = \frac{1}{2}, \quad \frac{0+1+2+3}{3+3+3+3} = \frac{6}{12} = \frac{1}{2},$$

and, in general, $$\frac{0+1+2+3+\ldots n}{n+n+n+n+\ldots n} = \frac{n(n+1)}{2n(n+1)} = \frac{1}{2}.$$

† *Opera* I, pp. 355–478.
‡ *Ibid.* I, p. 365.

Since a triangle may be considered as made up of an infinite number of lines whose lengths are in arithmetic progression and of which the maximum is the base,† then

$$\frac{\text{area of triangle}}{\text{area of parallelogram}} = \frac{1}{2}.$$

Similarly a conoid, being made up of an infinite number of planes with areas in arithmetic progression,‡ we have

$$\frac{\text{volume of conoid}}{\text{volume of circumscribing cylinder}} = \frac{1}{2}.$$

For the sums of the squares of the natural numbers,§ we have

$$\frac{0+1}{1+1} = \frac{1}{2} = \frac{1}{3}+\frac{1}{6}, \quad \frac{0+1+4}{4+4+4} = \frac{5}{12} = \frac{1}{3}+\frac{1}{12},$$

and, in general,

$$\frac{0+1+4+\ \dots\ +n^2}{n^2+n^2+n^2+\ \dots\ +n^2} = \frac{1}{3}+\frac{1}{6n}.$$

Hence,

$$\lim_{n\to\infty} \frac{\sum_0^n r^2}{\sum_0^n n^2} = \frac{1}{3},$$

and from this limit the area of the parabola, the areas of spiral sectors and the volumes of cones and pyramids can be determined. Extension to higher powers follows, i.e.

$$\lim_{n\to\infty} \frac{0^p+1^p+2^p+3^p+\ \dots\ +n^p}{n^p+n^p+n^p+n^p+\ \dots\ +n^p} = \frac{1}{p+1}.$$

This material was not, of course, new with Wallis, but it had not, in this form, been published before. What was, indeed, a new and exciting advance in the *Arithmetica* was the extension of these results to fractional, zero and negative exponents with the corresponding extension of the index notation to cover these cases.|| For the parabola (see Fig. 6.9) we have

$$\lim_{n\to\infty} \frac{0^2+1^2+2^2+\ \dots\ +n^2}{n^2+n^2+n^2+\ \dots\ +n^2} = \frac{1}{3} = \frac{\text{area } ABC}{\text{rectangle } ABCD},$$

† *Ibid.* I, p. 366.
‡ *Ibid.* I, p. 366.
§ *Ibid.* I, p. 373.
|| *Opera* I, pp. 389–411. For a full account of the stages in this development see Scott, J. F., *op. cit.*, pp. 35–44.

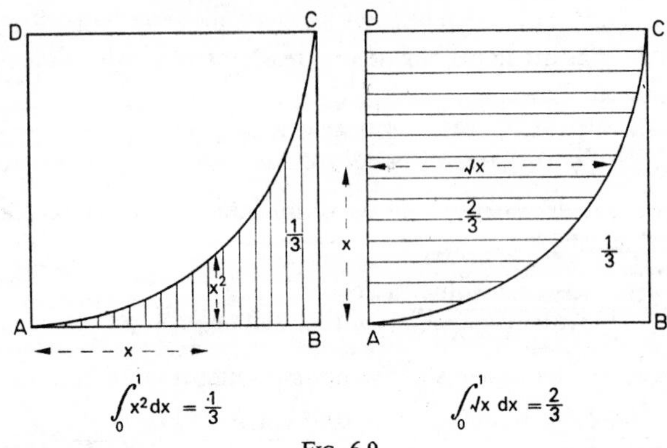

FIG. 6.9

$$\text{hence,}\quad \lim_{n\to\infty}\frac{\sqrt{0}+\sqrt{1}+\sqrt{2}+\ \dots\ +\sqrt{n}}{\sqrt{n}+\sqrt{n}+\sqrt{n}+\ \dots\ +\sqrt{n}} = \frac{\text{rectangle } ABCD-\text{area } ABC}{\text{rectangle } ABCD}$$

$$= \frac{2}{3}$$

$$= \frac{1}{\frac{3}{2}}$$

$$= \frac{1}{1+\frac{1}{2}}.$$

Wallis was thus led to conclude that $\sqrt{n}$ might be interpreted as $n^{\frac{1}{2}}$; similar considerations led to $\sqrt[3]{n} = n^{\frac{1}{3}}$, $1 = n^0$ *(Aequales)* and so on. Using what he termed the Principle of Interpolation Wallis was thus enabled to draw up a table of results demonstrating that, in general,

$$\lim_{n\to\infty}\frac{\sum_0^n r^p}{n^{p+1}} = \frac{1}{p+1},$$

where p can take any value, positive, negative or fractional, i.e. $\int_0^1 x^{p/q}\,dx = q/(p+q)$. Wallis was encouraged by his success in applying the same methods to binomial expansions of the form $(R^2-x^2)^p$, for p positive and integral, to attempt the quadrature of the circle by establishing a general integral of the form $\int_0^1 (1-x^2)^{\frac{1}{2}}\,dx$. In this he was only partially successful but after monumental labours by means of skilful interpolation succeeded in establishing the relation,†

$$\frac{4}{\pi} = \frac{3.\,3.\,5.\,5.\,7.\,7.\,9.\,9.\,11.\,11.\,\ldots}{2.\,4.\,4.\,6.\,6.\,8.\,8.\,10.\,10.\,12.\,\ldots},$$

continued to infinity.‡

Brouncker was able (some time in 1654) to give this the form of a continued fraction,§

$$\frac{\pi}{4} = \frac{1}{1+}\,\frac{1^2}{2+}\,\frac{3^2}{2+}\cdots$$

In the whole of the *Arithmetica* Wallis scarcely uses the word demonstrate. His avowed aim and purpose was to aid others and further the progress of mathematics by publishing his own method of investigation rather than by exhibiting proofs in the manner of the ancients.‖ This purpose was certainly fulfilled in that Newton on first reading the *Arithmetica* was inspired to begin an investigation on his own behalf and was able, by emulating the interpolation methods of Wallis, to obtain a general expansion for $\int_0^1 (1-x^2)^{\frac{1}{2}}\,dx$, which supplied the required quadrature of the circle.††

It is worth noting that whilst Wallis had taken a considerable step forward in freeing the concept of integral from its close association with surface area and volume, he was still obliged, in the case of trigonometric integrals, to link his work with a geometric model in the style of the *ungula* of Saint-Vincent. In the *Mechanica* (1669)‡‡ Wallis uses indivisibles freely and effectively in determining the centres of gravity of solids. In *De calculo centri gravitatis*,§§ by the standardisation of notation he shows himself deliberately striving to develop a primitive calculus: one which will enable him, without unnecessary repetition to pass from known results to more complex integral forms. In Prop. XVIII, for example,‖‖ he constructs a cylindrical section on a

† See Scott, J. F., *The Mathematical Works of John Wallis*, pp. 47–60. *Opera* I, Prop. 191, pp. 467–76.

‡ See also Whiteside, D. T., Patterns of mathematical thought, *Archive for History of Exact Sciences* I (1961), pp. 236–43.

§ *Arithmetica infinitorum*, Prop. 191; *Opera* I, p. 469.

‖ *Opera* I, p. 412.

†† See *The Correspondance of Isaac Newton* (ed. Turnbull, H. W.), Cambridge, 1959–, II, p. 111.

‡‡ *Opera* I, pp. 571–1063.

§§ *Ibid.* I, pp. 665–938.

‖‖ *Ibid.* I, pp. 766–80.

semicircular base, radius R and height πR (see Fig. 6.10). The sections of the cylinder parallel to the base are circular segments of area $= \frac{1}{2}aR - \frac{1}{2}s(R-v)$, where a is the arc $R\theta$, $s = R\sin\theta$ and v is the versine, i.e. $R(1-\cos\theta)$. Hence, summing, for all such sections,

$$\begin{aligned}\text{volume} &= \text{omn.}\left(\tfrac{1}{2}aR - \tfrac{1}{2}s(R-v)\right)\\ &= \text{omn.}\ \tfrac{1}{2}aR - \text{omn.}\ \tfrac{1}{2}sR + \text{omn.}\ \tfrac{1}{2}sv.\end{aligned}$$

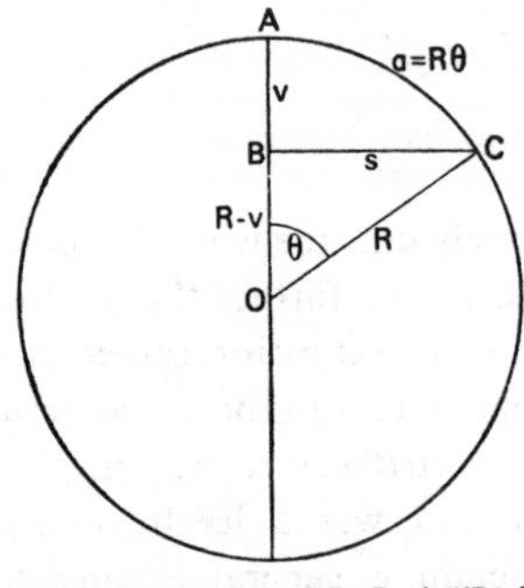

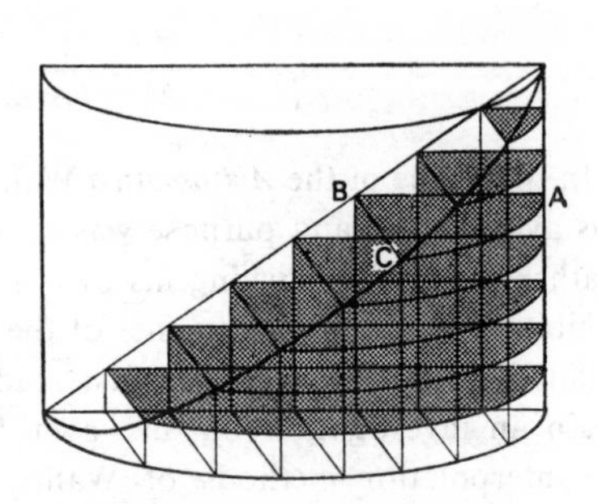

FIG. 6.10

Considering these sums separately, we have

$\text{omn.}\ \frac{1}{2}aR = \frac{1}{2}R\ \text{omn.}\ a = \frac{1}{4}RA^2$, where A is the maximum arc,

$\text{omn.}\ s = VR$ (where V is the maximum versine),

$\text{omn.}\ \frac{1}{2}sR = \frac{1}{2}VR^2$,

$\text{omn.}\ \frac{1}{2}sv = \frac{1}{2}VR^2 - \frac{1}{4}S^2R$ (where S is the maximum sine).

Hence,

$$\begin{aligned}\text{omn.}\left(\tfrac{1}{2}aR - \tfrac{1}{2}s(R-v)\right) &= \tfrac{1}{4}A^2R - \tfrac{1}{2}VR^2 + \tfrac{1}{2}VR^2 - \tfrac{1}{4}S^2R\\ &= \tfrac{1}{4}A^2R - \tfrac{1}{4}S^2R.\end{aligned}$$

In the notation of the calculus, we have

$$\frac{1}{2}\int_0^{\alpha} (R^2\theta - R^2\sin\theta\cos\theta)\, d(R\theta) = \tfrac{1}{4}R^3(\alpha^2 - \sin^2\alpha).$$

From this, by the usual process of moments, he was able to build up a further series of integrals. The weakness, of course, in all this work is that the whole is not founded on any standardised proof structure but taken more or less haphazard from any source and derived by any methods which yield results. Wallis never apparently felt any necessity to place these methods on a secure basis,† although a great many objections were justifiably raised against the style and manner in which his work was presented.‡ Wallis did his best to answer these objections, both in correspondence and later in the *Algebra* in which he enumerates the stages in the development of infinitesimal processes as follows:§

1. Method of Exhaustion (Archimedes).
2. Method of Indivisibles (Cavalieri).
3. Arithmetick of Infinites (Wallis).
4. Method of Infinite Series (Newton).

He emphasises once more that in the *Arithmetica* he had no intention of replacing the proof by Exhaustion by some other form of proof but supplied instead a profitable means of investigation "or finding out of things as yet unknown". He regarded the *Arithmetica* as the cornerstone of all subsequent development including the Method of Infinite Series of Newton and the rectification of the arcs of the semi-cubical parabola and the cycloid by Neile and Wren, respectively. In fact, although the *Arithmetica* acted as a powerful stimulant for a brief and important period in the development of infinitesimal methods, particularly in England, progress from 1660 onwards was so rapid and there were already so many different influences at work, that with Wallis the use of indivisibles as such virtually ceased. Barrow and Gregory, although duly impressed by the remarkable results published by Wallis in the *Arithmetica*, were themselves so strongly influenced by the methods they had learnt in Italy that their own work developed along entirely different lines. Newton's early mathematical work and some of his more important achievements were undoubtedly greatly influenced by the *Arithmetica* but there were also, even here, other factors at work, most notably the influence of his friend Isaac Barrow and that of a group of mathematicians in the Low Countries working under the guidance of Frans van Schooten.

† See Whiteside, D. T., *op. cit.*, pp. 323–4.

‡ See especially, Fermat to Digby: 15th Aug. 1657: *Oeuvres de Fermat* II, p. 343. Fermat says of the *Arithmetica*, "mais sa façon de démontrer, qui est fondée sur induction plutôt que sur un raisonnement à la mode d'Archimède, fera quelque peine aux novices, qui veulent des syllogismes démonstratifs depuis le commencement jusqu'à la fin. Ce n'est pas que je ne l'approuve; mais, toutes ses propositions pouvant être démontrées *viâ ordinariâ, legitimâ et Archimedeâ* en beaucoup moins de paroles que n'en contient son livre, je ne sais pas pourquoi il a préféré cette manière par notes algébriques à l'ancienne, qui est et plus convaincante et plus élégante, ainsi que j'espère lui faire voir à mon premier loisir."

§ Wallis, J., *A Treatise of Algebra*, London, 1685, pp. 280–355.

6.3. Infinitesimal Methods in the Low Countries

Frans van Schooten (1615–60), the son of a distinguished professor of the University of Leyden, was fortunate enough whilst still a student to be introduced to most of the important mathematical work in progress in France and England. In Leyden he met Descartes and saw the page proofs of *La Géométrie.* Through Mersenne in Paris Schooten made personal contact with Carcavy and Roberval and thus obtained access to the correspondence and work of Fermat which was at that time circulating, partly in the original and partly in drafts made by Carcavy and Mersenne.† On returning to Leyden Schooten determined to further by private teaching, public lectures and by publications the development of Cartesian methods. The group of young Netherlanders which he gathered round him and which included, for a brief period in 1646, Christiaan Huygens, was responsible for many original developments which Schooten, through his own publications, was able to make widely available.

In 1646 Schooten brought out a new edition of the works of Viète‡ replacing the unwieldy symbolism by the simpler notation of Descartes. In 1649 appeared the first Latin edition of Descartes' *La Géométrie* accompanied by De Beaune's *Notae breves* (1638) and a detailed commentary written by Schooten himself.§ Both De Beaune and Schooten stressed the importance for Descartes' tangent method of the construction of an imaginary circle to touch the curve and the consequent formation of an equation with equal roots. As a result of this influential publication interest in tangent methods intensified and in 1657 a significant correspondence developed between Huygens, Hudde, Schooten, Sluse and Heuraet on quadratures, tangents and centres of gravity.‖ This correspondence shows that all four had accepted the general idea of the use of a set of rules and a formula with appropriate notation in the determination of tangents to algebraic curves. Although Sluse appears to have been the first to draw up such a set of rules Hudde was the first to publish.

Born in Liège, René François de Sluse (1622–85) studied in Lyons and subsequently for ten years in Italy during which he was closely associated with Ricci, friend and pupil of Torricelli. In Italy he had ample opportunity to familiarise himself with the indivisible methods of Cavalieri and Galileo

† Hofmann, J. E., *Frans van Schooten der Jüngere,* Wiesbaden, 1962.
‡ Viète, F., *Opera mathematica* (ed. Schooten, F. van), Leyden, 1646.
§ Descartes, R., *Geometria* (ed. Schooten, F. van), Leyden, 1649.
‖ Huygens, C., *Oeuvres complètes,* The Hague, 1888–1950, II, *passim.*

and the mechanical tangent methods of Torricelli. Sluse appears to have tried out a variety of tangent methods but about 1655 he developed an analytic method of his own which was unfortunately not published until 1673.†

It was a matter of some concern to all those who involved themselves in the determination of tangents to try to reduce the amount of heavy algebra entailed in arriving at a solution. In Fermat's tangent method, for example, an approximate equation is formed:

$$f(x, y) \simeq f(x+e, y+ey/t),\ddagger \qquad (1)$$

where $f(x, y) = 0$ is the curve equation and t is the subtangent. Successive steps, being roughly equivalent to differentiation from first principles, lead to

$$f(x, y) \simeq f(x, y)+e\left[f_x(x, y)+\frac{y}{t}f_y(x, y)\right]+e^2\left[\ldots\right.$$

Neglecting terms above the first degree in e, we have

$$t = -\frac{yf_y(x, y)}{f_x(x, y)}. \qquad (2)$$

Much complicated and heavy algebra arose in passing from (1) to (2) and both Hudde and Sluse sought to find a way of moving immediately from the curve equation to the subtangent. Sluse had the distinction of being the first to find an algorithm of a general nature allowing direct determination of the tangent without calculation. That he should have appreciated the importance of establishing such an algorithm is perhaps of greater significance than the merit of the particular set of rules he devised. Unfortunately, owing to the severe pressure of church duties which he experienced after his return to Liège, Sluse made no attempt to publish his tangent method until he felt his priority challenged by Newton in 1673. Even at this juncture he provided only a brief note which was inserted in the *Philosophical Transactions* for 1672/3.§

The method is founded as follows:

$$\text{let } f(x, y) = \sum a_{pq}x^p y^q = 0.$$

† Le Paige, C., Correspondance de René François de Sluse, *Bull. di Bibl. e di Stor. delle Science* 17 (1884), pp. 427–554, 603–726. See also Rosenfeld, L., René-François de Sluse et le problème des tangentes, *Isis* 10 (1928), pp. 416–34.

‡ See pp. 168–70, above.

§ Sluse, R. F. de, A method of drawing tangents to all geometrical curves, *Philosophical Transactions* 7 (1672), No. 5143; 8 (1673), No. 6039.

For any point on the curve in the neighbourhood of (x_1, y_1) we have

$$f(x_1, y_1) - f(x, y) = 0,$$

and
$$\sum a_{pq}(x_1^p y_1^q - x^p y^q) = 0,$$

hence
$$\sum a_{pq}(x_1^p(y_1^q - y^q) + y^q(x_1^p - x^p)) = 0.$$

But
$$\frac{x_1^p - x^p}{x_1 - x} = x_1^{p-1} + x_1^{p-2}x + x_1^{p-3}x^2 + \ldots x^{p-1},$$

and similarly,
$$\frac{y_1^q - y^q}{y_1 - y} = y_1^{q-1} + y_1^{q-2}y + y_1^{q-3}y^2 + \ldots y^{q-1},$$

therefore,
$$x_1^p(y_1^q - y^q) = x_1^p(y_1 - y)(y_1^{q-1} + y_1^{q-2}y + \ldots y^{q-1})$$

and,
$$y^q(x_1^p - x^p) = y^q(x_1 - x)(x_1^{p-1} + x_1^{p-2}x + \ldots x^{p-1}).$$

Now, let $(x_1, y_1) \to (x, y)$, then

$$\frac{y_1 - y}{x_1 - x} \to \frac{y}{t},$$

where t is the subtangent, and since

$$\frac{y_1 - y}{x_1 - x} = -\frac{\sum a_{pq} y^q(x_1^{p-1} + x_1^{p-2}x + \ldots x^{p-1})}{\sum a_{pq} x_1^p(y_1^{q-1} + y_1^{q-2}y + \ldots y^{q-1})}$$

as $x_1 \to x$ and $y_1 \to y$, or more crudely if we put $x_1 = x$ and $y_1 = y$, then

$$\frac{y}{t} = -\frac{\sum p a_{pq} y^q x^{p-1}}{\sum q a_{pq} x^p y^{q-1}},$$

or, in the notation of the calculus,

$$\frac{dy}{dx} = -\frac{\partial f/\partial x}{\partial f/\partial y}.$$

Although it does not appear that Sluse ever established this method on a formal basis he made use of it from 1655 onwards. In the *Mesolabum*† he used Cavalierian methods in conjunction with algebraic notation to obtain the quadrature of curves by analytical transformations.‡ With Sluse, therefore, we

† Sluse, R. F. de, *Mesolabum*, Liège, 1659 (2nd ed. 1668).
‡ *Mesolabum* (2nd ed.); *Miscellanea*, pp. 100–6.

see a substantial advance towards the simplification of processes by the use of algebraic notation and substitution and by the development of formal rules for carrying out the processes now known as integration and differentiation. No specialised notation was introduced and, so far as we know, no deliberate effort made to establish the inverse relation between the two processes.

Johann Hudde (1628–1704) studied law in Leyden and had private lessons from Schooten in mathematics. In 1657 he participated in correspondence with Huygens and Sluse on tangents and quadratures. Unfortunately, however, although held in great esteem by his contemporaries, Hudde's concern for mathematics was of limited duration and, so far as we know, all his contributions were made during the period 1654–63. From 1663 until his death Hudde was fully engaged in the service of the city of Amsterdam, becoming Burgermaster twenty-one times in all.†

During his brief period of mathematical activity Hudde concerned himsel primarily with the improvement of algebraic methods; when he learnt through Schooten of Fermat's method of extreme values he set himself the task of developing a set of rules which, in the case of rational algebraic functions, would make possible a considerable simplification of the algebra.

An isolated example of *Hudde's method* appeared in the *Exercitationes mathematicae* of Schooten‡ (1657) and a short explanatory tract *De maximis et minimis* was included in the 1659 edition of the *Geometria.*§ A full account of the application of this method in the determination of tangents was given by Hudde in a letter dated, 21st November 1659, which was published in 1713 during the Newton–Leibniz controversy.||

Starting once again from Fermat's method of extreme values we have

$$f(x+e) \simeq f(x),$$

and, expanding in powers of e,

$$f(x)+ef'(x)+\frac{e^2}{2}f''(x)+\ldots \simeq f(x).$$

† See Haas, K., Die mathematischen Arbeiten von Johann Hudde, *Centaurus* 4 (1956), pp. 235–84.

‡ Schooten, F. van, *Exercitationes mathematicae*, Leyden, 1657; *Miscellaneae*, XXV and XXVI, pp. 493–9.

§ *Geometria à Renato Des Cartes* (ed. Schooten, F. van), Amsterdam, 1659, pp. 507–16: *Johannis Huddenii epistola secunda de maximis et minimis.*

|| *Journal literaire de la Haye* 1 (1713) (2nd ed. 1715), pp. 460–4: *Extrait d'une lettre de M. Hudde à M. van Schooten*, 21st Nov. 1659.

For an extreme value of the function $f(x)$, let $x = \alpha$, then $f'(\alpha) = 0$ and $f(x) = f(\alpha)$. The equation $f'(x) = 0$ has accordingly a root $x = \alpha$ and the equation $f(x) = f(\alpha)$ has a double root, $x = \alpha$.

The rule which Hudde formulated concerns the solution of equations with equal roots and runs as follows:†

> If, in an equation two roots are equal and, if it be multiplied by any arithmetical progression, i.e., the first term by the first term of the progression, the second by the second term of the progression, and so on: I say that the equation found by the sums of these products shall have a root in common with the original equation.

Let

$$f(x) = a_0+a_1x+a_2x^2+a_3x^3+\ldots\ a_nx^n = 0.$$

Multiplying by $p, p+q, p+2q, p+3q, \ldots$ we have

$$pf(x)+q(a_1x+2a_2x^2+3a_3x^3+\ldots\ na_nx^n) = 0,$$

$$pf(x)+qxf'(x) = 0.$$

Since, at an extreme value, $x = \alpha$, $f(\alpha) = 0$ and $f'(\alpha) = 0$, the above equation is satisfied by $x = \alpha$ and therefore has a root in common with the original equation.

Hudde demonstrates the rules in a number of cases where $f(x)$ is a polynomial in x and also shows how to apply it when the expression to be maximised is of the form, $f(x)/g(x)$.‡

To apply the rule in the determination of tangents Hudde instructs as follows:§

> Put all the terms on one side (= 0). Remove x, y, from the divisors. Arrange in descending powers of y and multiply each term by the corresponding term of any arithmetic progression whatsoever. Repeat this process for the terms containing x. Divide the first sum of products by the second. Multiply the quotient by $-x$ and this will give the subtangent.

Thus, if the expression $f(x, y)$ be arranged in descending powers of y and the terms multiplied by the corresponding terms of the progression, p, $p+q, p+2q, \ldots$, we have $k_1f(x, y)-qyf_y(x, y)$, and, if the same procedure be carried through for descending powers of x with the progression $r, r+s$, $r+2s, \ldots$, we have, $k_2f(x, y)-sxf_x(x, y)$. In practice, however, Hudde clearly intends that $q = s = -1$, so that,

$$t = -x\frac{k_1f(x, y)+yf_y(x, y)}{k_2f(x, y)+xf_x(x, y)} = -y\frac{f_y(x, y)}{f_x(x, y)}, \quad \text{since} \quad f(x, y) = 0.$$

† *Geometria* (1659), *De maximis et minimis*, p. 507.
‡ *Geometria* (1659), p. 512.
§ *Journal literaire*, pp. 460–1.

The essential simplicity of the operation is best illustrated by an example:[†]

let $$f(x, y) = ay^3+xy^3+b^2y^2-x^2y^2-\frac{x^3y^2}{2a}+2x^4-4ab^3.$$

Multiplying by	1	1	0	0	0	−2	−2
and by	0	1	0	2	3	4	0

we have

$$t = -x\frac{ay^3+xy^3-4x^4+8ab^3}{xy^3-2x^2y^2-\dfrac{3x^3y^2}{2a}+8x^4} = \frac{-axy^3-x^2y^3+4x^5-8ab^3x}{xy^3-2x^2y^2-\dfrac{3x^3y^2}{2a}+8x^4}.\ ‡$$

Thus, in the notation of the calculus,

$$t = -y\frac{f_y(x, y)}{f_x(x, y)}.$$

The importance of *Hudde's Rule* historically lay not in the process itself but in the publication of the general idea of standardising and simplifying tangent methods so that results could be achieved by the mechanical application of rules without the necessity for demonstration. Although the actual rules were limited in application to certain classes of rational algebraic functions others were inspired by Hudde's example to begin a systematic search for more general methods with wider fields of application.

Constantin Huygens (1596–1687) had connections with Descartes and Mersenne, and his son Christiaan Huygens (1629–95) was brought up in an atmosphere of scientific and mathematical endeavour which stimulated him to make contributions on his own account at an early age. As a pupil of Schooten Huygens was introduced to Cartesian methods and a piece of his work was included in the 1649 edition of the *Geometria.*[§] In 1651 Huygens found an error in Saint-Vincent's circle quadrature and himself investigated the quadrature of segments of a conic section by consideration of the centroid.[||] In

† *Ibid.*, p. 462. Hudde suggests that the progression be chosen in such a way as to eliminate the greatest number of terms.

‡ In this expression the numerator corresponds to $x(2f(x, y)-yf_y(x, y))$, and the denominator to $xf_x(x, y)$. Since, $f(x, y) = 0$, we have

$$\frac{t}{y} = -\frac{f_y(x, y)}{f_x(x, y)}, \quad \frac{dy}{dx} = -\frac{f_x(x, y)}{f_y(x, y)}.$$

§ *Geometria* (1649), pp. 203–5.

|| Huygens, C., *Theoremata de quadratura hyperboles, ellipsis et circuli*, Leyden, 1651. See *Oeuvres complètes* XI, pp. 281–337.

1654, using an idea from Viète, he simplified the quadrature of the circle by reducing the problem to the consideration of inscribed and circumscribed segments of a parabola.† Schooten's correspondence with Sluse and Hudde stimulated Huygens who, between the years 1655–65, experimented with analytic tangent methods, amending, combining and adapting the methods of Fermat and Descartes to his own use.‡ He developed a method of finding points of inflexion by applying Fermat's tangent method twice successively and subsequently adapted this method to the consideration of an equation with three equal roots.§ He used processes involving detached coefficients and in 1658 we find him applying Hudde's method (published in 1659).|| Subsequently he asserted in correspondence with Wallis that he had found the method before Hudde.†† In 1657, arising out of an examination of a faulty rectification of the parabolic arc by Hobbes,‡‡ he reduced the rectification of the parabola to the quadrature of the hyperbola.§§

The investigations which he began on the cycloid in response to Pascal's challenge and which led him to a full exploration of the theory of evolutes were conducted by means of indivisibles methods combined with skilful geometric transformations conducted on a cylindrical section.

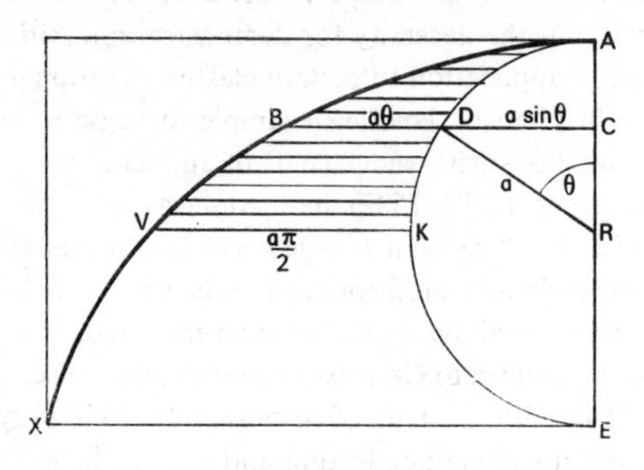

FIG. 6.11

For example, to obtain the quadrature of the general cycloid segment, *ABD* see (Fig. 6.11), Huygens represents the space *AVK* by the curved sur-

† Huygens, C., *De circuli magnitudine inventa*, Leyden, 1654; *Oeuvres* XII, pp. 113–81.
‡ *Oeuvres* XIV, pp. 181 *et seq.*
§ *Ibid.* XIV, p. 299; *Geometria* (1659), pp. 253–5.
|| *Oeuvres* XIV, p. 304.
†† *Ibid.* II, p. 417 (9th June 1659).
‡‡ Hobbes, T., *Elements of Philosophy*, London, 1656, pp. 199–202.
§§ *Oeuvres* XIV, pp. 234–70.

face of a cylindrical ungula with semicircular base, LMN, diameter $LM = AE$ ($= 2a$) and height $NF = AR$ ($= a$) (see Fig. 6.12). From the property of the cycloid, BD [$= a\theta$] $= ON = QP$ and the horizontal lines in the cylindrical surface FMN correspond with the lines in the cycloid space AVK, i.e. $AVK = FMN$, $ABD = FQP$.

To determine these surfaces† Huygens introduces a second cylindrical section or *ungula* (see Fig. 6.13) on the same base LMN, with height $NF' = LNM$ ($= \pi a$). The vertical lines in the curved surface $MNF'L$ correspond with the

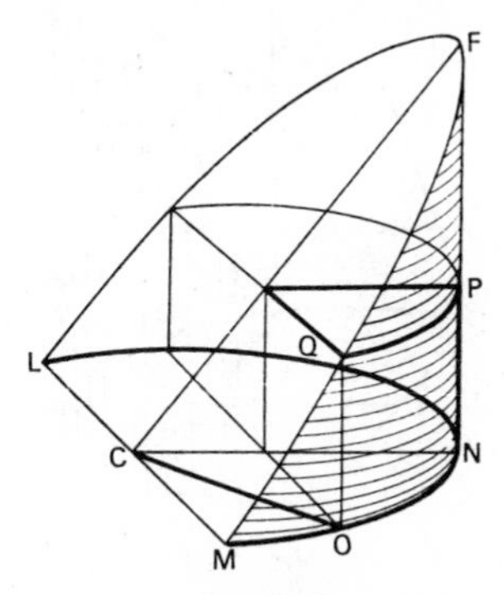

FIG. 6.12

lines in the surface of the hemisphere formed by rotating the semicircle LNM through two right angles about the diameter, LM (see Fig. 6.14). Thus, MNF' = surface formed by rotating arc MN about LM [$= \pi a^2$] and MOQ' = surface formed by rotating arc MO about LM [$= \pi a^2(1-\sin\theta)$]. Hence, since the surfaces MNF', MNF are in the ratios of their heights, space $MNF = a^2$, and space $MOQ = \frac{1}{2}MO^2$ [$= a^2(1-\sin\theta)$]. Thus FQP [$= ABD$, the required cycloidal space] $= MNF - MOQ - QPNO$ [$= a^2\big(1-(1-\sin\theta)-\theta\cos\theta\big) = a^2(\sin\theta-\theta\cos\theta)$, i.e. $a^2\int_0^\theta \theta\sin\theta\, d\theta$].

Although, by 1660 at any rate, Huygens was thoroughly conversant with the entire range of infinitesimal processes and in an excellent position for further advances he seems to have been constantly assailed by doubts as to the validity of the methods which he used with such facility. At an early stage he rejected Cavalierian indivisibles‡ and set himself to supply Archimedean

† *Oeuvres* XIV, pp. 347–50.
‡ See *Oeuvres* I, pp. 132–4; XI, p. 158; XIII, p. 753.

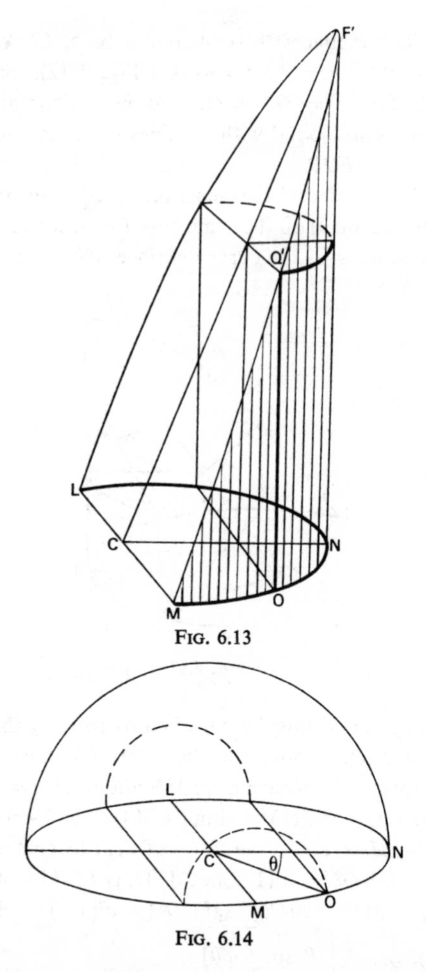

FIG. 6.13

FIG. 6.14

demonstrations for the quadrature of the higher parabolas and hyperbolas. In determining the tangents to these curves he used *reductio* proofs based on a lemma concerning the relation between arithmetic and geometric means† which is formally equivalent to an approximation, for fractional integral exponents, to the 1st term of the binomial expansion, i.e.

$$1 \pm pb/a \lessgtr (1 \pm nb/a)^{p/n}.$$

† *Oeuvres* XIV, pp. 283–4.

Similar lemmas were used by Torricelli, Fermat, Sluse and Barrow, in attempting to place the tangent constructions on a rigorous basis.†

Huygens had apparently in mind at this time the preparation of a comprehensive treatise containing all these results but eventually it seems the proofs became so lengthy and so tedious that he decided on a compromise, i.e.

> to furnish, not a rigorous demonstration but a demonstration of such a nature that those who examine it will not remain in any doubt as to the possibility of a rigorous demonstration.

At this stage he modified his attitude to Cavalierian methods:‡

> As to Cavalierian methods: one deceives oneself if one accepts their use as a demonstration but they are useful as a means of discovery preceding a demonstration. ... Nevertheless that which comes first and which matters most is the way in which the discovery has been made. It is this knowledge which gives most satisfaction and which one requires from the discoverers. It seems, therefore, preferable to supply the idea through which the result first came to light and through which it will be most readily understood. We will thereby save ourselves much labour and writing and the others the reading; it is necessary to bear in mind that mathematicians will never have enough time to read all the discoveries in Geometry (a quantity which is increasing from day to day and seems likely in this scientific age to develop to enormous proportions) if they continue to be presented in a rigorous form according to the manner of the ancients.

Eventually Huygens lost interest even in this limited goal and in the *Horologium oscillatorium* (1673) shelved the whole problem of proof structure and published only results.

When Leibniz came to Paris in 1672 it was to Huygens that he went for help and guidance in his mathematical studies. In this he was singularly fortunate for no one could have been better informed or more able to initiate him into the vast range of contemporary mathematics with such tact and understanding for the difficulties of a beginner.

6.4. The Rectification of Arcs

Following the publication, in 1659, of Hendrick van Heuraet's rectification of the arc of the semi-cubical parabola§ a violent controversy broke out on the whole issue of priority in the rectification of curved lines. In England, following a suggestion given by Wallis in the *Arithmetica*,‖ William Neile

† For Torricelli and Fermat (see pp. 186–7 and p. 162 above). For Sluse, see *Mesolabum* (2nd ed.), pp. 114–17. For Barrow, see pp. 246–7 below.

‡ *Oeuvres* XIV, p. 337.

§ *Geometria* (1659), pp. 517–20.

‖ *Opera* I, Prop. 38, Schol., pp. 380–1.

had rectified the semi-cubical parabola (in 1657) and Brouncker had improved the method in 1658,† whilst Wren had produced an elegant rectification of the general cycloid arc which was sent to Pascal in the same year. In 1657 Huygens had reduced the rectification of the parabolic arc to the quadrature of the hyperbola‡ but this was not disclosed until 1659.

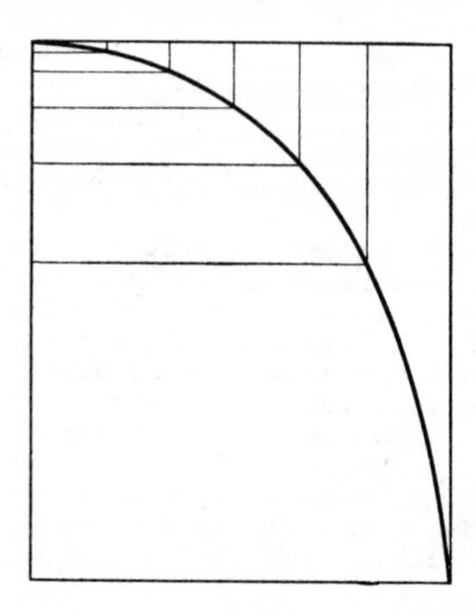

FIG. 6.15

Both Huygens and Wallis§ started from a consideration of the arithmetic differences between the ordinates of the curve resulting from equal divisions of the abscissa. Thus, for the parabolic arc ($y \propto x^2$), we have (see Fig. 6.15),

$$\begin{array}{llllllll} x \propto & 1 & 2 & 3 & 4 & 5 & 6 & 7\ldots n \\ y \propto & 1 & 4 & 9 & 16 & 25 & 36 & 49\ldots n^2 \\ \delta y \propto & 3 & 5 & 7 & 9 & 11 & 13 & 15\ldots (2n+1) \end{array}$$

Hence, successive chords obtained by joining the ends of ordinates (OT) are proportional to

$$\sqrt{(1+3^2)}, \quad \sqrt{(1+5^2)}, \quad \sqrt{(1+7^2)}, \quad \ldots, \quad \sqrt{(1+s^2)},$$

where $s = 2n+1$.

† For Wallis's account of these events see *De curvarum*, subjoined to *De cycloide* (1659): *Opera* I, pp. 550–69.

‡ See *Oeuvres* XIV, pp. 234–52 (27th Oct. 1657).

§ Huygens might have got the initial idea from reading the *Arithmetica* but the final method which he evolved is clearly independent of anything Wallis had said.

Thus, as n increases indefinitely, successive elements of the arc of the curve can be replaced by the ordinates of the hyperbola, $z^2 = 1+k^2x^2$, and the rectification of the parabola can be reduced to the quadrature of the hyperbola,
i.e.

$$\frac{ds}{dx} = \sqrt{(1+k^2x^2)}, \quad s = \int_0^x \sqrt{(1+k^2x^2)}\, dx = \int_0^x z\, dx.$$

Wallis suggested the extension of this method to other curves and Neile seized rapidly upon the idea of utilising a curve in which the small differences of the ordinates are proportional to the square roots of the abscissae; thus, if $\delta x = a$, $\delta y \propto a\sqrt{x}$ (i.e. $y \propto x^{3/2}$), $\delta s \propto a\sqrt{(1+x)}$, and the rectification of the semicubical parabola, $k_1y^2 = x^3$ reduces to the quadrature of the parabola, $z^2 = 1+k_2x$, where $k_2 = 9/4k_1$.

Both Neile and Huygens cast the original demonstration in geometrical form, constructing auxiliary curves to represent the transformations required. Heuraet[†] (who probably profited by a hint from Huygens) provided an unrigorous, yet general, algebraic treatment in which he made explicit use of a species of differential triangle to effect the necessary transformation.

Let ACE be a given curve, STR the differential triangle at a point C on the curve (see Fig. 6.16). $SR = \delta t \simeq \delta s$, $TR = \delta y$, $ST = \delta x$; CMQ is a triangle formed by the normal CQ, the subnormal MQ and the ordinate CM. From the similarity of the triangles, STR and CMQ, $ST/SR = MN/NC = MC/CQ$ $(= \delta x/\delta s)$. If MC be produced to I so that, $k/MI = ST/SR = MC/CQ$, then $MI.ST = k.SR$ and, summing for all such rectangles, $\sum MI.ST = k\sum SR$. Hence the quadrature of the segment $AGIM$ (see Fig. 6.16) of the second curve gives the arc of the first curve. If the curve ACE be the semi-cubical parabola, $y^2 = x^3/a$, then (using Descartes' tangent method) we have $AM = x$, $MC = y$, $CQ = v$, $AQ = s$, $QM = s-x$, and since $CQ^2 = QM^2+MC^2$, then $v^2 = (s-x)^2+y^2 = s^2+x^2-2sx+x^3/a$. Hence, using Hudde's rule and multiplying the terms of the equation

$$v^2 \;=\; s^2 \;+\; x^2 \;-\; 2sx \;+\; x^3/a$$

by

$$0 \quad\; 0 \quad\; 2 \quad\; 1 \quad\; 3$$

we have

$$0 = 2x^2-2sx+3x^3/a,$$

and

$$s-x = 3x^2/2a,$$
$$v^2 = 9x^4/4a^2+x^3/a.$$

† See *Epistola de transmutatione curvarum linearum in rectas* in *Geometria* (1659), pp. 517–20.

Hence,

$$ST/SR = MC/CQ = \sqrt{\left(\frac{x^3/a}{9x^4/4a^2+x^3/a}\right)} = \sqrt{\left(\frac{a^2/9}{ax/4+a^2/9}\right)} = k/MI.$$

Now, let $z = MI = \sqrt{(\frac{1}{4}ax+\frac{1}{9}a^2)}$,

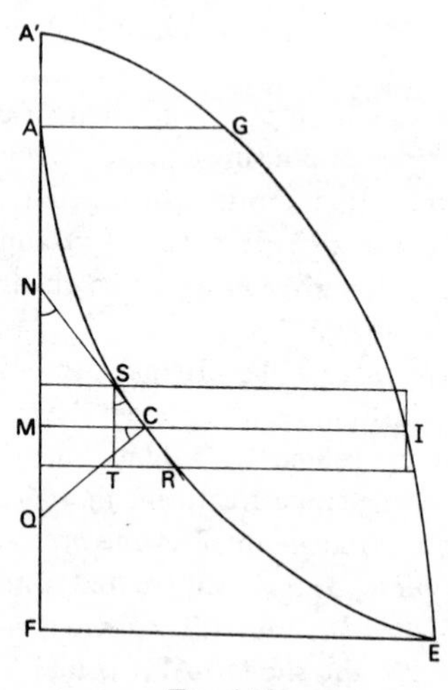

FIG. 6.16

then the curve *A′GIE* is a parabola (vertex *A′*) and the quadrature of this parabola gives the arc of the semi-cubical parabola, *ACE.*†

This was the kind of method which was to predominate in the latter part of the seventeenth century when the urge to develop new algorithmic processes took precedence over any kind of concern for rigorous proof structure. Further attempts to generalise the Exhaustion method to provide a basis for existence proofs in the field of infinitesimal methods were limited to a few isolated cases. The tract *De linearum curvarum cum lineis rectis comparatione*,‡ in which Fermat generalised and extended Wren's cycloid rectification, is a

† $y^2 = x^3/a$, $dy/dx = 3x^2/2ay$, $(dy/dx)^2 = 9x^4/4ax^3 = 9x/4a$,

$$s = \int_0^x \sqrt{\left(1+\frac{9x}{4a}\right)}\, dx = \frac{3}{a}\int_0^x \sqrt{\left(\frac{ax}{4}+\frac{a^2}{9}\right)}\, dx = \frac{3}{a}\int_0^x z\, dx,$$

$$\text{where} \quad z = \sqrt{\left(\frac{ax}{4}+\frac{a^2}{9}\right)}.$$

‡ *Oeuvres* I, pp. 211–54.

fine example of this type. The spirit and purpose of Fermat's proof is exemplified in his opening sentences:†

> To measure curved lines, I shall not make use of inscribed and circumscribed figures as did Archimedes, but only of circumscribed lines formed by the segments of tangents; I shall show, in fact, that there are two series of tangents, the one greater than the curve, the other less, and the analysts will see that the demonstration by the circumscribed figures only is much easier and more elegant.
>
> I say therefore that it is possible, following in the spirit of the method of Archimedes to circumscribe to any such curve two figures composed of lines and of which the one exceeds the curved line by a difference less than any given interval; the other, on the contrary, will be less than the curve by a difference less than any given interval.

Fermat bases his existence proof on two fundamental inequalities which he establishes by means of the convexity lemmas of Archimedes:‡ briefly, let

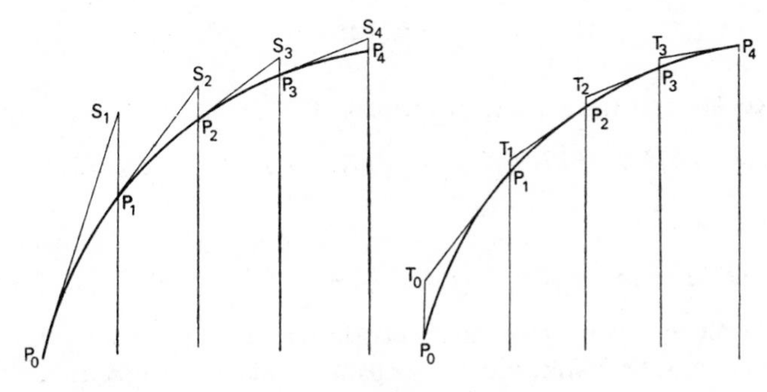

FIG. 6.17

$P_0, P_1, P_2, \ldots, P_n$ be points on a convex ascending curve situated at the ends of equidistant ordinates (see Fig. 6.17), and let the tangent at P_r meet the ordinates through P_{r-1} and P_{r+1} at T_{r-1} and S_{r+1} respectively. Finally, through S_{r+1} draw $S_{r+1}W$ to touch the curve at W§ (see Fig. 6.18).

Now, from the convexity lemmas,

$$T_{r-1}P_r < \text{chord } P_{r-1}P_r < \text{arc } P_{r-1}P_r, \qquad \text{(i)}$$

and

$$\text{arc } P_rP_{r+1} + \text{arc } P_{r+1}W < P_rS_{r+1} + S_{r+1}W,$$

† *Ibid.* I, pp. 213–14.

‡ See p. 37 above.

§ Fermat did not, apparently, remark that in the case of S_n (the last intercept) the point W will actually fall outside the curve. James Gregory provided an improved version of this existence proof in the *Geometria pars universalis*, Padua, 1668.

hence $$\text{arc } P_rP_{r+1} < P_rS_{r+1} - (\text{arc } P_{r+1}W - S_{r+1}W) < P_rS_{r+1}. \qquad \text{(ii)}$$

From (i) we have

$$T_0P_1 + T_1P_2 + T_2P_3 + \ldots T_{n-1}P_n < \text{arc } P_0P_n,$$

and from (ii) $$\text{arc } P_0P_n < P_0S_1 + P_1S_2 + P_2S_3 + \ldots P_{n-1}S_n.$$

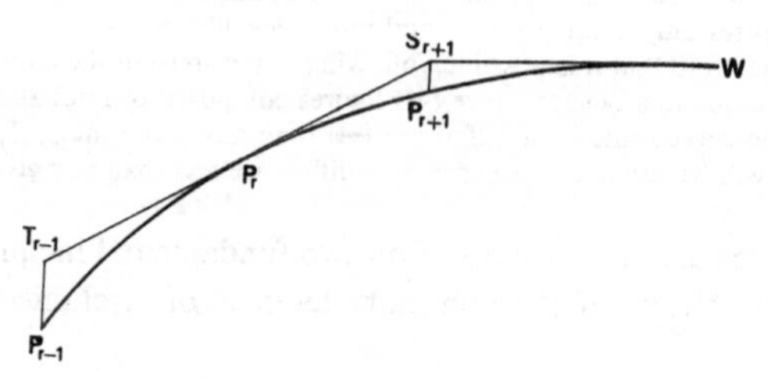

FIG. 6.18

Since, however (from equal intercepts),

$$P_1S_2 = T_0P_1, \quad P_2S_3 = T_1P_2, \quad P_3S_4 = T_2P_3, \ldots P_rS_{r+1} = T_{r-1}P_r,$$

we have

$$(P_0S_1 + P_1S_2 + P_2S_3 + \ldots P_{n-1}S_n) - \text{arc } P_0P_n < P_0S_1 - T_{n-1}P_n,$$

and the required conditions for the classic *exhaustion proof* are satisfied.

Fermat made no direct use of the characteristic triangle preferring to lean on the known tangent properties of the higher parabolas and to complete his proof by *reductio ad absurdum*.† In 1668 James Gregory gave an improved version of this existence proof‡ and incorporated in it a treatment of the differential or characteristic triangle. This work of Gregory's constitutes the finest, the most rigorous and at the same time the most general geometric integration method in the seventeenth century.§

6.5. JAMES GREGORY (1638–1675)

James Gregory was born into a scholarly family with established connections with mathematical and philosophic work. He received a sound mathema-

† *Oeuvres* I, pp. 226–7.

‡ Gregory, J., *Geometriae pars universalis*, Padua, 1668, pp. 1–3.

§ See Hofmann, J. E., *Die Entwicklungsgeschichte der Leibnizschen Mathematik*, pp. 67–8.

tical education and, at the age of 26, having already written and published his first work, the *Optica promota*, made contact with the Royal Society of London on his way to Italy, where he studied for four years (1664–8). Through Angeli, Gregory was introduced to the geometric indivisible methods of the Italians, Galileo, Cavalieri and Torricelli.† Although Torricelli's finest results remained unpublished his pupils and associates were able to continue his work and Ricci (1619–82) and Angeli (1623–97), besides publishing a great many small treatises on their own account,‡ were able to influence the development of mathematics through their contacts with Gregory and Sluse.

Gregory published two works during this period in Padua, the *Vera quadratura* (1667)§ and the *Geometriae pars universalis* (1668). In the *Vera quadratura* he was, to some extent, following up ideas gained from a study of Grégoire de Saint-Vincent and Pietro Mengoli (1625–86), Cavalieri's pupil and successor at Bologna.||

Gregory's main purpose, however, in the *Vera quadratura*†† was to make use of the idea of a convergent double sequence to define and determine as accurately as possible‡‡ such magnitudes as were not expressible in terms of a rational relation.§§ Through the skilful manipulation of inscribed and circumscribed polygons he was able to generate such a double sequence for the

† *James Gregory Tercentenary Memorial Volume* (ed. Turnbull, H. W.), London, 1939, pp. 1–15.

‡ Ricci, M., *Exercitatio geometrica*, Rome, 1666. Angeli, S. degli, *Problemata geometrica*, Venice, 1658; *De infinitis parabolis*, Venice, 1659–63; *Miscellaneum hyperbolicum et parabolicum*, Venice, 1659; *De infinitorum spiralium spatiorum mensura*, Venice, 1660; *De superficie ungulae et de quartis liliorum parabolicorum et cycloidalium*, Venice, 1661; *De infinitis cochleis*, Venice, 1661; *De infinitis spiralibus inversis infinitisque hyperbolis*, Venice, 1667.

§ Gregory, J., *Vera circuli et hyperbolae quadratura*, Padua, 1667.

|| Mengoli, P., *Via regia ad mathematicas per arithmeticam, algebram, speciosam*, Bologna, 1650; *Novae quadraturae arithmeticae*, Bologna, 1650; *Geometria speciosa*, Bologna, 1659; *Il problema della quadratura del circolo*, Bologna, 1672. In the *Geometria speciosa* Mengoli used inscribed and circumscribed figures to investigate the integral forms $\int x^p(1 - x^q)\,dx$ (Book VI, pp. 348–92) and calculated logarithms through the partial sums of harmonic series (Books IV, V, pp. 141–347). In the *Novae quadraturae* he developed further the work on series in order to throw light on the ultimate possibility of finding the sum of an infinite number of terms. He established that, in the case of the harmonic series $\Sigma 1/n$, the sums increase without limit. See Hofmann, J. E., *Geschichte der Mathematik* II, pp. 40–41. See also Whiteside, D. T., Patterns of mathematical thought, *op. cit.*, pp. 224–7.

†† See Scriba, C. J., *James Gregorys frühe Schriften zur Infinitesimalrechnung*, Giessen, 1957, pp. 13–28. Hofmann, J. E., Über Gregorys systematische Näherungen für den Sektor eines Mittelpunktkegelschnitte, *Centaurus* 1 (1950), pp. 24–37. Dehn, M. and Hellinger, E., On James Gregory's *Vera quadratura* in *Gregory Memorial Volume*, pp. 468–78.

‡‡ *Vera quadratura*, pp. 43–55.

§§ *Ibid.*, Preface, pp. 3–8.

sector of an ellipse, circle, or hyperbola, in the general form,

$$a_n = \sqrt{(a_{n-1}b_{n-1})},$$

$$b_n = \frac{2a_n b_{n-1}}{a_n + b_{n-1}},$$

where a_n, b_n, are the areas of the $(n+1)$th inscribed and circumscribed figures respectively† (see Fig. 6.19).

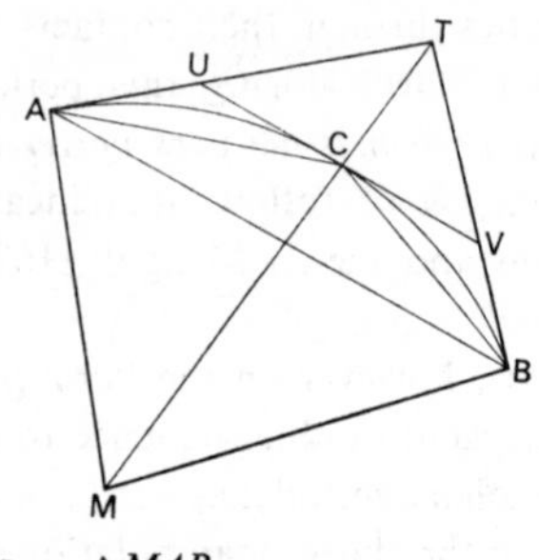

$$a_0 = \triangle MAB$$

$$b_0 = \text{quad. } MATB$$

$$a_1 = \sqrt{(a_0 b_0)} = \text{quad. } MACB$$

$$b_1 = \frac{2a_1 b_0}{a_1 + b_0} = \text{polygon } MAUVB$$

FIG. 6.19

After laying down the beginnings of a theory of convergence for such double sequences‡ (i.e. if $a_n = f(a_{n-1}, b_{n-1})$ and if $b_n = F(a_{n-1}, b_{n-1})$ then the double sequence (a_n, b_n) is convergent provided $|b_n - a_n| < |b_{n-1} - a_{n-1}|$ and, for any arbitrarily chosen small quantity ϵ_p, $|b_n - a_n| < \epsilon_p$, for $n > p$) Gregory attempted to establish the *impossibility* of rationally squaring the circle, ellipse or hyperbola by showing that no finite combination of the terms (a_n, b_n) of the above sequence could result in a rational function of (a_0, b_0). Although this proof was defective and in consequence rapidly incurred a storm of criticism§ it is to Gregory's credit that he was the first to formulate a proposition of this class.

† *Vera quadratura*, Props. I–V, pp. 11–14.
‡ *Ibid.*, Props. VIII–X, pp. 21–4.
§ See Huygens' criticism, 2nd July 1668, *Journal des Sçavans*, July 1668; given in *Oeuvres* VI, pp. 228–30; see also XX, pp. 303–7.

In the *Geometriae pars universalis* we have the first systematic attempt to compile a comprehensive treatise on all the processes for the determination of arc, tangent, area, volume and surface which would now find a place in an elementary textbook on the infinitesimal calculus. Gregory made no claim to originality† and was obviously using material from a wide variety of sources including Saint-Vincent and Tacquet as well as the Italian school. He rightly attached some importance to his contribution to method for he developed the valuable and time-saving idea of separating general from special proofs and carrying forward results already established. In this way he eliminated much of the necessity for constant repetition.

Most of the results which had been established piecemeal during the seventeenth century were reassembled by Gregory in the style and manner of a classical work. The treatment is throughout verbal and geometrical and the proof structure is based on a modified exhaustion proof.

In the first section‡ Gregory deals with the relations between curve, arc, tangent and area. For a monotonic ascending curve (*simplex* and *non sinuosa*) he sets up a system of inequalities on the basis of two sets of circumscribed tangents and establishes that the difference between the arc and the sum of either set of tangents can be made less than any arbitrarily chosen small quantity.§ He then makes use of the similarity of the normal triangle (formed from the normal (n), the ordinate (y) and the subnormal (s)) and the small triangle formed by the elements of tangent, ordinate and abscissa (δt, δy, δx)

† *Geometriae pars universalis*, Preface.
‡ *Ibid.*, Props. I–XI, pp. 1–29.
§ *Ibid.*, Prop. I, pp. 1–3. See pp. 227–8 above. Instead of introducing the point W (as did Fermat) Gregory uses:

$$\text{arc } P_rP_{r+1} < P_rQ+QP_{r+1} < P_rQ+QS_{r+1} < P_rS_{r+1}.$$

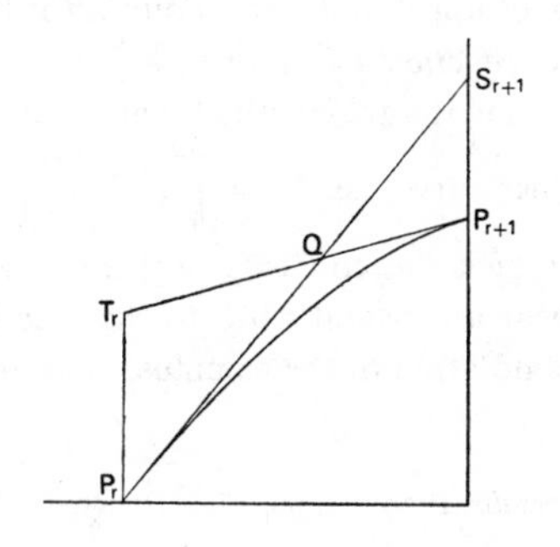

to establish the relation

$$\frac{\delta x}{\delta t} = \frac{y}{n} \quad \left[= \frac{k}{z}\right] \qquad \text{(see Fig. 6.20).}$$

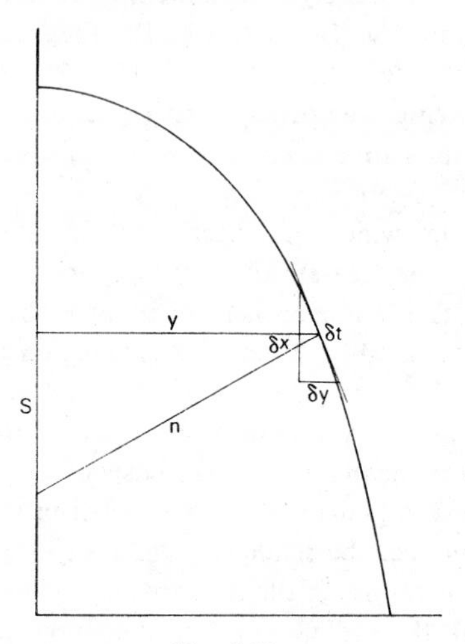

FIG. 6.20

Two auxiliary curves are now constructed, $n = f(x)$, $z = g(x)$: for the first of these we have $\sum n\ \delta x = \sum y\ \delta t$, and the quadrature of $n = f(x)$ gives the surface area of the cylinder formed by rotating the original curve about the x-axis. For $z = g(x)$, we have $\sum z\ \delta x = k \sum \delta t$, and the quadrature of this curve gives the arc length of the original curve. Nowhere does Gregory approximate arc to tangent or inscribed rectangle to area under curve, but in each case he sets up a system of upper and lower bounds for the required sums and completes the proof by *reductio ad absurdum*.

Having shown how to find the arc length of a given curve by replacing it by the quadrature of another curve, i.e. $ks = \int_0^x z\, dx = \int_0^x \sqrt{(1+(dy/dx)^2)}\, dx$, Gregory goes on to attack the converse problem i.e. to find a curve, $u = g(x)$, whose arc shall bear a constant ratio to the area under a given curve, $y = f(x)$.† Thus, in the notation of the calculus, it is required to determine

† *Geometriae pars universalis*, Prop. VI, pp. 17–20.

$u = g(x)$, such that

$$ks = k\int_0^x \sqrt{(1+(du/dx)^2)}\, dx = \int_0^x y\, dx,$$

or

$$k^2(1+(du/dx)^2) = y^2,$$
$$k^2(du/dx)^2 = y^2-k^2 = z^2,$$
$$k\, du/dx = z,$$

hence

$$ku = \int_0^x z\, dx.$$

Gregory constructs two auxiliary curves,

(i) $z = g_1(x) = \sqrt{[(y^2-k^2)]}$

(ii) $ku = \int_0^x z\, dx.$

He then completes the proof by constructing the tangent to the latter curve, i.e. by marking off the subtangent, $t = uk/z$ $(du/dx = u/t = z/k)$. It is in this step, subsequently carefully justified on the basis of a *reductio* proof, that Gregory shows himself completely aware of the inverse relation between tangents and quadratures—or differentiation and integration—in the sense that he is able to pass at once from $ku = \int_0^x z\, dx$, to $k\, du/dx = z$.

Proposition VI of the *Geometriae* must therefore rank as the first published statement, in geometrical form, of what is now termed the fundamental theorem of the calculus.†

The concept of *evolution* and *involution* was developed by Gregory‡ as a means of applying results obtained by the use of rectangular coordinates for the higher parabolas $(y/a)^m = (x/b)^n$, to the spirals $(r/a)^m = (\theta/\alpha)^n$. If a curve, given in the form $y = f(x)$, be replaced by a curve $r = F(\theta)$, where $r = y$ and $(dr)^2+r^2(d\theta)^2 = (dx)^2+(dy)^2 = (ds)^2$, then the first curve is termed the *evolute* and the second curve the *involute*. The transformation can thus be expressed in the form, $r = y$, $x = \int_0^\theta r\, d\theta$, $\theta = \int_0^x \frac{dx}{y}$, $y = r$.

† It must nevertheless be borne in mind that for us differentiation and integration are inverse operations, applied to abstract functions, whereas for Gregory they were applied in a purely geometric context between tangents, arcs and areas.

‡ *Ibid.*, Props. XII–XVIII, pp. 29–41. The general idea was used by Torricelli, Fermat and others. Wallis discusses it in the *Tract. duo* (1659).

Representing these transformations in geometrical form, we have

$$A_1 = \int_0^x y\,dx = \int_0^\theta r^2\,d\theta = 2A_2,$$

where A_1 and A_2 represent the areas of the first and second curves, respectively (see Fig. 6.21): also, since $dy/dx = 1/r\ (dr/d\theta)$, the angle which the tangent to the evolute makes with the ordinate is equal to the angle which the tangent to the involute makes with the radius vector. The arc of the involute is, accordingly, equal to the arc of the evolute. Gregory had thus, in the theory of involutes and evolutes, an extremely useful tool which considerably increased the power of, what he termed, his *transmutations.*

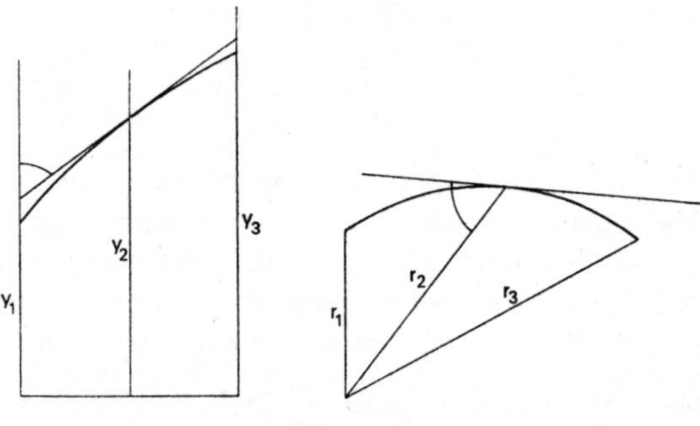

FIG. 6.21

The fundamental mode of attack in all of Gregory's fascinating collection of theorems is, indeed, a progression through successive *transmutations*† (or geometrical transformations) until a final form is reached which is capable of direct resolution. The reader, picking his way delicately through a complex network of auxiliary curves and constructions can, in the end, only feel profound awe and admiration for the masterly way in which Gregory handles this particular tool. As with Archimedes, the indirect proof structure frequently obscures the method of approach and, in consequence, the emergence of the final result is often accompanied by a sensation of surprised wonder.

One of the finest examples of Gregory's method of geometrical integration appears in a later work, the *Exercitationes geometricae,*‡ which Gregory wrote after his return to England partly to answer some of the criticisms

† *Methodus transmutandi.*

‡ Gregory, J., *Exercitationes geometricae*, London, 1668.

directed against the *Vera quadratura* and partly to develop geometrically Mercator's quadrature of the rectangular hyperbola by expansion in series.† Although the complex sequence of geometric transformations which Gregory develops to establish the equivalent of $\int_0^\alpha \sec\theta\, d\theta$ may appear tedious and involved to those accustomed to integration processes based on algebraic symbolism and analytical transformations, in a sense they epitomise the contribution made, throughout the seventeenth century, by geometric imagery in the development of integration methods. For this reason I am unable to resist the temptation to give a brief sketch of Gregory's solution to this problem.‡

1. *Representation of the Problem in Geometric Terms*

At each point P, on a circular arc, BPQ, centre O, let a line PT' be drawn perpendicular to the plane of the sector and equal in length to OT, where

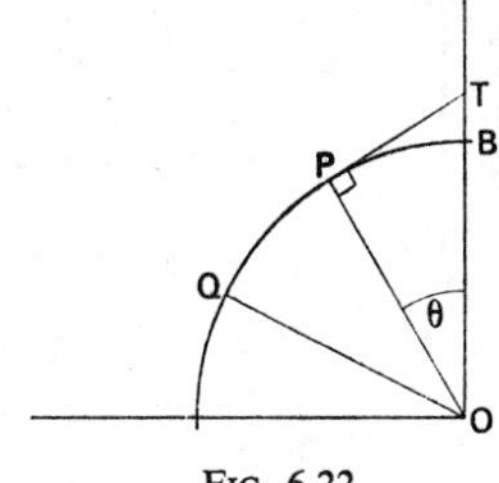

FIG. 6.22

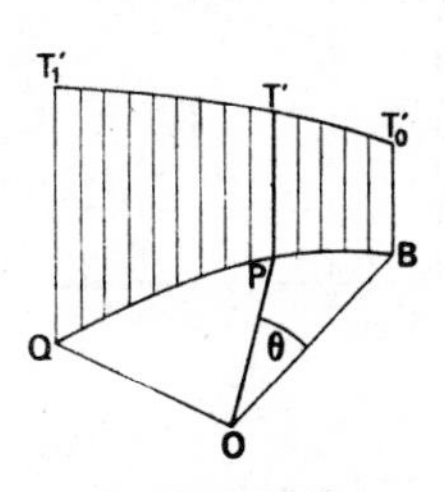

FIG. 6.23

PT is the tangent at P. Consider the area of the cylindrical surface, $QT_1'T_0'B$ $\left[\text{i.e. if } OP = OB = OQ = a,\ \widehat{QOB} = \alpha,\ \widehat{POB} = \theta,\ PT' = a\sec\theta,\ I = \int_0^\alpha a\sec\theta\, d(a\theta)\right]$.

2. *First Transformation*

Through each point P on the curve QPB, draw the line NPX, parallel to OT, and such that $NX/OT = OT/OP$ [$= \delta s/\delta x$].

† Mercator, N., *Logarithmotechnia*, London, 1668.
‡ *Exercitationes geometricae*, Props. I–II, pp. 14–17.

Consider the surface area X_1N_1BO. $\left[NX = a\sec^2\theta,\ \sum NX\,\delta x = \sum OT\,\delta s,\right.$

$\left. I = \int_0^\alpha a\sec\theta\, d(a\theta) = \int_0^\alpha a\sec^2\theta\, d(a\sin\theta) = \text{area } X_1N_1BO\right]$. (Fig. 6.24.)

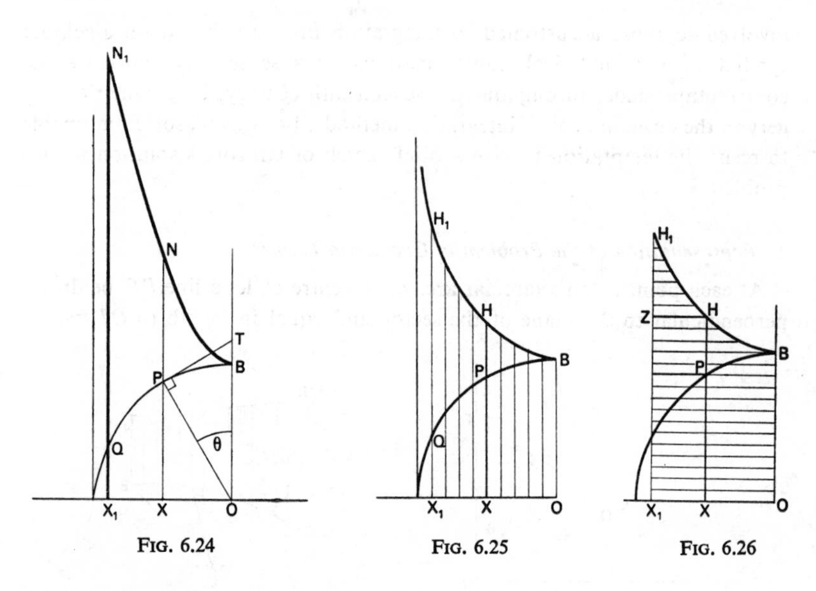

FIG. 6.24 FIG. 6.25 FIG. 6.26

3. *Second Transformation*

Through each point X on the line OX, draw XH, parallel to OB, such that, $HX^2 = OP.\,NX$ $(HX = a\sec\theta)$.

Consider the volume of the solid formed by rotating X_1H_1BO about X_1O.

$\left[\text{We have } a\sum NX.\,\delta x = \sum HX^2.\,\delta x, \text{ volume of solid} = \pi\int_0^\alpha a^2\sec^2\theta\, d(a\sin\theta) = \pi a I.\right]$ (Fig. 6.25.)

4. *Third Transformation*

This solid can be considered as made up of cylindrical elements formed by rotation of line segments ZH, and so on, about X_1O. $\left[\text{Since } ZH = a(\sin\alpha - \sin\theta) \text{ then } aI = 2\int_0^\alpha a(\sin\alpha - \sin\theta)a\sec\theta\, d(a\sec\theta) + a^3\sin\alpha.\right]$ (Fig.6.26.)

5. *Fourth Transformation*

This can be represented (in the manner of Saint-Vincent) by constructing an isosceles triangle (right-angled), $X_1\,AH_1$, with $AH_1 = X_1H_1$. Then,

$$2\times \textit{ductus}\ H_1AX_1\ \text{in}\ H_1X_1OB = 2\sum X_1Z.ZH = aI. \quad \text{(Fig. 6.27.)}$$

It is now necessary to transform this volume integral into a quadrature by expressing the product $X_1Z.ZH$ as a single term [i.e. $X_1Z.ZH = (a \sin\alpha - a\sin\theta)\, a\sec\theta$].

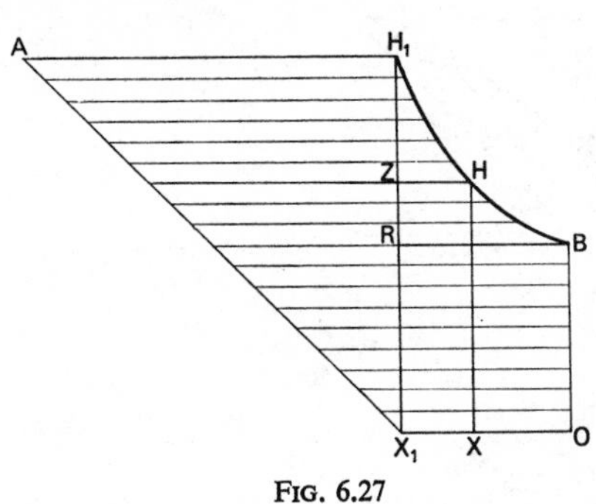

FIG. 6.27

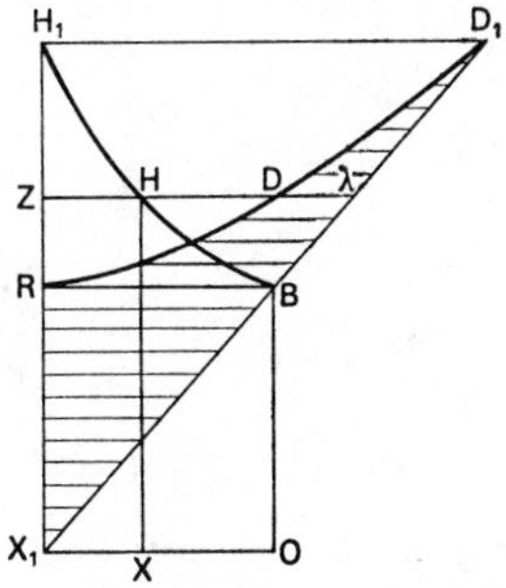

FIG. 6.28

6. *Fifth Transformation*

Construct a rectangular hyperbola, vertex R, such that

$$H_1X_1^2 - H_1D_1^2 = a^2 = ZX_1^2 - ZD^2. \quad \text{(Fig. 6.28.)}$$

Join X_1D_1 and let ZH produced cut X_1D_1 in λ. [Since $H_1X_1 = a\sec\alpha$, $ZX_1 = HX = a\sec\theta$, $H_1D_1 = a\tan\alpha$, $ZD = a\tan\theta$, then

$$Z\lambda = \frac{a^2\tan\alpha\sec\theta}{a\sec\alpha}$$

$$= a\sin\alpha\sec\theta,$$

$D\lambda = Z\lambda - ZD = a\sec\theta\,(\sin\alpha - \sin\theta)$.]

Hence,

$2\times \textit{ductus}\ H_1AX_1$ in $H_1X_1OB = 2a\times$sector $RDD_1\lambda X_1$, and $I = 2\times$sector RDD_1X_1.

7. Sixth Transformation

Draw RR', DD', D_1D_1', perpendicular to the asymptote X_1Y, then, sector RDD_1X_1 = segment $RDD_1D_1'R'$ [since sector $RX_1D_1 + \Delta X_1D_1D_1'$ = segment $RR'D_1'D_1 + \Delta RR'X_1$, and since the curve is a rectangular hyperbola, $X_1R' . RR' = X_1D_1' . D_1D_1'$, i.e. $\Delta X_1D_1D_1' = \Delta RR'X_1$]. (Fig. 6.29.)

Hence, $I = 2\times$hyperbolic segment $RR'D_1'D_1 = a^2 \log (RR'/D_1D_1')$
$(= a^2 \log (\sec \alpha + \tan \alpha))$.†

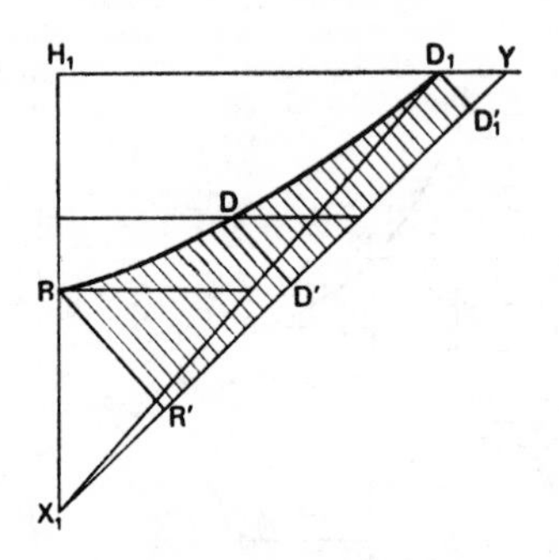

FIG. 6.29

Fascinating as these processes are, it can only be with a certain measure of relief that the end is finally reached and a result achieved which, in terms of analytical transformations, can be arrived at in a couple of lines. With Gregory and his contemporary, Isaac Barrow, we reach the end of this particular line of development, although Gregory himself, always hampered by a lack of facility in the use of Cartesian notation, went on to play his part in the development of the new analysis.

The algebraic tangent method which Gregory included in the *Geometriae* (Prop. VII)‡ was an adaptation of Fermat's method applied to the curve now known as the "Pearls of Sluse".§ Historically the main interest in this example lies in the publication by Gregory of the adaptation made by Beaugrand to

† $D_1D_1' = \dfrac{a(\sec \alpha - \tan \alpha)}{\sqrt{2}}$, $RR' = a/\sqrt{2}$.

‡ *Geometriae*, pp. 20–2.

§ The curves, $y^p = x^q(x-a)^r$, were well known to Sluse, Huygens and Schooten and occupied a prominent place in the correspondence between them on tangent methods. The curves are discussed in some detail by Ricci in the *Exercitatio geometrica* (Rome, 1666).

Fermat's notation† in which he used a small '*o*' instead of the '*e*' of Fermat. The transcription made by Beaugrand of Fermat's method circulated widely and presumably Gregory saw a copy in Italy. The use of this suggestive but misleading sign by Gregory and subsequently more extensively by Isaac Newton has led to much speculation regarding a possible connection with any of these sources.‡

6.6. Isaac Barrow (1630–1677)

Isaac Barrow, first Lucasian Professor of Mathematics at Cambridge from 1663, was in process of preparing his *Lectiones geometricae*§ for the press when he received a copy of the *Geometriae pars universalis* through John Collins, mathematical correspondent of the Royal Society. The similarity in content of the two works should, however, be attributed rather to a common background of study in Italy than to any direct influence of Gregory on Barrow.

Barrow was educated at Charterhouse, Felsted, and Trinity College, Cambridge, where he graduated and subsequently obtained a fellowship. Later, for religious reasons, he left Cambridge and travelled abroad for three years during which time he visited Paris, Florence, Smyrna, Constantinople and Venice returning home through Germany and the Low Countries. The letters which he wrote to the University at this time indicate the diversity and breadth of his interests rather than any concentration on mathematical studies, but there can be no doubt that he welcomed the opportunities afforded him of establishing communication with eminent mathematicians in France and Italy.||

Immediately on his return to England Barrow was ordained and, a year later, was appointed to the Regius Professorship of Greek in the University of Cambridge. The stipend for this appointment was small and Barrow took the opportunity to augment it by accepting a Gresham Professorship of Geometry. During the period, 1662–3, he lectured at Gresham College in Astronomy and Geometry and, less frequently, at Cambridge in Greek. The Royal Society met at Gresham College soon after Barrow's appointment and his name appears among the first list of Fellows elected after the Society's incorporation. He served on various committees and, although his member-

† See Chapter V above, pp. 172–3.

‡ The word subsequently scarcely applies to Newton for he appears to have been privately making use of the symbol from 1663 onwards. It appears in the *De analysi* (1669).

§ Barrow, I., *Lectiones geometricae*, London, 1670.

|| Osmond, P. H., *Isaac Barrow*, London, 1944.

ship was not, apparently, a particularly active one, he corresponded with John Collins from 1663 and accepted his assistance in the publication of his lectures.

In 1663 Barrow was appointed first Lucasian Professor of Mathematics at Cambridge and held this appointment until 1669 when he relinquished it in favour of Isaac Newton. In 1670 he was appointed Doctor of Divinity by Royal mandate and became Master of Trinity College in 1672.

From this brief account it is clear that the close and intimate relationship which developed between Barrow and Newton probably dates from 1663, when Barrow gave up the Gresham Professorship and began to give mathematical lectures regularly in Cambridge. Besides guiding Newton's studies, lending him books and advising on reading Barrow gave a series of lectures during the session, 1664–5, which Newton almost certainly attended. These lectures were of a philosophic nature and dealt with general concepts of space, time and motion.† At this time Barrow had already begun to consider the study of curves as the paths of moving points and the first five lectures which he included in the *Lectiones geometricae* probably date from this period.‡

In the first lecture Barrow discusses the nature of Time and Motion and the geometric representation of such magnitudes in the manner of Oresme and Galileo. Time and motion he says, naturally measure each other. Time is made up either of the simple addition of rising moments, or of the continual flux of one moment. A line, being the trace of a point moving forward, may be conceived as the trace of a moment continually flowing. Time can thus be represented by a uniform right line.

Divisibility is always possible, infinitely or indefinitely. Instants are infinitely small parts of time in which infinitely small spaces are moved through. Any degree of velocity can be generated starting from rest, or the lowest degree of velocity, by increase, or continual flowing, in the same manner as a line may be produced by the apposition of points or by motion, or time by the succession of instants. Velocities can, accordingly, be represented by straight lines at right angles to the line representing time; the space enclosed represents the aggregate distance covered during the motion (see Fig. 6.30).

Barrow goes on to consider the generation of curves and surfaces by com-

† Barrow, I., *Lectiones mathematicae*, London, 1683.

‡ See Rigaud, S. P., *Correspondence of Scientific Men of the 17th Century*, Oxford, 1841, II, p. 34: Barrow to Collins (probably Dec. 1663). See also *Lectiones geometricae*, Preface. J. M. Child (in *The Geometrical Lectures of Isaac Barrow*, Chicago, 1916) discusses the dating of Barrow's lectures in some detail.

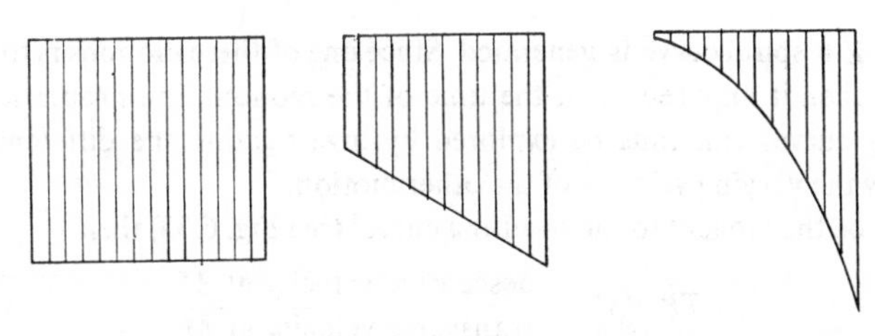

Fig. 6.30

pound motion.† A curved line is generated by the motion of a point along a right line whilst, at the same time, the line moves parallel to itself (see Fig. 6.30): a cone is generated by the rotation of a line segment about one end: a circle results if the line segment rotates in a plane (see Fig. 6.31). The Cavalierian concept of indivisibles is linked with these ideas of motion through the *apposition* of points. Any curved line may be considered as made up of innumerable such lines, indefinitely small: the space of a circle is made up of concentric circular peripheries, as many in number as there are points in the radius: the surface of a cone is made up of an infinity of triangles. In effect, Barrow says: "I speak in the phrase of the atomists, for the sake of ease, brevity and perspicuity; and I make no scruple to use their method because of its truth. The method of indivisibles is the most expeditious of any, and no less certain and infallible when rightly applied."

In Lecture III Barrow goes on to generate curves systematically by the composition of rectilinear and parallel or rotational motions. If a straight line AZ (see Fig. 6.32) move transversely and uniformly parallel to itself, whilst at the same time a point M moves along the line from A to Z, then the successive positions of the point M generate a curve. Similarly, if the line AZ rotate uniformly about A, whilst a point M moves along the line from A

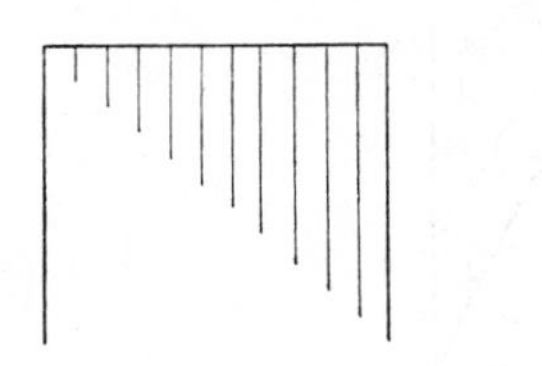

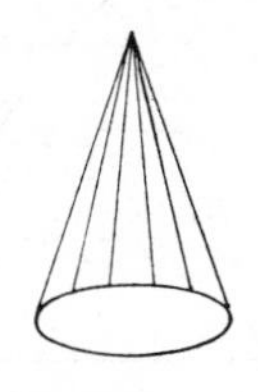

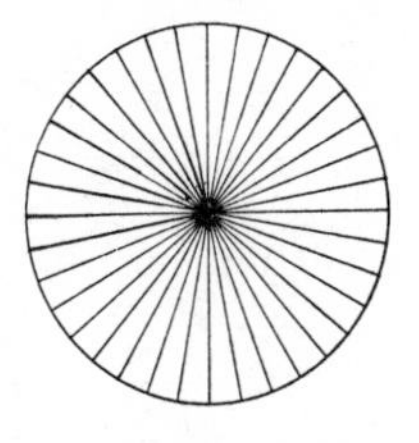

Fig. 6.31

† Lecture II.

towards Z a spiral curve is generated. Since one of these motions is to be kept uniform then it may represent the time of the motion. The properties of the resulting curves can thus be explored by investigating the differences produced by the varying velocity of the other motion.

If TM be the tangent to the resulting curve† (see Fig. 6.33) then

$$TP/PM = \frac{\text{descensive velocity at } M}{\text{transverse velocity at } M}$$

$$= (v/u = vt/ut).$$

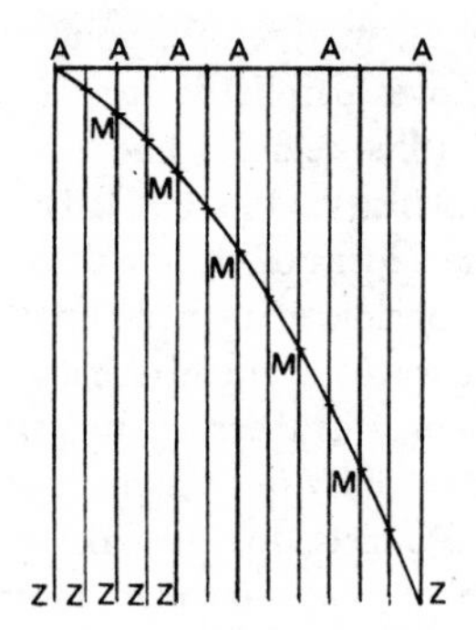

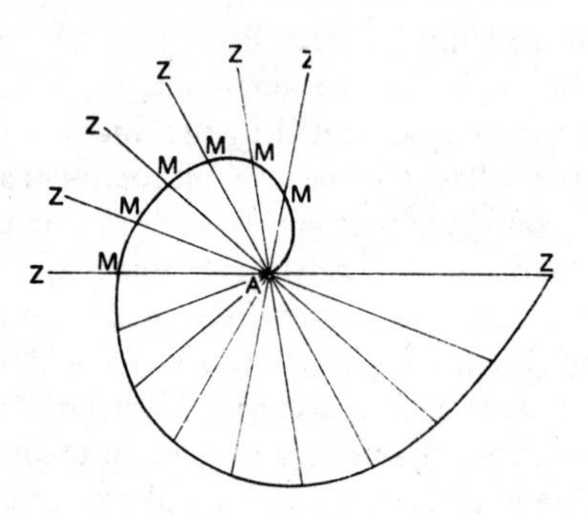

FIG. 6.32

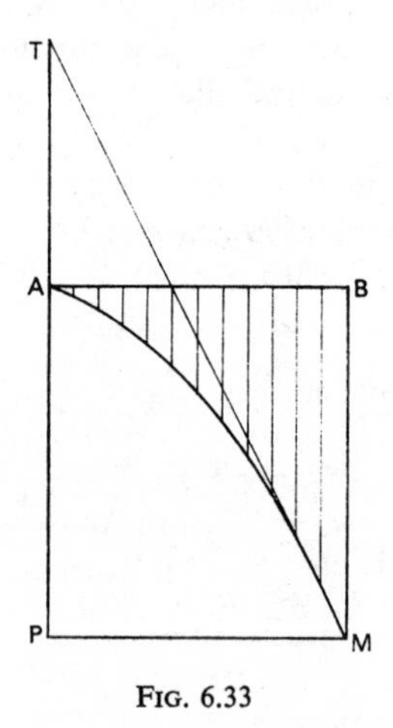

FIG. 6.33

† Lecture IV.

Hence, if PM represent the distance covered under uniform transverse velocity in time t, TP represents the distance covered in the same time with velocity equal to the final descensive velocity at M.

Now, let a curve *am* be constructed, with abscissa $ab \propto t$, and such that the space *abm* (see Fig. 6.34) = BM = descensive motion to M (in time t). Since this is a space–time graph, the ordinate *bm*, drawn at any point *m* on the curve, in general represents the velocity (descensive) at each instant of time, i.e. if $\int_0^t z\,dt = y$, $z = dy/dt$.

Hence $$\frac{\text{space } abm}{\text{rectangle } ab\,.\,bm} = \frac{BM}{TP} = \frac{AP}{TP}.$$

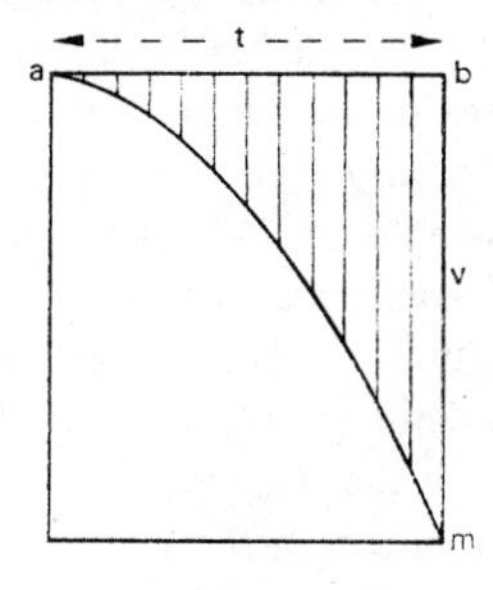

Fig. 6.34

Thus, from a knowledge of the tangent to one curve it is possible to pass to the construction and quadrature of another, and conversely: if, for example, the curve AM be of the form, $y = kx^n$, ($AB = x$, $BM = y$),

then $$TP/PA = n, \quad \frac{\text{space } abm}{\text{rect. } ab\,.\,bm} = \frac{1}{n},$$

$bm = dy/dt = z = v = ny/x = nkx^{n-1}$.†

It is clear that out of the mediaeval concepts of time and motion, linked with the indivisibles of Galileo and Torricelli, Barrow had been able to build for himself an intuitive understanding of the inverse relation between the differential and integral processes. These ideas were not, of course, new or original with Barrow, but had a long historical development from Aristotle

† If the curve AM is of the form $y = kx^n$, $dy/dx = nkx^{n-1} = ny/x = TP/PM = vt/ut$, $vt = ny$, $x = ut$. Space $abm = y = vt/n = \int_0^t z\,dt$, $dy/dt = bm = v = z$.

through the mediaevalists to Galileo, Cavalieri, Torricelli and Roberval. In particular the unpublished manuscripts of Torricelli contained most of the material discussed by Barrow.† None the less, it must be regarded as of some significance in the historical development of the calculus that Barrow should have rediscussed these ideas in such a fundamental way at a time when Newton was already beginning his own researches in this field. There were, of course, a great many factors influencing the development of Newton's early mathematical work but one of the strongest and the most potent was certainly the application of the concepts of time and motion to the generation of curves.

For Barrow himself these concepts formed only an introduction to the study of curves. In his later work he attached much more importance to the geometrical proofs which he supplied in Lectures VII–X. In these lectures Barrow develops systematically the fundamental theorems relating to tangent, arc, area and surface, giving each theorem both in the axial and in the polar form. The ground covered by Barrow is very much the same as that covered by Gregory in the *Geometriae pars universalis* but Barrow goes more deeply, both into the foundations and into further developments.

The proofs depend, for the most part, on the application of certain inequalities used in conjunction with the tangent properties of the curve. Barrow's concept of tangent was antiquated, in the sense that for him, as for Fermat, a tangent line meets the curve in one point only and remains without the curve at all other points. If, therefore, a tangent can be constructed at a point P on a convex curve, according to some proposed rule (see Fig. 6.35) it is only necessary to validate the construction by showing that, for points P' and P'' above and below P,

$$T'X' > P'X', \quad T''X'' > P''X''$$

In a sense, therefore, all of Barrow's methods are concealed, for the proposed constructions appear from no very obvious source and are confirmed by the systematic application of established inequalities. The construction of tangents proceeds, accordingly, by means of auxiliary curves, for which the tangents are already available. For example,‡ if EFE and HGH are two curves, such that, for corresponding points on the curves, the ordinates are in a given ratio, i.e. $IH/IE = XG/XF$ (see Fig. 6.36), then, if TG be the tangent to HGH at G, TF is the tangent to EFE at F: since TG is tangent to the curve

† See pp. 190–3 above.
‡ Lecture VIII, Prop. 6.

HGH at G, $IL > IH$ and $IL/IE > IH/IE$: thus, $IL/IE > XG/XF$ and, since $XG/XF = IL/IK$, then $IL/IE > IL/IK$ and $IK > IE$. Hence, K lies outside the curve EFE and FT is a tangent, i.e. $d(ky)/dx = k\, dy/dx$.

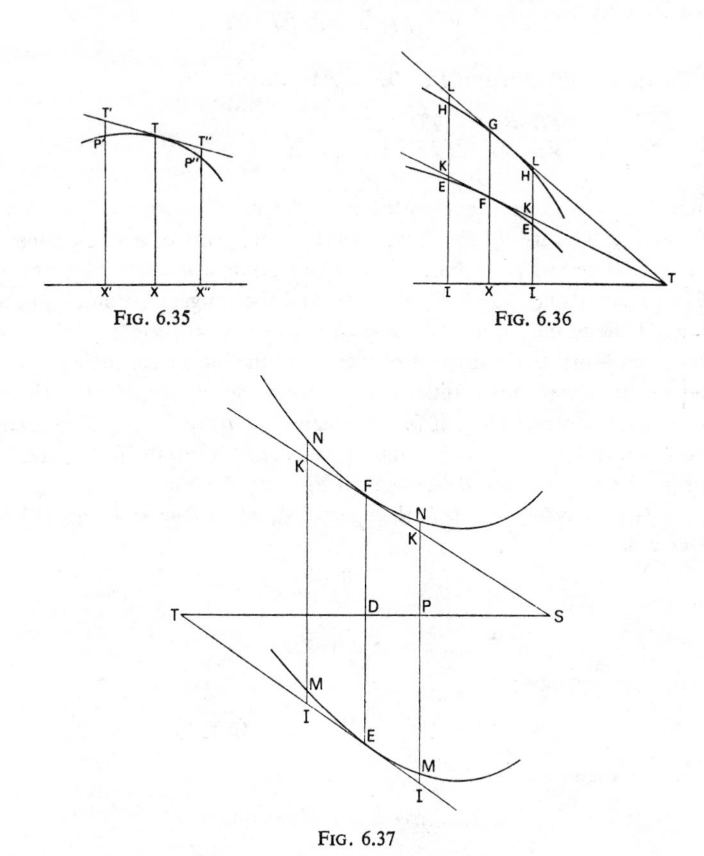

FIG. 6.35

FIG. 6.36

FIG. 6.37

A fruitful and general proposition on the same lines† relates the tangents to curves NFN and MEM, whose ordinates FD and DE are reciprocals, i.e. $FD.DE = NP.PM$ (see Fig. 6.37): let TE be the tangent to the curve MEM at E, then SF is tangent to NFN at F, provided $SD = DT$.

† Lecture VIII, Prop. 9.

Let TE and SF meet the ordinate MPN at I and K respectively: we have, $TP/PM > TP/PI\,(= TD/DE)$ and $SP/PK = SD/DF$.

Hence, (multiplying) $$\frac{TP.SP}{PM.PK} > \frac{TD.SD}{DE.DF}.$$

But, $TD.SD > TP.SP$, and $DE.DF = PM.PN$,

therefore, $$\frac{TD.SD}{PM.PK} > \frac{TD.SD}{PM.PN}, \quad \text{and} \quad PN > PK:$$

the line FS, accordingly, touches the curve NFN.

By varying the form of the curve MEM the tangents to a whole range of curves can be determined: thus, if MEM be a straight line, say $y = kx$, then NFN is a rectangular hyperbola, $y = k/x$, and the tangent becomes immediately available in the form, $t = -x\,(dy/dx = -y/x = -k/x^2)$.

To develop further the tangent properties of the higher parabolas, Barrow makes use of certain inequalities derived from the properties of arithmetic and geometric means.† Thus, if $(n-1)$ means $(a_1, a_2, a_3, \ldots, a_{n-1})$ be placed between a and b and the same number of geometric means $(g_1, g_2, g_3, \ldots)$ between a and c, Barrow shows that, if $g_1 \geqslant a_1$, then in general, $g_m > a_m$ and $c > b$. Conversely, if $c \leqslant b$, then $g_m < a_m$, where $0 < m < (n-1)$.‡

If we write,

$$a_m = a + \frac{m}{n}(b-a),$$

$$g_m = a(c/a)^{m/n},$$

the condition becomes,

$$a(c/a)^{m/n} \gtreqless a + \frac{m}{n}(b-a), \qquad (c \gtreqless b).$$

If $c = b$, then

$$a(b/a)^{m/n} < a + \frac{m}{n}(b-a),$$

† Lecture VII, Props. 11–13.

‡ The argument runs as follows: if the two series be given in the form

$$a, \quad a+d, \quad a+2d, \quad a+3d, \quad \ldots, \quad a+(n-1)d$$
$$a \quad ar \quad ar^2 \quad ar^3, \quad \ldots, \quad ar^{n-1},$$

then, if $ar > a+d$, since $a+ar^2 > 2ar$ $[(r-1)^2 > 0]$, and $a+ar^2 > 2(a+d)$, we have $ar^2 > a+2d$. Repeating the process, since $a+ar^3 > ar+ar^2$, $[(1-r)(1-r^2) > 0]$, then, $a+ar^3 > 2a+3d$, $ar^3 > a+3d$. Barrow concludes: "And you may reason on in the same manner as long as you please."

and, putting $b-a = x$, we have

$$a(1+x/a)^{m/n} < a+mx/n,$$

and $$(1+x/a)^{m/n} < 1+mx/na \qquad (0 < m < n-1).$$

This inequality, formally equivalent to an approximation to the first term of the binomial expansion for fractional exponents was, of course, used in some form or other by all those who sought to establish the tangent properties of the higher parabolas and hyperbolas on a rigorous basis. As $x \to 0$,

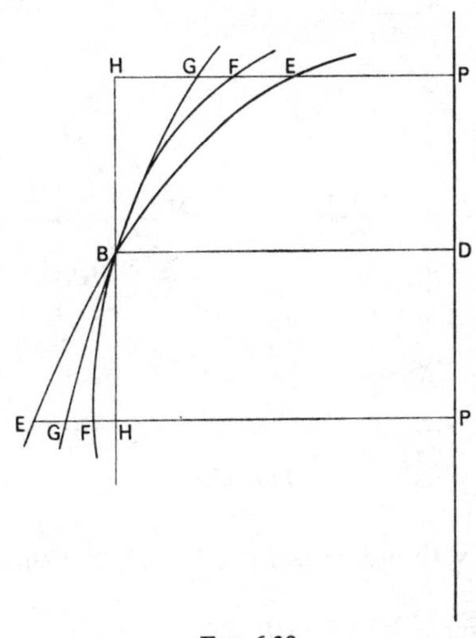

FIG. 6.38

$(1+x/a)^{m/n} \to 1+mx/na$, and the application to tangent methods is readily made:† let BD be a given straight line, and let HP, drawn parallel and equal to BD, cut the lines EBE, FBF, GBG (intersecting in B) in E, F, G, respectively, such that (see Fig. 6.38),

(i) PG is an arithmetic mean between PH and PE,
(ii) PF is a geometric mean between PH and PE (PF and PG are means of the same order).

GBG and *FBF* will touch at *B* for it is evident that the line *GBG* falls entirely outside the line *FBF*.

† Lecture VII, Prop. 17.

Analytically, if $PE = z = f(x)$, $PH = a$, $PF = y$, $PG = u$, we have, $u = a+m/n\,(z-a)$, $y = a(z/a)^{m/n}$. Hence the curves *FBF*, *GBG*, have a common tangent at *B*. If, however, *EBE* becomes a straight line, then *GBG* becomes the actual tangent to *FBF* at $B\,(z = a+kx,\ y = a(1+kx/a)^{m/n}$, $u = a+m/n(z-a) = a+(m/n)kx)$.

The fundamental theorem of the calculus appears in the following guise:† let *ZGEG* be an ascending curve (i.e. the ordinates *VZ*, *PG*, *DE*, *MG*, ... increase continually from the first, *VZ*) with axis *VD*, and let a second curve,

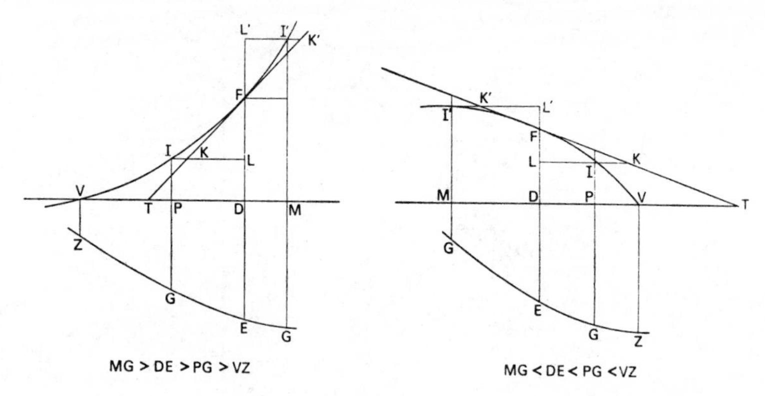

Fig. 6.39

VIF, be constructed with ordinates *PI*, *DF*, *MI'*, ... and such that (see Fig. 6.39)

$$\text{space } VDEZ = R.FD,$$

$$\text{space } VPGZ = R.IP,$$

then

$$\text{space } PDEG = R(FD-LD) = R.FL.$$

Since, however,

$$\text{space } PDEG < \text{rectangle } PD.DE \qquad PG < DE$$

$$R.FL < PD.DE,$$

and

$$R/DE < PD/FL. \tag{i}$$

Thus, if a line *FT* be drawn through *F*, meeting the axis, *VD* in *T* and such that, $TD/DF = R/DE$, and if *IL* meet *FT* in *K* then $KL/LF = TD/DF = R/DE$ and also, by virtue of (i) above, $KL/LF < PD/LF$, $KL < PD$, i.e. $KL < IL$.

† Lecture X, Prop. 11.

Hence, the point K lies outside the curve. In order to complete the proof it is necessary to go through the reasoning again for points G, I on the other side of E, F.

The same theorem is given in polar form at Lect. X, Prop. 13, and the converse appears in Lect. XI, Prop. 19.

Although this material, because of its manner of presentation, scarcely makes exciting reading, Barrow was making a systematic effort to include within a single volume the entire range of infinitesimal processes known to him and he therefore thought it worth while to include an algebraic tangent method. He did so, he says, on the advice of a friend and this friend was Isaac Newton who has confirmed that a discussion on tangent methods took place between himself and Barrow at this time. As for the method, it was unquestionably Barrow's own modification of Fermat's tangent method for, by this time, Newton's own work had developed along quite different lines.

In the algebraic examples given by Barrow in the *Lectiones*† he makes no distinction between the indefinitely small portions of arc and tangent but thinks it reasonable to identify these lines on the grounds of the "indefinite smallness" of the part of the curve. The method, as used by Barrow, accordingly consists of forming an *exact* equation between the coordinates of neighbouring points on the curve and subsequently rejecting in this equation any terms containing second order (or greater) infinitesimals. He then constructs two similar triangles, the one being an infinitesimal triangle containing the small portion of arc (or tangent), ordinate and abscissa, the other containing the intercept of the tangent on some line or lines given in the figure. In practice Barrow works mainly in terms of rectangular axes but is equally at home with polar coordinates. Although apparently he saw no particular point in Cartesian notation (x, y) he introduced one useful feature in that he gave separate letters (a, e) to two infinitesimals $(\delta x, \delta y)$. A simple example will suffice to make the process clear;‡ let the coordinates of two points, $M(f, m)$ and $N(f+e, m-a)$ on a curve MN be connected by the relation

$$f^3+m^3 = (f+e)^3+(m-a)^3 \quad \text{(see Fig. 6.40).}$$

Discarding terms containing powers of a and e above the first, we have

$$0 = 3ef^2-3am^2,$$

and

$$e/a = m^2/f^2.$$

† Lecture X.
‡ Lecture X, Example 2.

Thus, if t denote the subtangent (TP in the figure),

$$t/m = e/a = m^2/f^2.$$

The diagram bears the familiar appearance of an elementary textbook in the calculus today, save for the important blurring of the relation between tangent and arc. Because of his antiquated concept of tangent Barrow was unable to see the tangent as the limiting position of the chord and he can

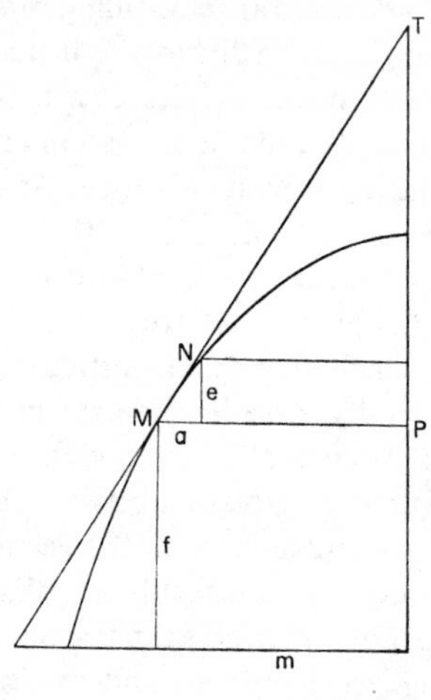

FIG. 6.40

scarcely have found much satisfaction in this crude identification of tangent and arc. Nevertheless he was adept at constructing and using infinitesimal triangles and this provided him with some neat tangent constructions.

For the trigonometric functions he constructs two differential triangles, one for the curve itself and one for the circular quadrant from which it is derived.† Thus, let $BEDC$ be a circular quadrant, centre C (see Fig. 6.41) and let $CB = r$, $CK = f$, $KE = g$, arc $EF = e$; let AMO be the curve of tangents (see Fig. 6.42), AQ = arc BF, AP = arc BE, $QP = EF = e$, $PM = BG$ $(= r \tan \theta) = m$, $MR = GH = a$, $NQ = m-a = BH$.

Now, from similar triangles, $r/g = e/LK$, $LK = eg/r$, $CL = f+eg/r$, $LF = \sqrt{(r^2-CL^2)}$. Since $CL/LF = CB/BH = CB/NQ$, we have, $CL^2/LF^2 = CB^2/BH^2 = CB^2/NQ^2$; substituting and omitting terms in a^2, e^2, ae (ac-

† Lecture X, Example 5.

cording to the rules) we have,

$$\frac{f^2+2fge/r}{g^2-2fge/r} = \frac{r^2}{m^2-2ma}$$

and finally, since $m/r = g/f$, $rfma = gr^2e+gm^2e$

and $$\frac{a}{e} = \frac{g(r^2+m^2)}{rfm} = \frac{m^2+r^2}{r^2}.\dagger$$

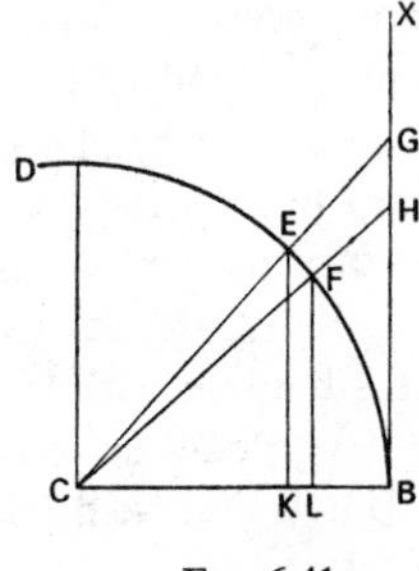

FIG. 6.41

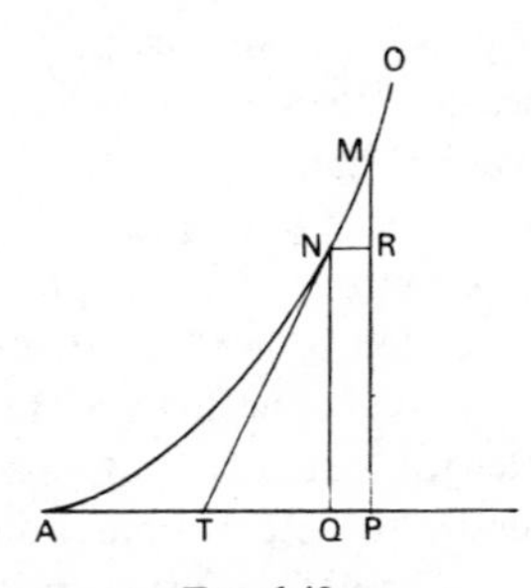

FIG. 6.42

In conclusion it is perhaps worth saying once again that Barrow's *Geometrical Lectures* should be viewed, not as an isolated study, but as the culmination of all the seventeenth-century geometrical investigations leading to the calculus. In this context the work represents the most detailed and systematic treatment of these properties of curves such as tangents, arcs, areas and so on, which, in the hands of Newton and Leibniz, led so rapidly to the invention of the calculus. By the use of modern notation, it is, of course, possible to transform the geometrical results arrived at by Barrow into standard differentials and definite integrals and Child has drawn up a formidable array of results which he obtained by so doing.‡ Moreover, Barrow was able to integrate the concepts of time and motion with those of space in the manner suggested by Torricelli, Galileo and Roberval, and thus to move nearer to Newton's fluxions.

Mathematical invention is a process of continuous change and development rather than something which takes place at a given point in time, but if it be considered necessary to draw a line between those mathematicians of

† If $y = r \tan \theta$, we have $dy/d(r\theta) = 1+\tan^2 \theta = \sec^2 \theta$.

‡ See Child, J. M., *The Geometrical Lectures of Isaac Barrow*, pp. 30–2.

the seventeenth century who "had the calculus" and those who "had not" the line would inevitably exclude Barrow on the grounds that he exhibited no calcular rules and used no specialised notation or symbolism. The claim made by Child† that Barrow privately made use of notation, rules and symbols and that he turned these over to Newton whilst preferring to publish his own work in purely geometrical language, cannot be considered seriously. Barrow was a skilful geometer, not only in the purely formal sense, but also in his intuitive appreciation, through the concepts of time and motion, of the properties of curves, tangents and areas. His approach to the study of curves was made possible by the acceptance, in his thinking processes, of Cavalierian indivisibles, and there is no evidence that he evolved, or indeed felt any need for, any kind of analytical procedure. He made a noteworthy effort to relate the entire range of results to the only satisfactory proof structure he knew, and it was something of a disappointment to him that his work failed to make the impact he felt it merited. Whether his resignation of the Lucasian Chair was in any sense connected with the failure of his book or with his recognition of Newton's superior powers, or whether it was due, quite simply, to his own desire to obtain promotion within the University, we do not know. At all events, he did no further mathematical work.

† *The Geometrical Lectures of Isaac Barrow*, pp. 28–9.

CHAPTER 7

Epilogue: Newton and Leibniz

7.0. INTRODUCTION

The concepts and techniques of the infinitesimal calculus are the result of a long line of mathematical development stretching almost unbroken from antiquity to the present day and the work of all those who contributed in one way or another should be fully acknowledged. Above all, it is fitting that, in this line of advance, the achievements of Newton in extending and unifying the range of processes and of Leibniz in connecting them with a new calculus and new analytical operations should be especially recognised.

This is not the place for any lengthy discussion of the Leibniz MSS., at present in the Lower Saxony State Library, Hanover, but, in any case, these papers have been catalogued by Bodemann† and by Rivaud,‡ and intensively studied by a succession of eminent German historians, most notably Gerhardt,§ Wieleitner and Mahnke.|| All this work has recently been finalised in the meticulous researches of Dr. J. E. Hofmann.†† Although the Leibniz

† *Die Leibniz-Handschriften der Königlichen Öffentlichen Bibliothek zu Hannover* (described by Bodemann, E.), Hanover, 1895; henceforth referred to as *Bodemann.*

‡ *Catalogue critique des manuscrits de Leibniz*, Vol. II (ed. Rivaud, A.), Poitiers, 1914–24, henceforth referred to as *Cat. crit.*).

§ Gerhardt, C. I., *Die Entdeckung der Differentialrechnung durch Leibniz*, Salzwedel, 1848. Leibniz, G. W., *Der Briefwechsel mit Mathematikern* (ed. Gerhardt, C. I.), Berlin, 1899 (henceforth referred to as *Briefwechsel*). Leibniz, G. W., *Die philosophischen Schriften* (ed. Gerhardt, C. I.), Berlin, 1875–90, 7 vols. Leibniz, G. W., *Mathematische Schriften* (ed. Gerhardt, C. I.), Berlin, 1849–63, 7 vols. Leibniz, G. W., *Sämtliche Schriften und Briefe* (ed. Preuss. Akad. d. Wiss. Berlin), Darmstadt–Leipzig, 1924– (6 vols. to date).

|| Mahnke, D., Zur Keimesgeschichte der Leibnizschen Differentialrechnung, *Sitz. d. Gesell. z. Beförd. d. Ges. Naturw. zu Marburg* 67, (1932), pp. 31–69; Neue Einblicke in die Entdeckungsgeschichte der höheren Analysis, *Abh. d. Preuss. Akad. d. Wiss.*, 1925, Phys.-math. Klasse, 1, Berlin, 1926, pp. 1–64.

†† Hofmann, J. E., Erste Versuche Leibnizens und Tschirnhausens eine algebraische Funktion zu integrieren, *Archiv. f. Geschichte d. Math. d. Naturw. u. d. Techn.* 13 (1931), pp. 277–92 (with Wieleitner, H.); Die Differenzenrechnung bei Leibniz, *Sitz. d. Preuss. Akad. d. Wiss.*, 1931, Phys.-math. Klasse, 26, Berlin, 1931, pp. 562–600 (with Wieleitner

papers, being in the nature of study exercises are not suitable for publication in full, the resources of the University of Pennsylvania Library† have ensured that virtually all of this material is available in microfilm for those who have time and patience enough to decipher it. In consequence, it is now possible to get a clear picture of the background and circumstances of the invention of the notation of the differential and integral calculus by Leibniz during the years 1673–6.

On the other hand, the development by Newton in Cambridge of the fluxional calculus and the method of infinite series during the years 1663–71 has up to the present been quite inadequately studied.‡ A beginning was made in 1850 by Edleston§ who attempted to establish the essential facts about the period of Newton's residence in Cambridge and at Woolsthorpe during the Plague years. Rigaud published one or two extracts from Newton's early papers and a general description of some others.‖ A printed catalogue of the Portsmouth Collection†† of Newtonian manuscripts, now for the most part in the University Library, Cambridge, has been available since 1888. Until recently, these papers attracted very little attention. At last, however, the initial researches of the late Professor H. W. Turnbull, followed up by the painstaking editorial work of Dr. J. F. Scott, have made it possible to embark on the publication of the whole of the voluminous Newton Correspondence.‡‡ In the meantime Dr. D. T. Whiteside who has, for some years now, devoted himself to the detailed study of the mathematical papers of Newton in the Cambridge University Library, hopes soon to publish, with

H. and Mahnke, D.); *Die Entwicklungsgeschichte der Leibnizschen Mathematik während des Aufenthaltes in Paris* (1672–6), Munich, 1949.

† A collection of thirty-five reels of microfilm of the letters and manuscripts of G. W. Leibniz, at present in the Lower Saxony State Library, Hanover: selected by Paul Schrecker on the basis of the Bodemann catalogue, and Concordance of Leibniz Microfilm with this catalogue prepared by Sterling, C. G., on behalf of the University of Pennsylvania Library, Philadelphia, 1955.

‡ See Whiteside, D. T., The expanding world of Newtonian research, *History of Science* 1 (1962), pp. 16–29. In this article Whiteside surveys the field of Newtonian research currently in progress.

§ Edleston, J., Synoptical view of Newton's life, in *The Correspondence of Sir Isaac Newton and Professor Cotes*, Cambridge, 1850.

‖ Rigaud, S. P., *Historical Essay on the First Publication of Sir Isaac Newton's Principia*, Oxford, 1838.

†† This printed catalogue was the work of a Syndicate of the University of Cambridge and included initially J. C. Adams and G. G. Stokes. See footnote 2, p. 27 in Whiteside, D. T., The expanding world, *op. cit.*

‡‡ See, Turnbull, H. W., *The Mathematical Discoveries of Newton*, London, 1945. *The Correspondence of Isaac Newton* (ed. Turnbull, H. W.), Cambridge, 1959–. Vols. I–III have now appeared. Dr. J. F. Scott has been appointed editor of Vol. IV onwards.

the assistance of Dr. M. A. Hoskin, an "exhaustive collected edition of Newton's mathematical work both printed and unprinted".[†]

Whilst any final judgement on the origins and development of Newton's mathematical work must await the publication of all this material[‡] it would seem both necessary and desirable that something should be said at this stage as to the studies which took place in Cambridge from 1663 to 1671 and in Paris from 1673 to 1676. These studies led both to the invention of the infinitesimal calculus and to the infamous Leibniz–Newton controversy which has clouded the history of mathematics for 250 years. The main purpose of this brief account which I am including here is to establish the background and source material for the early investigations of Newton and Leibniz. In no sense is it intended as a detailed analysis of the work itself.

7.1. Isaac Newton (1642–1727)

Isaac Newton was born at Woolsthorpe, near Grantham, on 25th December 1642. He was educated at the King's School, Grantham, and at Trinity College, Cambridge, where he was awarded the B.A. degree in 1665.

Although there are in existence a great many contradictory anecdotes regarding Newton's early mathematical education, or lack of it, we have little direct information available on the subject until we come to the years 1663–5, which were marked by certain significant developments.[||] The College Record shows that Newton's first tutor was Benjamin Pulleyn, subsequently Regius Professor of Greek in the University. From 1663 onwards, however, Newton must have come into increasing contact with Barrow, an established figure within the University, undoubtedly one of the most outstanding mathematicians of the day and an early member of the Royal Society. Although many stories have been told of Newton's ignorance of geometry

† Since this book went to press the first volume of the edition referred to above has appeared: *The Mathematical Papers of Isaac Newton* I (ed. Whiteside, D. T.), Cambridge, 1967. I regret that it has not been possible to make use of this scholarly and highly relevant work.

‡ It is worth noting that Hofmann, who has done so much to establish the details of the mathematical work of Leibniz, also contributed a valuable paper (based on published material) on Newton's early work in this field; see Hofmann, J. E., Studien zur Vorgeschichte des Prioritätstreites zwischen Leibniz und Newton um die Entdeckung der höheren Analysis, I: Materialien zur ersten Schaffensperiode Newtons (1665–75), *Abh. d. Preuss. Akad. d. Wiss.*, 1943, Math.-naturwiss. Kl. 2, Berlin, 1943. See also *Unpublished Scientific Papers of Isaac Newton* (ed. Hall, A. R. and Hall, M. B.), Cambridge, 1962. An initial selection of mathematical material is included in this work, which is primarily concerned with Newton's ideas on matter.

§ Whiteside, D. T., The expanding world, *op. cit.*, pp. 24–5.

|| For biographical details of Newton's life see More, L. T., *Isaac Newton*, London, 1934.

when Barrow tested him in the scholarship examination in 1664, it seems most unlikely that he did not study Euclid effectively during his first two years at Cambridge. The annotations in Newton's personal copy of Barrow's *Euclid*, a work especially written for beginners, show that this book, at any rate, was thoroughly mastered.†

In 1663 Newton read Oughtred's *Clavis*‡ and this probably represented his introduction to algebra. It was, however, from Viète§ that Newton gained his knowledge of the sophisticated and elegant techniques for handling equations of degree higher than the second. At about the same time he read the *Arithmetica*|| of Wallis and the *Geometria*†† of Descartes.

The extent of Newton's debt to Wallis is well known. He has himself described how his own early investigations in connection with infinite series arose from his study of Wallis's *Arithmetica* in the year 1664‡‡. The general binomial theorem, the quadrature of the circle and the rectangular hyperbola, were the immediate results of the interpolation procedures Newton embarked on in following up Wallis's work on the squaring of the circle.§§

During the centuries since the death of Newton the nature of the particular edition of the *Geometria* which he studied has been the subject of protracted controversy. Newton did not read French so that it must have been either the

† Barrow, I., *Euclidis elementa*, Cambridge, 1655.

‡ Oughtred, W., *Arithmeticae in numeris et speciebus institutio*, London, 1631. Newton probably read the 3rd Latin edition, *Clavis mathematicae*, Oxford, 1652.

§ Viète, F., *Opera mathematica* (ed. Schooten, F. van), Leyden, 1646. In this edition Schooten effected a substantial modernisation of Viète's notation.

|| C.U.L. (Add. 4000, fol. 14ʳ). Entry in Newton's hand dated 4th July 1699: "By consulting an Accompt of my expenses at Cambridge in ye years 1663 & 1664, I find yt in ye year 1664 a little before Christmas, I being then Senior Sophister, I bought Schootens Miscellanies, & Cartes's Geometry, (having read this geometry & Oughtred's Clavis above 1/2 a year before) and borrow'd Wallis's Works, & by consequence made these Annotations out of Schootens & Wallis, in the Winter between ye years 1664 & 1665. At which time I found the method of Infinite Series. And in Summer 1665, being forced from Cambridge by ye Plague, I computed the Area of the Hyperbola, at Boothby in Lincolnshire, to two & fifty figures, by the same method."

Add. 4000 contains notes made by Newton from the *Clavis* (on extraction of roots), from Viète (extraction of roots in affected powers, sundry geometric propositions and notes on angular sections), from Schooten (miscellaneous propositions) and various annotations from Wallis.

†† This could have been either the 1st Latin edition (1649) or the 2nd Latin edition (1659).

‡‡ C.U.L., Add. 4000, fol. 14ᵛ. See also Newton to Oldenburg (24th Oct. 1676) in *The Correspondence of Isaac Newton* II, pp. 110–29.

§§ C.U.L., Add. 3958.3, fols. 70–2; Add. 4000, fols. 18–19ᵛ. See Whiteside, D T., Newton's discovery of the General Binomial Theorem, *Math. Gaz.* 45 (1961), pp. 175–80. See also Whiteside, D. T., Henry Briggs: the Binomial Theorem anticipated, *ibid.*, pp. 9–12. A paper in which Newton formalised this theorem is printed in Vol. II of the *Correspondence* (pp. 168–70).

first (1649) or the second (1659) Latin edition, published by Schooten. The second edition, it will be recalled, contained probably the most important collection of material available at the time, including a valuable commentary by Schooten in which he discussed the tangent method of Fermat and Huygens' method of determining points of inflexion as well as Hudde's *Methodus de maximis et minimis* and Heuraet's rectification of the arc of the semi-cubical parabola.† All this material was new and much discussed in England at the time.‡

The existence in the Cambridge University Library of an edition of the *Geometria* dated 1649, and purporting to be Newton's own book,§ has often been accepted as evidence that Newton had not the advantage of using the 1659 edition. I have been unable to discover what evidence there is that this particular book ever belonged to Newton but, in any case, from a detailed study of the development, page by page, of Newton's mathematical work during the years 1664–6 there can be no conceivable doubt that he was using the 1659 edition, and that he had continuous access to all the new material made available by Schooten.

In the first stages of Newton's investigations into the properties of curved lines he relied mainly on Descartes' tangent method. He took over Descartes' characteristic symbolism (v, s) as well as the method of equal roots.|| This he incorporated with *Hudde's Rule* (as given in Schooten's *Exercitationes mathematicae*, Leyden, 1657, Book V, *Sectiones Miscellaneae*, XXV and XXVI). Thus (Add. 4000, fol. 93^{v}) we have $ae = x$, $ab = y$, $ed = a$, $sb = s$, $se = v$ (see Fig. 7.1),

$$s^2 = y^2+(v-x)^2.$$

Now, if the curve be given by

$$a^2x+a^3 = y^2x,$$

then

$$a^2x+a^3 = xs^2-x(v-x)^2 = xs^2-xv^2+2vx^2-x^3,$$

and

$$x^3-2vx^2+v^2x-s^2x+a^2x+a^3 = 0.$$

Multiplying (by Hudde's Rule)

$$2 \quad 1 \quad 0 \quad 0 \quad 0 \quad -1,$$

† See pp. 217–19, 225–6.

‡ See Rigaud, S. P., *Correspondence of Scientific Men of the 17th Century* I, p. 127; II, p. 304; I, p. 132. The controversy between Wallis and Huygens regarding the rectification of curved lines also brought this material into prominence. Wallis's account was given in *De cycloide* (1659); see pp. 223–5 above.

§ C.U.L., Adv. d. 39.1.

|| See pp. 165–6 above.

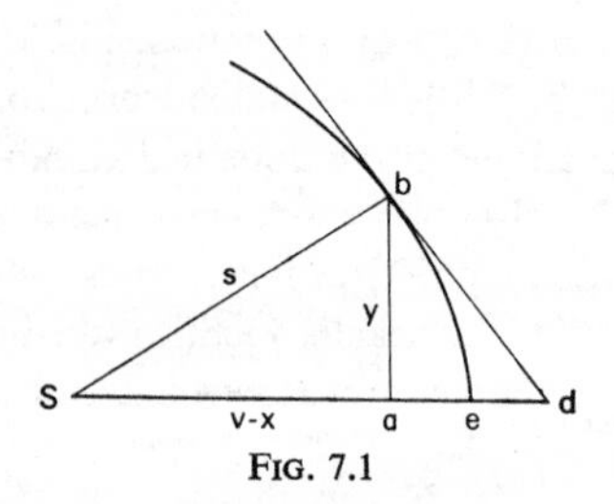

FIG. 7.1

we have

$$2x^3 - 2vx^2 - a^3 = 0, \quad \text{and} \quad v = \frac{2x^3 - a^3}{2x^2}.\dagger$$

For the hyperbola, $xy = a^2$, we have, $v = \dfrac{x^4 - a^4}{x^3}$,‡ and

$$v - x = -a^4/x^3.$$

Now, let a second curve be constructed with ordinate *og* (see Fig. 7.2) such that $og = z$ and $y/(x-v) = a/z$ ($z = -a\,dy/dx$). Thus, as *og* moves over the surface *okhg*, *ne* moves over the surface *neqm* (see Fig. 7.2), i.e.

$$\int_{x_1}^{x_2} z\,dx = a\int_{y_2}^{y_1} dy = a(y_1 - y_2), \quad \text{or} \quad \int_{x_1}^{x_2} (a^3/x^2)\,dx = a^3/x_1 - a^3/x_2.$$

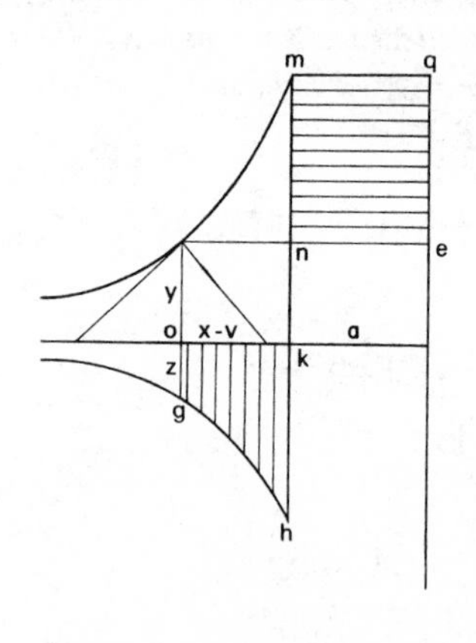

FIG. 7.2

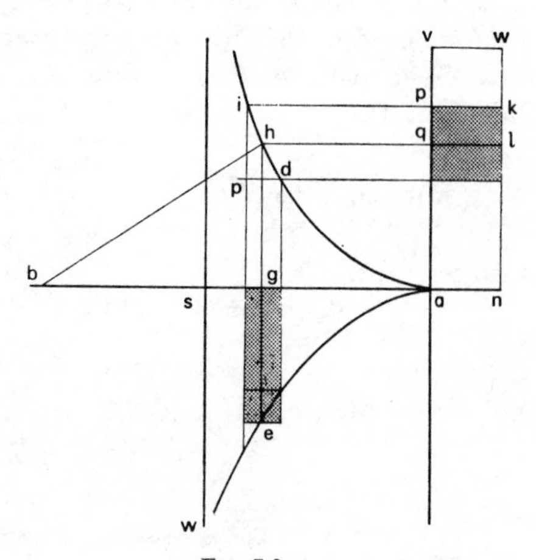

FIG. 7.3

† $v = x + y\,dy/dx = x - a^3/2x^2$.
‡ $v - x = y\,dy/dx = -a^4/x^3$.

If $yx^2 = a^3$, $y = a^3/x^2$, $z = 2a^4/x^3$, and so on. This theorem is, of course, general, and relates to the inverse nature of integration and differentiation (or tangents and quadratures). This approach was subsequently formalised by Newton† in a paper entitled "A Method whereby to square such crooked lines as may be squared". Thus, with the same construction, we have (see Fig. 7.3)

$$hg/bg = pd/ip = an/ge.$$

$$pd \, . \, ge = ip \, . \, an.$$

Hence, adding for all such rectangles "wch if they be infinite are equall to superficies $saw = vwna$".

In fact,

$$\int_{x_2}^{x_1} z \, dx = a \int_{y_1}^{y_2} dy = \Big[ay \Big]_{y_1}^{y_2}.$$

In the *Waste Book* (Add. 4004) Newton develops further his work on tangent methods, curvature and points of inflexion. The work in this book for the most part runs consecutively and sundry dates are given (1664–6). Although Hudde's tangent method‡ was not specifically given in Schooten it was a straightforward enough matter for Newton to draw his own conclusions from the available material and from his investigations.§ In September 1664 he formulated a calcular rule‖ for the determination of the subnormal (now called v), e.g. if $x^4 - y^2x^2 + a^2yx - y^4 = 0$, then

$$v = \frac{4yx^3 - 2y^3x + a^2y^2}{4y^3 - a^2x + 2yx^2} \quad (= y \, dy/dx).$$

The rule is on the same lines as *Hudde's Rule* and runs as follows: multiply each term of the equation by so many units as x has dimensions in that term, divide it by x and multiply it by y for a numerator. Again, multiply each term of the equation by so many units as y has dimensions in that term and divide it by $-y$ for a denominator.

† Given in the *Correspondence* II, pp. 164–7. (Add. 4000, fols. 134^v–135^v.)

‡ Compare the wording of Hudde's tangent rule, given in a letter to Schooten of 21st Nov. 1659, and printed in the *Journal literaire de la Haye*, 1713. See Chap. VI, pp. 217–19.

§ C.U.L., Add. 4004, fol. 8^v. The first curve which Newton investigates ($x^3 + y^3 = a\,xy$), the folium of Descartes, was given by Hudde as an example of the application of his method (to find the maximum breadth of the curve). This example appeared in Schooten's *Exercitationes mathematicae* (see p. 217 above) which Newton purchased in 1664 (see p. 256 above).

‖ $f(x, y) = 0$, $dy/dx = -f_x/f_y$.
$f_x = 4x^3 - 2y^2x + a^2y$, $f_y = -2yx^2 + a^2x - 4y^3$, $v = -yf_x/f_y = y \, dy/dx$.

The tangent problem being temporarily resolved† Newton turns to the study of curvature, which he approaches on the basis of the intersection of consecutive normals. In the first example which he carries through he finds the subnormal (v) by the above calcular rule and then applies the condition that a line through (ξ, η) with gradient $-y/v$ shall cut the curve in coincident points (using Hudde's method for equal roots).‡

In another method§ Newton applies the condition that a circle, centre (ξ, η) shall cut the curve in three coincident points.‖ In this case the result is arrived at by applying Hudde's method twice successively.††

† The small quantity "o" appears (Add. 4004, fol. 30) where Newton makes some notes on Fermat's tangent method "blotting out" identical terms, dividing the rest by o, then supposing $o = bc$ to vanish.

‡ In the notation of the calculus, if a curve be given in the form, $y = f(x)$, then the normal at (x, y) can be written, $(X-x)+(Y-y)f'(x) = 0$, where (X, Y) is a point on the normal. This intersects the curve where $(X-x)+(Y-f(x))f'(x) = 0$. If this equation has equal roots, then differentiating with respect to x, we have, $-1+(Y-y)f''(x)-(f'(x))^2 = 0$, and $Y-y = \dfrac{1+(f'(x))^2}{f''(x)}$. The point (X, Y), found in this way, is (ξ, η), the centre of curvature Add. 4004, fol. 30v).

§ Add. 4004, fol. 33. (The date is given as Feb. 1664—probably in error for 1665.)

‖ Newton had available Schooten's comments on the determination of points of inflexion by the use of three equal roots and comparison with the equation $(x-e)^3 = 0$ (given with a reference to Huygens). See *Geometria* (1659), pp. 253–5.

†† Let $(\xi-x)^2+(\eta-y)^2 = s^2$, i.e. $\xi^2+x^2-2x\xi+\eta^2+y^2-2y\eta = s^2$, and if $y = f(x)$, differentiating with respect to x, we have $2x-2\xi+2f(x)f'(x)-2\eta f'(x) = 0$, and again,

$$1+(f'(x))^2+f(x)f''(x)-\eta f''(x) = 0,$$

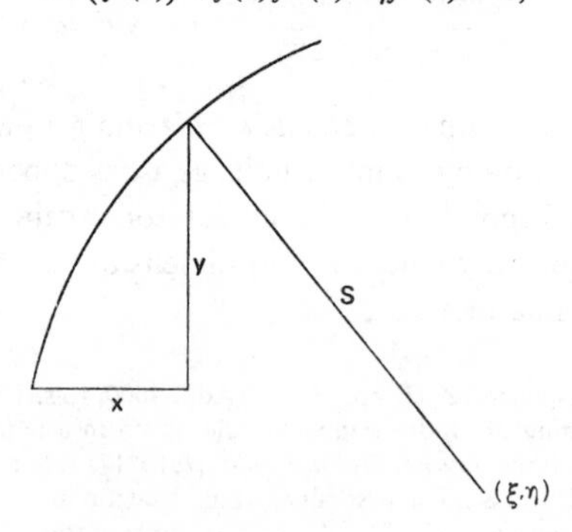

giving,

$$\eta-y = \frac{1+(f'(x))^2}{f''(x)}.$$

In both of these methods Newton determines, as an example, the centre of curvature of the parabola, $y^2 = rx$.

In May 1665 Newton returns to another variant of Descartes' tangent method,† when he forms the equation of a circle, centre $(d, 0)$, which cuts the curve, $y = f(x)$, at (x, y) and $(x+o, z)$. Hence, we have (see Fig. 7.4)

$$s^2 = v^2+y^2 = (v-o)^2+z^2,$$
$$y^2 = -2vo+o^2+z^2,$$

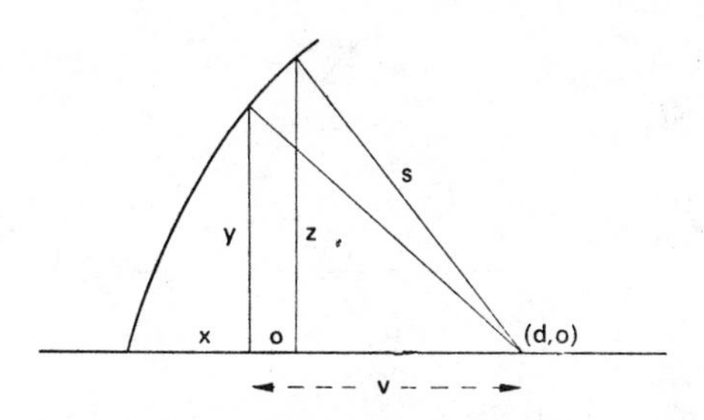

FIG. 7.4

and, letting o "vanish into nothing", v can be determined. Thus, if $y = f(x)$, $\left(f(x+o)^2\right)-\left(f(x)\right)^2 = 2vo$,

$$f(x)\frac{\left(f(x+o)-f(x)\right)}{o} = v$$

$$(\text{or } v = y\,dy/dx).$$

The extension of this method to curves of the form $f(x, y) = 0$ enables Newton to confirm the calcular rule already formulated and, at this stage, he makes the first of a variety of suggestions for symbolising the process. If, for example, the equation of a curve be written in the form, $ay^3+by^2+cy+d = 0$, where a, b, c, d are functions of x, then

$$v = -\frac{\ddot{a}y^3+\ddot{b}y^2+\ddot{c}y}{3axy^2+2byx+cx}\left[= -y\frac{f_x}{f_y}\right].$$

If $a = x^m$, $\ddot{a} = mx^m$, and so on.‡

In another paper (C.U.L., Add. 3960. 12) we find Newton tirelessly building up the differential process into a series of organised techniques (notation, A, $\mathit{A\!\!\!-}$, B, $\mathit{B\!\!\!-}$, v). Integration (i.e. given $v = f(x)$, to determine y) is tackled

† C.U.L. Add. 4004, fol. 47. Compare with Descartes' interpretation of Fermat's tangent method given in a letter to Hardy (see pp. 170–2 above).

‡ For example, if $a = x^3$, $\ddot{a} = 3x^3$, $b = x^2$, $\ddot{b} = 2x^2$.

by transforming the function to be integrated into a result previously obtained by a differential process.

The next step is to polish the method still further by using the characteristic triangle and avoiding the clumsy manipulation involved in the construc-

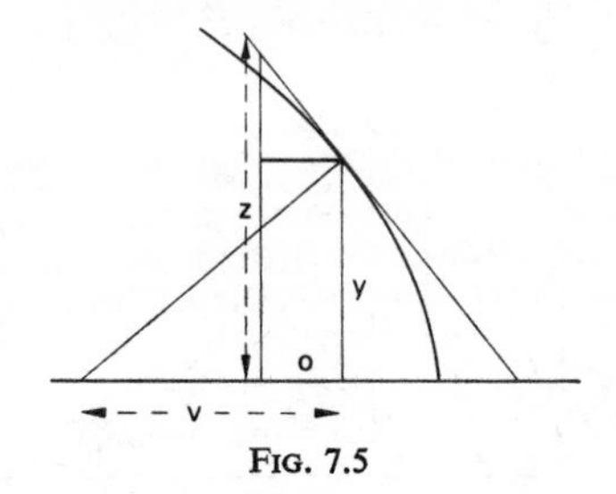

Fig. 7.5

tion of circles.† From similar triangles (see Fig. 7.5), we have

$$y/v = o/(z-y),$$

and

$$z = y+ov/y.$$

Hence, from the curve equation,

$$f(x, y) = f(x+o, y+ov/y),$$

$$0 = f_x+\frac{v}{y}f_y,$$

and, once again,

$$v = -y\frac{f_x}{f_y}.$$

Applying the same approach to the problem of curvature‡ by the direct use of small increments Newton gives the radius of curvature in the form

$$\rho = \frac{(\dot{x}^2y^2+\dot{x}^2x^2)^{3/2}}{(2\dot{x}\ \dot{x}\ \dot{x}-\dot{x}\ \dot{x}\ \dot{x}-\dot{x}\ \dot{x}\ \dot{x})xy}$$

where

$$x=f(x,y),\ \dot{x}=xf_x,\ \dot{x}=xyf_{xy}$$
$$\dot{x}=yf_y,\ \dot{x}=x^2f_{xx},\ \dot{x}=y^2f_{yy}$$

† Add. 4004, fol. 47ᵛ.
‡ Add. 4004, fol. 48ᵛ.

This rule corresponds to

$$\varrho = \frac{(f_x^2+f_y^2)^{3/2}}{2f_{xy}f_xf_y-f_{yy}f_x^2-f_{xx}f_y^2}.\dagger$$

Work now commences on the determination of tangents to mechanical curves by motions‡ and this culminates in the small tract, *To resolve problems by motion* (1666).§ This collection of material represents a first synthesis of all the work Newton had been developing over the previous two years and in it he deliberately elects to make the concept of motion the fundamental basis for the determination of tangents, curvature, points of inflexion, quadratures and arcs and also a means of relating these processes effectively to each other.

The differential process is defined as follows: let $f(x, y, z) = 0$ and let p, q, r be the motions of x, y and z respectively (i.e. $dx/dt, dy/dt, dz/dt$); then, if o be the infinitely little time in which each motion takes place, we have

$$f(x, y, z) = f(x+po, y+qo, z+ro).$$

Thus

$$0 = pf_x+qf_y+rf_z,$$

where f_x, f_y, f_z are obtained by ordering the terms in $f(x, y, z)$ according to the powers of x, y, z, respectively, multiplying each term by the corresponding exponent of x, and dividing by x, y and z, respectively, i.e. f_x, f_y, f_z, correspond to $\partial f/\partial x, \partial f/\partial y, \partial f/\partial z$.

Newton then turns to the consideration of problems of the form, given $f(p/q, x) = 0$, to find $y = F(x)$, i.e. the inverse tangent problem (or differential equations). If, for example, $q/p = ax^{m/n}$, then

$$y = \frac{an}{m+n}x^{(m+n)/n}.$$

In this case the only difficulty arises when $m/n = -1$: y is then the logarithm of a number and this case will be dealt with separately. The division method follows and the symbol □ is used to represent a quadrature of this type.

† Regarding what D. T. Whiteside calls Newton's "dottage" (i.e. the use of pricked letters to denote differential operations on $f(x, y)$) he experimented a great deal with a variety of forms.

‡ Add. 4004 (8th Nov. 1665), fols. 50^v, 51^r ff. Schooten discusses the determination of the tangents to mechanical curves by "motions", referring to Torricelli, Galileo, Roberval (*Geometria*, pp. 264–70).

§ Add. 3958.3. Another draft: *De Solutione problematum per motum.*

Further examples follow in which algebraic transformations are used to effect quadratures, particularly those expressed in terms of surd quantities.

The fundamental operations having been established (differentiation and integration) these are then applied to certain basic problems:

1. The finding of tangents to curved lines; i.e. if $f(x, y) = 0$, then the subtangent, $t = py/q$, where $pf_x + qf_y = 0$.
2. Determination of the radius of curvature by an ingenious method depending on rotation about the centre of curvature, i.e. the point of the normal in *least* motion.
3. Determination of points of inflexion. The radius of curvature is, in this case, infinite and we have

$$2\ddot{x}\;\dot{x}\;\dot{x} - \dddot{x}\;\dot{x}\;\dot{x} - \dddot{x}\;\ddot{x}\;\ddot{x} = 0$$

4. Finding the points at which lines are most, or least, crooked. At such points the radius of curvature is a maximum or a minimum.
5. To find the curves whose areas (y) are expressed by a given relation, i.e. given $f(x, y) = 0$, to find the ordinate. The surfaces, *abc* and *abde* (see Fig. 7.6) are considered as described by the moving ordinate, *ebc*. Hence

$$\frac{\text{increase of } abc}{\text{increase of } abde} = \frac{bc}{be} = \frac{q}{p}.$$

But,

$$q/p = -\frac{xy}{xx} \quad (= -f_x/f_y)$$

and

$$q = -pf_x/f_y.$$

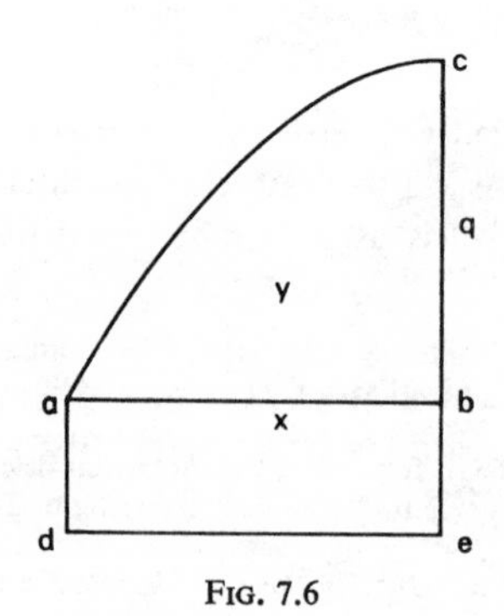

FIG. 7.6

Here, p is recognised as an independent variable and can, accordingly, be put equal to unity.†

This, of course, contains the germ of the idea of *moment*, which Newton was to develop in the *De analysi* (1669).

6. To rectify curved lines. As the normal rotates about the centre of curvature each point on the normal traces out the arc of a curve (see Fig. 7.7). These arcs are: (i) perpendicular to the normal, (ii) parallel to each other. A curve $\beta\delta m$ can be drawn which the normal always touches. As each point of the normal moves perpendicular to the curve $\beta\delta m$ there will be no sliding. Hence, the line γm, will *measure* the curve $\beta\delta m$.

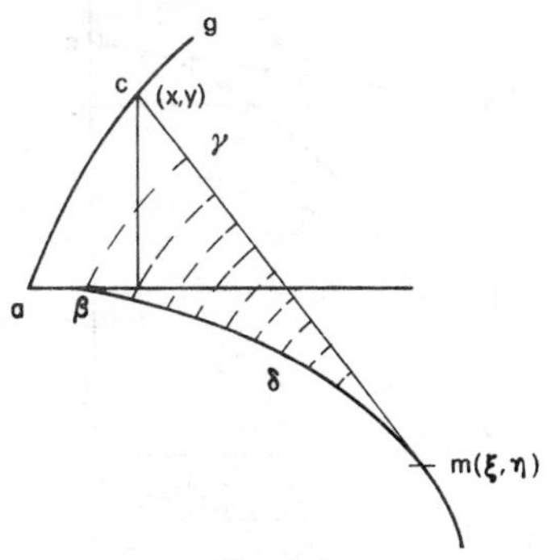

FIG. 7.7

For example, if acg is the parabola, $y^2 = rx$, from previous work,

$$cm = \varrho = \frac{(r+4x)^{3/2}}{2\sqrt{r}}.$$

The position of $m(\xi, \eta)$ is given by $\xi = 3x+r/2$, $\eta = -4y^3/r^2$; hence, $\beta\delta m$ is the curve,

$$(\xi-r/2)^3 = \tfrac{27}{16}\eta^2 r \text{ (the evolute of the normal).}$$

If $x = 0$, $y = 0$, $a\beta = r/2$; and

$$\beta\delta m = \frac{(r+4x)^{3/2}}{2\sqrt{r}} - \frac{r}{2}$$

and the semi-cubical parabola is rectified.

† For example, let $9y^2-4rx^3 = 0$, $q = -\dfrac{\varkappa y}{\varkappa x} = \dfrac{12rx^3y}{18y^2x} = \dfrac{2}{3}rx^2/y = \sqrt{rx}$, $q^2 = rx$, and the curve is a parabola. Conversely, if $y^2 = rx$, $\int_0^x y\,dx = \int_0^x \sqrt{rx}\,dx = \dfrac{2}{3}\sqrt{rx^3}$, $9A^2-4rx^3 = 0$.

Further, in a number of fascinating examples Newton explores, by means of a differential triangle representing the *motions* of a point on the curve, all the essential relations between tangent, area and arc.

For example, if x, y, z represent abscissa, ordinate and arc, and, if p, q, r be their respective motions, then, if $f(z, x) = 0$, we have $pf_x + rf_z = 0$, and $p/r = f_z/f_x$. From the figure (see Fig. 7.8), we have $c\partial/r = \partial e/p = c\epsilon/q$, and,

$$\frac{\sqrt{(r^2-p^2)}}{p} = \frac{q}{p}.$$

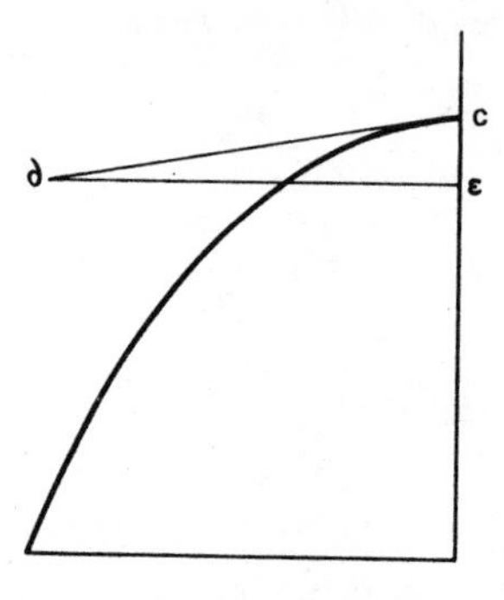

FIG. 7.8

Thus, from r/p (ds/dx) we can find q/p (dy/dx) and hence (by integration) y, i.e. given $s = f(x)$, $ds/dx = f'(x)$, $dy/dx = \sqrt{((ds/dx)^2-1)}$, and $y = \int_0^x \sqrt{((f'(x))^2-1)}\, dx$.

In this brief account of the development of Newton's work on curves during the years 1663–6, I have tried particularly to indicate the extent to which he was drawing for his initial ideas on material from Schooten and from Wallis. By 1666 he had unified his work and the synthesis he had effected was characteristically his own. The basic principles having been established we find him concerned, at each stage, to develop the necessary techniques on an effective working basis and there are innumerable examples of pages in which he has drawn up tables of results.

As for notation, although by 1666 he had tried out a variety of forms, he ultimately made more use of the (x, y, z, p, q, r) notation than of any of the others. No particular symbol was required for the integral, although he used occasionally a simple rectangle $\Box$. This was, of course, in fairly general use in the seventeenth century to denote a quadrature.

In 1668 the publication of Mercator's *Logarithmotechnia*,† which contained the quadrature of the hyperbola in the form

$$\int_0^x \frac{dt}{1+t} = x - x^2/2 + x^3/3 - x^4/4 \ldots = \log(1+x),$$

excited much interest. It was clear that Mercator had taken a considerable step forward and there were rumours that he was on the point of bringing forward "series on the circle".‡ In the autumn of 1668 James Gregory gave, in his *Exercitationes geometricae*, a rigorous geometrical demonstration of the quadrature of the hyperbola and also presented the series in an improved form,

$$\int_{-x}^{+x} \frac{dt}{a+t} = \int_{-x}^{+x} \frac{a\,dt}{a^2-t^2} = \log \frac{a+x}{a-x}.$$

When news of this publication reached Cambridge Newton apparently felt that he had lost the priority in the method of division and in the quadrature of the hyperbola. Nevertheless, perhaps spurred on by the fear that Mercator was about to publish further series on the circle, Newton decided to put forward his own results, and, in 1669, he submitted his *Analysis per aequationes numero terminorum infinitas*, through Barrow to Collins at the Royal Society.§ For Newton this paper represented the introduction to a powerful general method which he went on to develop in a further treatise.

The *De analysi* contained the foundations for Newton's method of infinite series. These are handled by straightforward operations in which the basic processes are division and root extraction: the binomial theorem is neither stated nor used. The methods of interpolation which Newton had learnt from Wallis, and which he had already put to so many exciting uses, are scarcely touched on in this tract.

Although many of the ideas and concepts which Newton developed in detail in the treatise on Fluxions|| emerge for the first time in the concept of *Moment*, the words *fluxion* and *fluent* do not appear, nor is there any use of pricked-letter notation. Nevertheless, perhaps the most interesting advance lies in the application of the idea of motion in the determination of surfaces,

† Mercator, N., *Logarithmotechnia*, London, 1668.

‡ Collins to Gregory, 15th Mar. 1669: *James Gregory Tercentenary Memorial Volume*, p. 71.

§ *De analysi per aequationes numero terminorum infinitas*, printed in *Commercium Epistolicum D. Johannis Collins et aliorum de analysi promota*, London, 1712, pp. 3–20.

|| C.U.L., Add. 3960.14 (1671).

volumes and arcs. The Wallisian formula, the equivalent of $\int_0^1 x^n\,dx = 1/(n+1)$, was presented by Wallis himself as an inductive process but was derived initially through summation. Indeed, it would be true to say that, until Newton, integration techniques had rested ultimately on some process through which elemental triangles or rectangles were added together. This is evidenced in the published work of Wallis, Neile, Heuraet, Mercator, Gregory and many others. By considering the moving ordinate as the *moment* of the surface Newton established the essential generality and inverse nature of the operations now known as differentiation and integration.

The *De analysi* contains no tangent method and no curvature; we know, however, that Newton had already developed a fully operational rule for finding the tangents to algebraic curves and this was the rule he quoted in the well-known tangent letter of 1672.†

Through the study of the finest mathematical work of the time Newton had been led to make important new syntheses. In order to develop these fully he had acquired a mastery of analytical techniques unsurpassed in the seventeenth century. He was thereby enabled to derive important results by processes which, compared with the laborious work of his contemporaries, were simple and general. Newton thought analytically in the modern sense and he thereby gained an enormous advantage over his many distinguished contemporaries. The Herculean labours of Barrow and of Gregory inspire awe and admiration but demand long and exacting study before recognisable results emerge. Newton, on the other hand, obtained easily and in greater generality results which they could only reach after endless difficult and complicated geometrical transformations.

7.2. Gottfried Wilhelm Leibniz (1646–1716)

Although, according to his own account, Leibniz always loved mathematics, he had little opportunity to study it in any depth, either at school or at the University.‡ His initial university studies were primarily in the fields of

† Newton to Collins, 10th Dec. 1672: *Correspondence* I, pp. 247–8.

‡ *Historia et origo calculi differentialis (1714)*, in Leibniz, G. W., *Mathematische Schriften* (ed. Gerhardt, C. I.), Berlin–Halle (1849–63), 7 vols.; reprinted, Hildesheim, 1962, V, pp. 392–410. See also draft of postscript of letter to James Bernoulli (1703) in *Mathematische Schriften* III, pp. 66–73. The evidence in support of these statements lies, however, mainly in the Leibniz MSS., which show that, even in later years, Leibniz never achieved any great competence in elementary processes. A simple geometric demonstration, or a lengthy piece of algebra, alike lead to frequent errors. Leibniz never succeeded in gaining any great proficiency in any techniques other than those which he himself had forged.

philosophy, logic and law. In his doctoral dissertation† (1666), which was already directed towards his lifelong project of a universal characteristic, or an algebra of thought, he displays a certain superficial knowledge of mathematical literature‡ and draws on examples from elementary number theory (permutations and combinations).§ At that time he appears to have been entirely ignorant of the more recent mathematical works and quotes Seth Ward, instead of John Wallis, as the author of the *Arithmetica infinitorum*.‖ In Nuremberg (1667) Leibniz studied the writings of the alchemists, became associated with the Rosicrucians, and ultimately, through Boineburg, gained an introduction to the Elector of Mainz, whose service he subsequently entered. During this period at Nuremberg, Leibniz recalls that he saw copies of two mathematical works, Léotaud's *Examen circuli quadraturae* and Cavalieri's *Geometria indivisibilibus*.†† His multifarious activities must have rendered it impossible for him to do much more than glance at these works, for when he wrote the *Hypothesis physica nova*‡‡ (1671) his views on indivisibles were still extremely vague. He drew largely on the *Elementa philosophiae* of Thomas Hobbes§§ and he certainly had no knowledge whatsoever of the uses of indivisible techniques in mathematics.

In the spring of 1672 Leibniz went to Paris as a man of the world, a diplomat and a politician, but also a scholar with wide interests in natural philosophy, theology and law as well as mathematics. He met everyone, Jansenists, Jesuits, courtiers, statesmen, philosophers and scientists, and it was not until the autumn that he found time to get in touch with Huygens, then at the height of his powers and the leading mathematician and natural philosopher in Europe.‖‖ Through Huygens came the impulse which led Leibniz to embark on his first mathematical investigations and thereafter to pursue mathematics

† *Disputatio arithmetica de complexionibus*, later printed as *Dissertatio de arte combinatoria: Mathematische Schriften* V, pp. 7–79.

‡ Amongst the works briefly mentioned are the *Geometria* of Descartes, Schooten's *Principia matheseos universalis*, Barrow's *Euclid*. A great many examples are taken directly from Schwentor–Harsdörffer's *Erquickstunden*.

§ Tables of binomial coefficients: see *Mathematische Schriften* V, p. 17.

‖ *Ibid.* V, pp. 18–19.

†† *Historia et origo: Mathematische Schriften* V, p. 398.

‡‡ *Mathematische Schriften* VI, pp. 17–59. A copy of the *Philosophical Transactions* having come into the hands of Leibniz whilst he was in Nuremberg, he dedicated this tract to the Royal Society and sent a copy to London. It was discussed at a meeting of the Society on 23rd Mar. 1671 and given into the hands of Boyle, Wallis, Wren and Hooke to read and pass judgement on. See Birch, T., *The History of the Royal Society of London* (4 vols.), London, 1756–7, II, p. 475. This was Leibniz' first contact with the Royal Society.

§§ Hobbes, T., *Elementa philosophiae*, London, 1655–8.

‖‖ Probably October 1672.

in a fever of haste and mounting enthusiasm for the whole of his time in Paris.[†]

Leibniz had prepared the ground carefully for his introduction to the learned members of the French Academy and the Royal Society of London[‡] and he had developed a calculating machine which he subsequently exhibited both in Paris and in London. He hoped rapidly to gain acceptance in these circles and was ambitious to play a leading part in the development of mathematics and natural philosophy. At this stage he appears to have been completely unaware of the extent of his own ignorance, both of contemporary mathematical literature and of techniques and, in consequence, he over-estimated the possibilities open to him. It was only through the good-humoured guidance afforded him through his continued association with Huygens that he was able, at least to some extent, to make good these deficiencies.

The first successes obtained by Leibniz in Paris were obtained by means of "difference" series and he has himself described how he came to develop the concepts of differences, differences of differences and so forth, which came to play such a fundamental role in all his mathematical thinking.

In logic, he asserted that the ultimate analysis of truths reduced to two things, definitions and identical truths.[§] In order to demonstrate that profitable results could emerge through the consideration of identical truths alone, he developed some of the properties of elementary number series.[‖] Thus, starting from $A = A$, or $A-A = 0$, we have

$$A-A+B-B+C-C+D-D+E-E = 0,$$

and $$A-(A-B)-(B-C)-(C-D)-(D-E)-E = 0.$$

Putting $$A-B = L, \quad B-C = M, \quad C-D = N, \quad D-E = 0,$$

we have $$A-L-M-N-0-E = 0$$

and $$L+M+N+0 = A-E.$$

Thus, *the sum of the differences equals the difference of the first and the last terms.*

† See Hofmann, J. E., *Die Entwicklungsgeschichte der Leibnizschen Mathematik, op. cit.* For the whole of this section I am deeply indebted to Dr. Hofmann, not only for the printed word, but also for the invaluable help and guidance he has so freely offered me in personal correspondence and in private conversation.

‡ The first part of Leibniz' treatise, *Hypothesis physica nova,* was dedicated (and despatched) to the Royal Society, the second part, *Theoria motus abstracti,* to the French Academy. *Mathematische Schriften* VI, pp. 17–59 and 61–80.

§ See Körner, S., *The Philosophy of Mathematics*, London, 1960, pp. 21–5.

‖ *Historia et origo: Mathematische Schriften* V, pp. 395–6.

For the series,	0		1		4		9		16		25
with differences		1		3		5		7		9,	

we have

$$1+3+5+7+9 = 25-0,$$

and so on for any number of terms.

Now, if in a series, the terms, $A, B, C, D, \ldots$, diminish to zero, the possibility clearly arises of summing an infinite series by the same means. Leibniz was thus able to solve a problem given him by Huygens,† to find the sum of the infinite series, $1/1+1/3+1/6+1/10+\ldots$, i.e. the series whose terms are the reciprocals of the triangular numbers.‡

Briefly, if the terms are	1		1/2		1/3		1/4 ...
and the differences		1/2		1/6		1/12 ...	

we have

$$0 = 1-1+1/2-1/2+1/3-1/3+1/4-1/4+\ldots$$
$$0 = 1-(1-1/2)-(1/2-1/3)-(1/3-1/4)-(\ldots)$$
$$1 = 1/2+1/6+1/12+1/20+\ldots$$
$$2 = 1+1/3+1/6+1/10+\ldots$$

At the same time Leibniz was able to integrate this method with the summation of geometric series,§ thus,

$$0 = 1-(1-1/2)-(1/2-1/4)-(1/4-1/8)-\ldots$$
$$1 = 1/2+1/4+1/8+1/16+\ldots$$
$$0 = 1-(1-1/3)-(1/3-1/9)-(1/9-1/27)\ldots$$
$$1 = 2/3+2/9+2/27+2/81+\ldots$$

Immensely pleased with these methods, Leibniz assembled them in a variety of forms, finally evolving the idea of the harmonic triangle (Table 1).

† Autumn 1672. Huygens himself had explored this problem in discussions with Hudde (1665) on probability in games of chance (see *Oeuvres* XIV, pp. 144–52.

‡ See *De summis serierum fractionum*: Bodemann XXXV. 3B. 10, fols. 2–3. See also *Leibniz for Galloys* (end of 1672): Bodemann XXXV. 3A. 32, fols. 1–10; *Accessio ad arithmeticam infinitorum*, published in Leibniz, G. W., *Sämtliche Schriften und Briefe* II, 1, pp. 222–9.

§ Huygens probably recommended Leibniz to read Saint-Vincent and Wallis. It is certain that he examined the relevant sections in the *Opus geometricum* but even so his approach is typically his own. He may have looked into Wallis's *Arithmetica* but he clearly got no detailed knowledge from it. The idea of tabulating numerical differences he already had. At this stage, Leibniz had neither time, patience, nor understanding enough to study any of these mathematical works in detail.

TABLE 1

(1) Units.	(2) Natural nos.	(3) Triangular nos.	(4) Pyramidal nos.	(5) Triangular–Triangular nos.	(6) Triangular–Pyramidal nos.
1/1	1/1	1/1	1/1	1/1	1/1
1/1	1/2	1/3	1/4	1/5	1/6
1/1	1/3	1/6	1/10	1/15	1/21
1/1	1/4	1/10	1/20	1/35	1/56
1/1	1/5	1/15	1/35	1/70	1/126
1/1	1/6	1/21	1/56	1/126	1/252
0/0	1/0	2/1	3/2	4/3	5/4

The sums, for the columns (3), (4), (5), ..., were obtained directly: the entries in columns (1) and (2), arrived at by a leftward extension of the table, bear a peculiar appearance but need not necessarily imply a misconception on the part of Leibniz. Everything depends on the interpretation he gave to the quotients, 0/0, 1/0.†

In the spring of 1673‡ Leibniz paid his first visit to London. Although the primary purpose of the visit was diplomatic and Leibniz was consequently committed to the fulfilment of many formal duties, he naturally took the opportunity to make the personal acquaintance of Oldenburg, with whom he had corresponded since 1670. He attended two meetings of the Royal Society, at the first of which he showed his calculating machine.§ On the second occasion he heard Sluse's tangent letter read.‖ These formal meetings were followed by informal private visiting with some members of the Royal Society. Moray introduced Leibniz to Morland,†† who had similar interests in calculating machines, and it was at a meeting at Boyle's house that Leibniz found the opportunity to discuss his mathematical activities with Pell.‡‡ Leibniz spoke of his general method of summing series by differences and

† See Hofmann, J. E., *Die Entwicklungsgeschichte*, pp. 12–13.

‡ 14th Jan. 1673. Schönborn and Leibniz arrived in London.

§ 22nd Jan. 1673. See Birch, T., *The History of the Royal Society of London* III, p. 73.

‖ 29th Jan. 1673. The letter was sent by Sluse to Oldenburg (17th Jan. 1673 (N.S.)) and printed in the *Phil. Trans.* 7, pp. 5143–7 (20th Jan. 1673). Published in Leibniz, G. W., *Briefwechsel*, pp. 232–3.

†† See Oldenburg to Leibniz, 30th Jan. 1673. Given in *Briefwechsel*, pp. 73–4.

‡‡ *Ibid.* Leibniz to Oldenburg, 3rd Feb. 1673: pp. 74–78.

Pell replied that this material was all available in a contemporary work of Mouton, published in France, and given as a result of Regnauld.†

Leibniz, distressed by the situation, and wholly ignorant of the literature in this field, either in England or elsewhere, consulted the book in the library of the Royal Society and found Pell's statement substantially correct. Anxious to avoid a charge of plagiarism he drafted a hasty explanatory letter which he submitted through Oldenburg to the Royal Society. This letter was not regarded as particularly satisfactory. Leibniz attended no further meetings of the Society and at subsequent meetings Hooke showed himself extremely critical of the calculating machine and expressed himself about it in no uncertain terms. Soon after, the diplomatic purpose of the visit having failed, Leibniz and Schönborn left hurriedly for Paris.

On 19th February, at a further meeting of the Society, Leibniz was formally accepted as a member and thereafter benefited by regular correspondence with the Society through Collins.

Although the visit had brought with it a certain measure of success there can be no question that the impression made by Leibniz in London, fostered particularly by Hooke and by Pell, was that of an over-ambitious amateur, promising more than he was able to achieve and not above using the ideas of others to further his own ends. On this occasion, Leibniz did not, apparently, have the opportunity of meeting Collins, mathematical correspondent of the Royal Society, who alone could, at this time, have supplied information on the development of the mathematical work of Newton and Gregory on infinite series. The notes which Leibniz made relating to mathematical and scientific matters during this visit do not suggest that he gained any insight into the nature of the important mathematical work which was going forward in Cambridge and in St. Andrews at that time. However, his attention was drawn, through discussion, probably centred round Sluse's letter and Barrow's book, to the tangent problem. He returned to Paris taking with him a copy of the *Philosophical Transactions* for Huygens: we know also that he purchased Mercator's *Logarithmotechnia* and Barrow's works (the *Optical* and *Geometrical* Lectures in one volume). Much has been made of this latter purchase‡ by those seeking to establish Barrow as the inventor of the calculus and to represent the labours of Newton and Leibniz merely as a translation of Barrow's work into algebraic form with, in the case of Leibniz especially, the development of new notation.

† Mouton, G., *Observationes diametrorum solis et lunae apparentium*, Lyons, 1670.

‡ Child, J. M., *The Early Mathematical Manuscripts of Leibniz*, Chicago, 1920. See also Barrow, I., *The Geometrical Lectures* (ed. Child, J. M.).

After his return to Paris, Leibniz received fairly regular letters from Collins and, through these, he became more fully aware of the extent of his own ignorance and of the necessity for further study. News of developments in England was, at any rate, at first, given only in the most general terms. Collins was quite unselective in the information he provided, partly because he had no clear idea of Leibniz' interests and also because he was, himself, an uncertain judge of the merits of the material he was handling.

Once again in contact with Huygens, Leibniz was presented with a copy of the newly published *Horologium oscillatorium*. The contents of this work were well beyond Leibniz' comprehension[†] but when he asked some naïve questions Huygens gave sound advice on reading. Through the spring and summer of 1673 Leibniz read Pascal, Fabri, Saint-Vincent, Descartes and Sluse.[‡] Although the notes he made show that he was studying seriously he never seems to have aimed at mastery except on the odd occasion when the ideas struck a chord which harmonised with his own thinking. He used the books as an inspiration and a starting point for his own enquiries and made no effort to follow through any detailed range of techniques.

According to his own account, repeated many times in later life, it was from a reading of Pascal[§] that Leibniz gained his first important insight which enabled him, in the summer of 1673, to develop a general transformation through which he expressed the area of a circular quadrant as an infinite series. Pascal's work probably made easier reading for a beginner than most of the other published works on infinitesimal methods available at this time and, from the *Letters of Dettonville*, Leibniz speedily learnt of the use in geometry of the principle of moments and its application in the determination of centres of gravity, solids of revolution and so forth.[||] Pascal's simple, triangular and pyramidal sums were well understood by Leibniz and interpreted in terms of his own developing ideas on differences, sums and sums of sums.

Interested particularly in the concept of moment Leibniz noted that the

[†] Leibniz to James Bernoulli, April 1703 (draft of an intended postscript): *Mathematische Schriften* III, pp. 66–73.

[‡] *Cat. crit.* II: Ref. nos. 497, 500–1, 503–5, 510–64. All these papers contain notes on the works he was reading. Nos. 544–6 refer to Pascal.

[§] Bodemann XXXV. 15. 1, fols. 18–23. *Ex d'Ettenville seu Pascalii geometricis, excerpta cum additamentis.* See also Gerhardt, C. I., Leibniz und Pascal, *Sitz. d. König. Preuss. Akad. d. Wiss. z. Berlin*, 1891, pp. 1053–68. Translated and criticised by Child, J. M., in *The Early Mathematical Manuscripts of Leibniz*, pp. 196–227.

At the point at which these notes were made Leibniz was prepared to speak of sums and differences, sums of sums, and so on, but used no notation, either algebraic or operational.

[||] See Chap. VI, pp. 200–1.

theorem in which Pascal established the relation $\int_0^{a\pi/2} y\,ds = a\int_0^a dx = a^2$ for the quadrant of a circle (Fig. 7.9) was not, in fact, restricted to the circle but applied in general. Thus, if n be the normal (Fig. 7.10), we have $\int_{s_1}^{s_2} y\,ds = \int_{x_1}^{x_2} n\,dx$. This result was, of course, far from new,[†] but the satisfaction derived from the achievement encouraged Leibniz to go further. Seizing on the general concept of the division of a plane surface into elements, he explored the use of

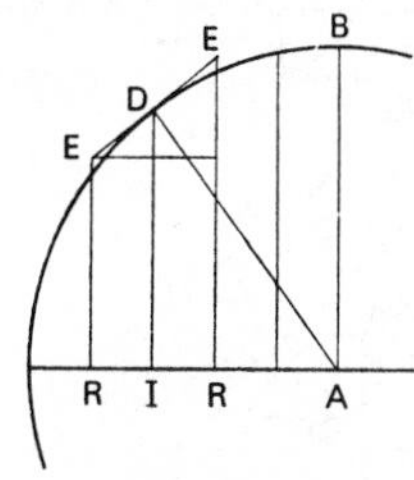

FIG. 7.9.

Pascal uses $\sum DI.EE = AB\sum RR$,

i.e. $\int_0^{a\pi/2} y\,ds = a\int_0^a dx.$

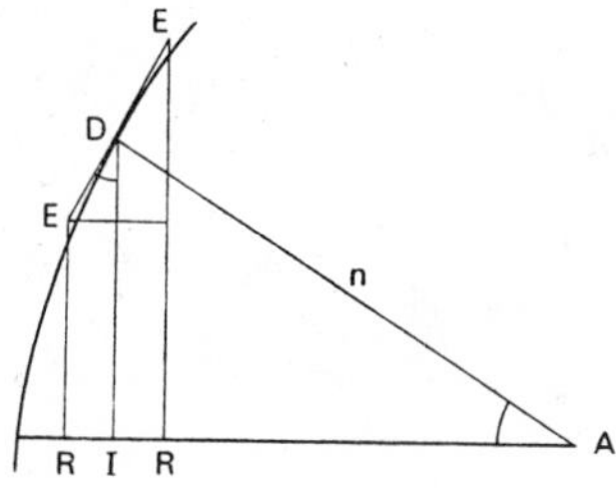

FIG. 7.10.

In general $\sum DI.EE = \sum DA.RR$

or $\int_{s_1}^{s_2} y\,ds = \int_{x_1}^{x_2} n\,dx.$

triangular elements instead of the usual trapezoidal forms. This led to the famous *transmutation* of which he gave so many accounts over the years.

Briefly, if OPQ be the arc of a given curve (see Fig. 7.11) and if PQ be an indefinitely small element of the arc (or tangent), we have, segment $OPQ = \sum\triangle OPQ$. Now, let QP be produced to cut the axis OY in T and let OX be drawn perpendicular to $PT(=p)$. The triangle OTX, so constructed, is similar to the differential triangle, PQR, and we have, $OX/TO = PR/PQ$; $OX.PQ = PR.TO$. But, $OX.PQ = 2\triangle POQ$, hence, $\sum OX.PQ = 2\sum\triangle OPQ = 2$ segment $OPQ = \sum PR.TO$. In general, however, segment $OPQ = \sum PA.AB - \frac{1}{2}OB.BQ$, so that $\sum TO.PR = 2\sum PA.AB - OB.BQ$.

In the notation of the calculus,

$$\int_0^x t\,dx = 2\int_0^x y\,dx - xy \qquad (\text{where } t = y - x\,dy/dx).$$

† Huygens, himself, had used this relation to obtain the surface area of solids of revolution: see *Oeuvres* XIV, p. 234.

The importance of this approach lay in linking the problem of quadrature with that of tangent construction. Thus, for the higher parabolas, $(y/a)^m = (x/b)^n$, we have, by the well-known tangent construction,

$$dy/dx = ny/mx, \quad t = y - ny/m = \left(\frac{m-n}{m}\right) y.$$

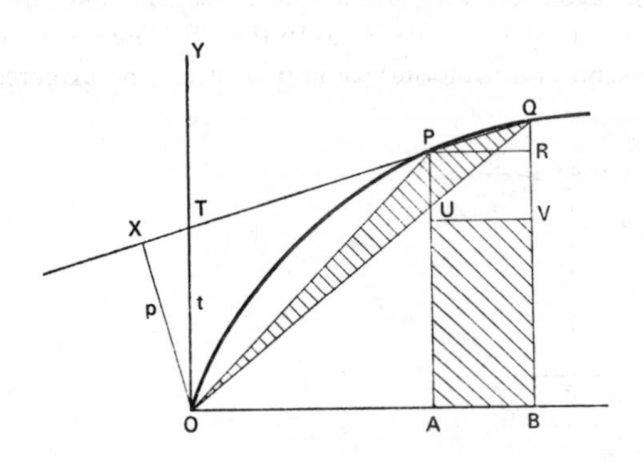

FIG. 7.11

Hence,
$$\int_0^x t\,dx = \frac{m-n}{m} \int_0^x y\,dx = 2 \int_0^x y\,dx - xy,$$

and
$$\int_0^x y\,dx = \frac{m}{m+n} xy.$$

Similar results apply for the hyperbolas $(y/a)^m (x/b)^n = 1$. For the circle we have, from geometrical considerations (see Fig. 7.12),

$$t/a = y/(2a-x) = x/y = \sqrt{(x/(2a-x))},$$
$$t^2/a^2 = x/(2a-x),$$
$$x = 2at^2/(a^2+t^2).$$

Now, applying the transformation directly (see Fig. 7.12), we have

circular segment $OQP = \frac{1}{2}$ shaded area OSD,
circular sector $OQPC = \frac{1}{2}$ shaded area $OSD + \frac{1}{2} OC.PD$,

thus circular sector $OQPC = \frac{1}{2} \int_0^x t\,dx + \frac{1}{2} ay.$

But,
$$\int_0^x t\,dx = tx - \int_0^t x\,dt,$$

hence,

$$\text{circular sector } OQPC = \frac{1}{2}(tx+ay) - \frac{1}{2}\int_0^t x\,dt$$

$$= at - a\int_0^t \frac{t^2}{a^2+t^2}\,dt.$$

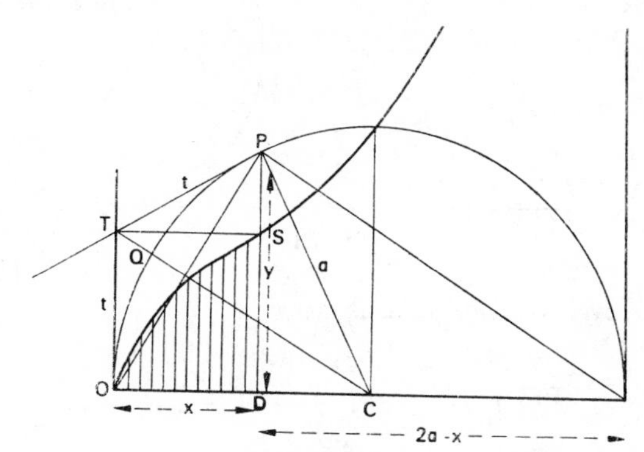

FIG. 7.12

From Mercator and from Wallis, Leibniz had learnt how to deal with this by long division and term by term integration, so that the result is readily obtainable in the form of an infinite series, i.e.

$$\text{circular sector} = a(t - t^3/3a^2 + t^5/5a^4 - t^7/7a^6 + \ldots).$$

For $t = a = 1$ we have

$$\pi/4 = 1 - 1/3 + 1/5 - 1/7 + \ldots$$

This well-known series has been associated with the name of Leibniz ever since, except in this country where it is known as Gregory's series. Actually James Gregory developed the series in February 1671, as part of a general expansion (now known as a Taylor expansion).†

For the cycloid (see Fig. 7.13) segment, we have, $y = u+s$, $u^2 = x(2a-x)$, and, from similar triangles,

$$\frac{\delta x}{\delta u} = \frac{u}{a-x}, \quad \frac{\delta x}{\delta s} = \frac{u}{a}, \quad \text{thus} \quad \frac{\delta x}{u} = \frac{\delta u}{a-x} = \frac{\delta s}{a} = \frac{\delta u + \delta s}{2a-x}.$$

† *Gregory Memorial Volume*, pp. 168–72.

Now $$t = y-x\,dy/dx = (u+s)-\frac{x(2a-x)}{u} = s.$$

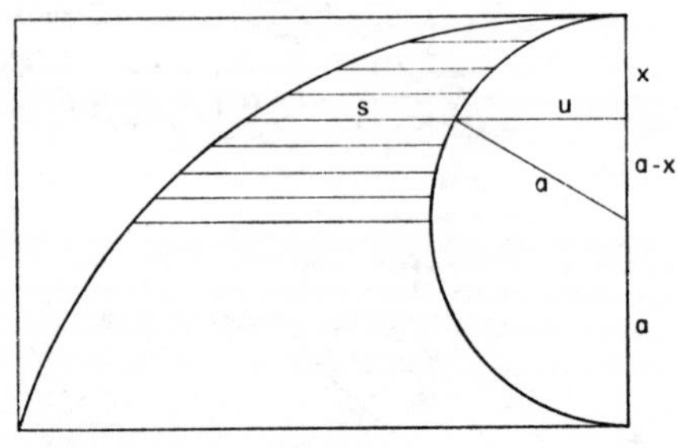

FIG. 7.13

Hence, for the cycloid segment, we have

$$\int_0^x s\,dx = \int_0^x t\,dx = 2\int_0^x y\,dx-xy$$
$$= 2\int_0^x (u+s)\,dx-x(u+s)$$
$$= xu+xs-2\int_0^x u\,dx^{\dagger}$$
$$= xu+xs-((sa-u(a-x)) = ua-s(a-x).^{\ddagger}$$

† Since s is the circular arc and u the chord

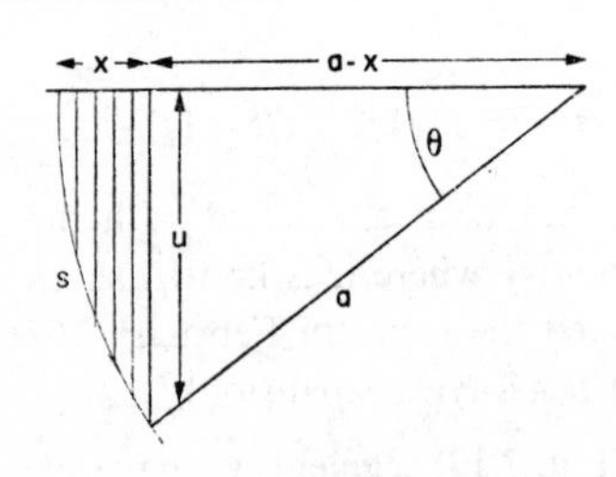

$$\int_0^x u\,dx = \text{sector}-\text{triangle} = \tfrac{1}{2}\,\text{as}\,-\tfrac{1}{2}\,u(a-x).$$

‡ $$\int_0^x s\,dx = a^2\int_0^\theta \theta\sin\theta\,d\theta = a^2[-\theta\cos\theta]_0^\theta+a^2\int_0^\theta \cos\theta\,d\theta$$
$$= -a^2\theta\cos\theta+a^2\sin\theta = -s(a-x)+ua.$$

I must admit that, although I have studied with some care the evidence presented by Hofmann and others in this matter,† I am not satisfied that I have fully grasped the reasons for dating this work as early as the summer of 1673.‡ The first references made by Leibniz to these matters occur in the summer of 1674 when he was quoting results to Huygens§ and to Oldenburg.‖ Once again we find him inclined to over-estimate the power of his method and to promise somewhat more than he had actually achieved.††

These letters of Leibniz had, apparently, the effect of putting Oldenburg on his guard and the reply he makes is cool, withdrawn and critical.‡‡ From this point onwards the correspondence is conducted on a more serious note for, although Oldenburg still doubts the ability of Leibniz to compete with the English in this field, he sees him, perhaps for the first time, as a serious contender. The subsequent Leibniz–Oldenburg correspondence, conducted over the years 1674–6 and culminating in the exchanges with Newton in 1676, is, of course, of profound mathematical significance and has been scrutinised frequently over the last 250 years as part of the history of the Leibniz–Newton controversy.§§ Hofmann has done much to clarify the position of Leibniz in relation to this correspondence and has dealt in an enlightened and sympathetic manner which is comparatively rare in the history of mathematics

† Hofmann, J. E., *Die Entwicklungsgeschichte*, Chap. V, pp. 27–36.

‡ *Ibid.*, p. 36.

§ Bodemann XXXV. 11. 7, fol. 1: *Inventa aliquot mea geometrica*: draft intended for Huygens (summer 1674). Acknowledged in a letter from Huygens to Leibniz (6th Nov. 1674); see *Oeuvres* VII, pp. 393–5.

‖ Leibniz to Oldenburg (15th July 1674): see *The Correspondence of Isaac Newton* I, pp. 322–5.

†† In the letter to Oldenburg (16th Oct. 1674) Leibniz indicates that he can find the arc, the sine being given, i.e. the inverse sine series ($\sin^{-1} x$). I have seen no evidence that he was able to achieve this. At this time he was not able to produce an infinite series by root extraction (the device adopted by Newton in the *De analysi*) nor had he any understanding of the use of binomial expansions for this purpose. In fact, although Leibniz had indeed hit on a powerful transformation he was not able to develop the method as fully as he would have liked. The inverse sine series was obtained by Newton in 1663/4 (included in the *De analysi*) and was obtained independently by Gregory in the autumn of 1670 (see *Gregory Memorial Volume*, pp. 148–52).

‡‡ Oldenburg to Leibniz (8th Dec. 1674): given in *Briefwechsel*, pp. 108–10.

§§ *Commercium epistolicum D. Johannis Collins et aliorum de analysi promota*, London, 1712. *Commercium epistolicum D. Johannis Collins et aliorum de analysi promota*, London, 1722. *Commercium epistolicum J. Collins et aliorum de analysi promota etc. ou Correspondance de J. Collins et d'autres savants célèbres du XVII*^e^ *siècle* (ed. Biot, J. B. and Lefort, F.), Paris 1856.

De Morgan drew attention to the dominant part played by Newton in the earlier investigations of the Royal Society into the origins of the calculus. De Morgan, A., *Essays on the Life and Work of Newton* (ed. Jourdain, P.), London, 1914.

with the human problems arising from this conflict between two men of genius; Turnbull has thrown a good deal of light on the lonely struggles of Gregory in St Andrews, connected only by a single slender link, his correspondence with Collins, with the outside world. Publication of most of the relevant correspondence should now make possible a final survey of the entire controversy in which, one hopes, both mathematical and human problems, could be fully explored.

This, however, is not the place for such a study and after this brief reference I must return to the central theme, which is the development by Leibniz, during the years 1674–6 in a series of supremely important papers, of the notation of the differential and integral calculus. Whilst it is possible to appreciate the magnitude of his achievements by reference only to the papers themselves, the circumstances under which he was labouring must be constantly borne in mind; the insecurity of his personal position and the ever-present desire so to establish himself that he could look forward to an indefinite stay in Paris; the consequent necessity for him to follow up a great many of his varying interests in the expectation that one of them might bring the hoped return; the constant stimulus of the correspondence with Oldenburg leading him at times to unwarrented optimism and at others to equally unjustified despair. In the history of mathematics there is perhaps no finer example of an individual contribution made in the face of such tremendous external pressures.

During the summer of 1673 Leibniz greatly extended his knowledge of contemporary mathematical work and, although his background of technical proficiency was still extremely weak, he at least gained some insight into the significance for mathematical development of certain outstanding problems. Huygens, who still hoped for the analytical quadrature of the circle, introduced him to Gregory's *Vera circuli et hyperbolae quadratura,* the work which had given rise to his own bitter controversy with Gregory. Leibniz may possibly have read the *Geometriae pars universalis* at this time. Whatever the motivation, we find that by August he had settled down to an exploration of the inverse tangent problem apparently fully recognising the relation between inverse tangents and quadratures.

At this time he was using Cartesian notation with a certain facility, understood the general principles of Descartes' tangent method and made constant use of a characteristic triangle, often in association with specific symbols denoting δx and δy respectively (see Fig. 7.14).

Thus, we have, $GW/WL = g/w = TS/SL$. For the parabola, $y^2 = ax$, $ST = 2x$, $SL = \sqrt{(ax)}$, hence, $g/w = 2x/\sqrt{(ax)} = c/v$ (where $v = c\,dy/dx$), and $v = c\sqrt{a}/2\sqrt{x}$, $4xv^2 = c^2a$. Thus, the quadrature of the curve, $4xv^2 = c^2a$, gives the original parabola, $y^2 = ax$, and, in general, the quadrature of the curve $\varphi(x, v) = 0$, gives the curve $f(x, y) = 0$; the tangent relation for $f(x, y) = 0$, gives the curve $\varphi(x, dy/dx) = 0$.† All this does not come out

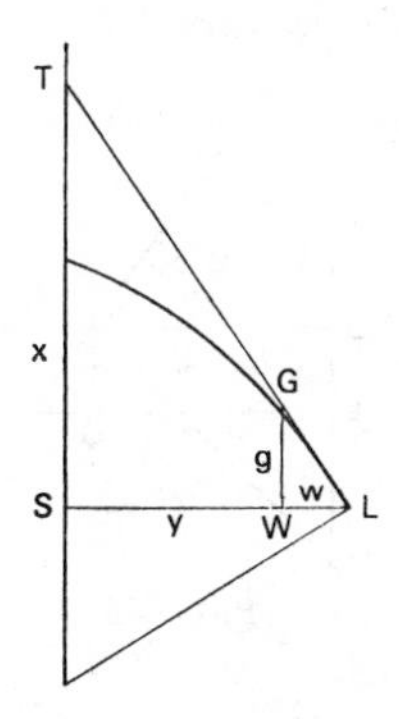

FIG. 7.14

very successfully for, in the process of squaring equations, inadequate use of index notation, defective algebra and the necessity to keep the equations homogeneous by introducing new constants, no very definite pattern emerges.‡

To determine the tangents to algebraic curves Leibniz made use of Sluse's Rules *(regula Slusiana)* without, apparently, any interest in deriving them for himself. In later papers,§ however, we find some limited exploration of the use of small differences. Thus, if $y^2/2a = x$ (see Fig. 7.15), and if $CG = \beta$,

then $$GH = \frac{(y+\beta)^2}{2a} - \frac{y^2}{2a} = \frac{\beta y}{a}.$$

† $4xv^2 = c^2a$, $v^2 = c^2a/4x$, $v = c\sqrt{a}/2\sqrt{x}$, $cy = \int v\,dx = \frac{c\sqrt{a}}{2}\int \frac{dx}{\sqrt{x}}$ hence $y^2 = ax$.

‡ Bodemann XXXV. 8. 3, fols. 1–8: *Methodus tangentium inversa seu de functionibus* (Aug. 1673). On this see Scriba, C. J., The Inverse Method of Tangents: a dialogue between Leibniz and Newton (1675–7), *Archive for History of Exact Sciences* 2 (1964), pp. 113–37.

§ Bodemann XXXV. 5. 1: *Schediasma de methodo tangentium inversa ad circulum applicata* (Oct. 1674), *Cat. crit.* II, no. 791.

Much of this work was unprofitable, and it was not till October 1675 that Leibniz fell on a more fruitful line of development. In a notable paper, the *Analysis tetragonistica ex centrobarycis*,† he turns once again to examine the application of the principle of moments to the elements of curves and areas.

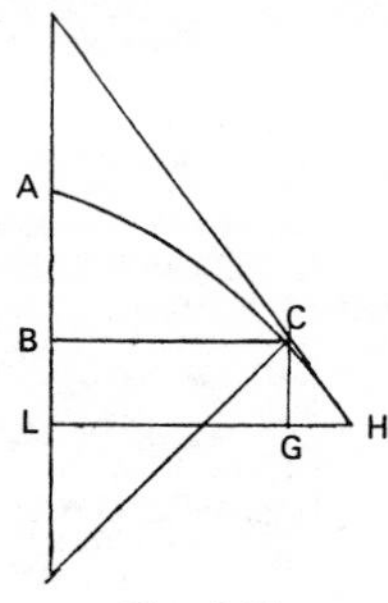

FIG. 7.15

Although there are no new discoveries here, through the development of the symbols and their operational use the work takes on an increasingly analytical appearance. Results arise from substitution in general formulae and diagrams become increasingly redundant.

Taking moments about AD (see Fig. 7.16), we have,

moment of $ABCEA$ = moment of rectangle $ABCD$ − moment of $ADCEA$.

Now, if the curve AEC be of the form, $y = f(x)$, and if $AB = b$, $BC = c$, then $\int_0^b xy\,dx = b^2c/2 - \int_0^c (x^2/2)\,dy$.

Leibniz gives this in the form

$$\text{omn. } \overline{yx \text{ ad } x} \sqcap b^2c/2 - \text{omn. } \overline{x^2/2 \text{ ad } y}\,.$$

Hence, given $f(x, y) = 0$, by substituting $xy = z$, we obtain $\varphi_1(x, z) = 0$, and substituting $x^2 = 2w$, $\varphi_2(x, w) = 0$. The quadratures of $\varphi_1(x, z) = 0$ and $\varphi_2(x, w) = 0$ are thus mutually dependent, the one being the complement of the other.

Developing these ideas further (26th October 1675) by considering the

† Bodemann XXXV. 8. 18 (25th, 26th, 29th Oct., 1st Nov. 1675). *Briefwechsel*, pp. 147–60.

moments of the differences (w) about a line perpendicular to the axis we have (see Fig. 7.17),

$$\text{omn. } \overline{xw} \sqcap \text{ult. } x\, \overline{\text{omn. } w} - \text{omn. } \overline{\text{omn. } w},$$

where ult. x is the last x, or the upper limit and $y = \text{omn. } w$. This Leibniz recognises as a *general* statement, applying, not only to the differences (w) but also to the terms (y).

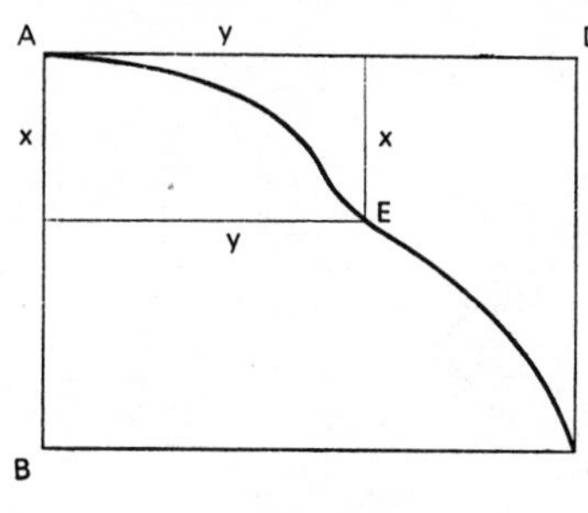

Fig. 7.16

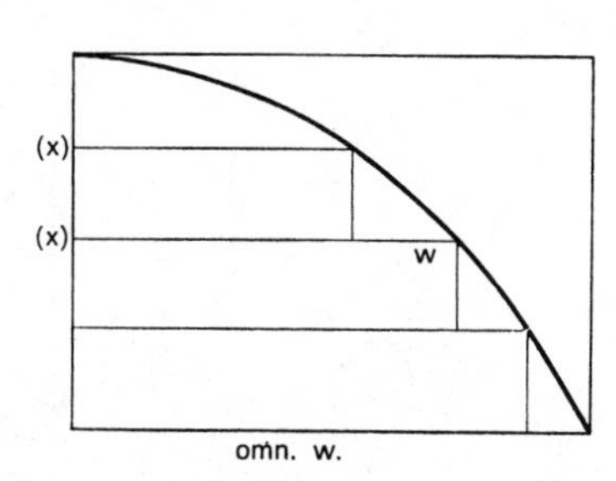

Fig. 7.17

For the terms, we have

$$\text{omn. } \overline{xy} \sqcap \text{ult. } x\, \overline{\text{omn. } y} - \text{omn. } \overline{\text{omn. } y}.$$

Thus, in the notation of the calculus,

$$\int_0^a xy\, dx = a \int_0^a y\, dx - \int_0^a dx \int_0^x y\, dx.$$

It is now possible to proceed by purely analytical substitutions. For example, if $xw = a$, then

$$\text{omn. } a \sqcap \text{ult. } x\, \overline{\text{omn. } a/x} - \text{omn. } \overline{\text{omn. } a/x},$$

and the sum of the logarithms (omn. $\overline{\text{omn. } a/x}$) is given in terms of the quadrature of the hyperbola ($\overline{\text{omn. } a/x}$). From this point onwards the entire direction of the work changes and becomes wholly an exploration of symbolism.

Returning to inverse tangents (see Fig. 7.18), if $BL = y$, $WL = l$, $BP = p$, $GW = a$, then, $p/y = l/a$, $p = ly/a$, $p = (y/a)l$. But, from previous results,†

† $p = y\, dy/dx, \quad \int p\, dx = \int y\, dy.$

$\text{omn.}\, p = \dfrac{\overline{\text{omn.}\, l}^{\,2}}{2}$. Since, however, $y = \text{omn.}\, l$, then $p = \dfrac{\overline{\text{omn.}\, l}}{a}\, l$,

and
$$\text{omn.}\, p = \text{omn.}\, \frac{\overline{\text{omn.}\, l}}{a}\, l$$
$$= \frac{\overline{\text{omn.}\, l}^{\,2}}{2}.$$

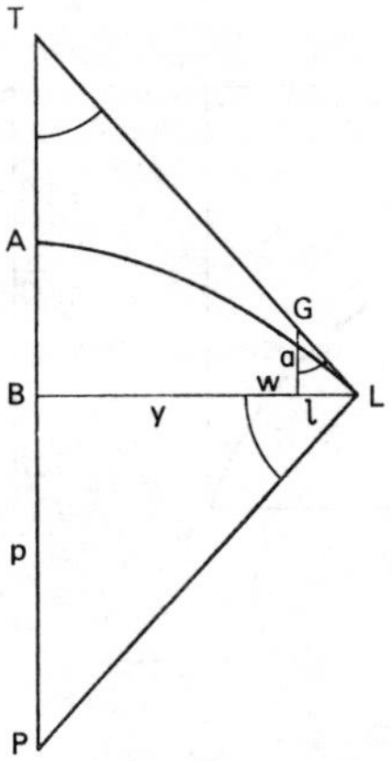

FIG. 7.18

Impressed by this result Leibniz† introduced the symbol $\int$ (i.e. a long *S*, the first letter of the word *summa*) for omn. and $\int l$ for omn. l, so that he can now write,

$$\frac{\overline{\int l}^{\,2}}{2} = \overline{\int \overline{\int l}\, \frac{l}{a}},$$

and
$$\int \overline{xl} = x \int l - \overline{\int \overline{\int l}}.$$

Regarding the operational rules for this new symbol, it is clear that, if $\int l$ be given analytically, then so also is l: the converse, however, is not true.

† "Ergo habemus theorema quod mihi videtur admirabile, et novo huic calculo magni adjumenti loco futurum ...", *Analysis tetragonistica* (Part II) given in *Briefwechsel*, p. 154.

Also $\int x = x^2/2$ and $\int x^2 = x^3/3$. In general, if a/b is a given constant,

$$\int \frac{a}{b}\, l = \frac{a}{b} \int l.$$

If only one variable is to be considered, then the differences (l) can be taken as constant and $\int l = x$.

The example which follows is of special interest because it brings into prominence some of the difficulties with which Leibniz is struggling; to determine e in the following equation,

$$a^3e \sqcap \int \frac{c}{a} \overline{\int l}^{\,2} + ba^2 + \overline{\int l}^{\,3} + \int l^3.$$

The form of the equation is dictated by dimensional considerations which clearly also enter into the solution.

We have, $\quad l = a[=1], \quad \int l = x, \quad \overline{\int l}^{\,2} = x^2,$

$$\int \overline{\int l}^{\,2} = x^3/3, \quad \overline{\int l}^{\,3} = x^3, \quad \int \overline{\int l}^{\,3} = x^4/4,$$

$$l^3 = a^3, \quad \int l^3 = a^3x.\dagger$$

Thus,

$$a^3e = \frac{cx^3}{3} + ba^2x + \frac{x^4}{4} + xa^3 \ddagger$$

(1) (2) (3) (4)

† $\int l = \int dx = x, \ \overline{\int l}^{\,2} = \left(\int dx\right)^2 = x^2, \ \int \overline{\int l}^{\,2} = \int x^2\, dx = x^3/3, \ \overline{\int l}^{\,3} = \left(\int dx\right)^3 = x^3,$
$\int \overline{\int l}^{\,3} = \int x^3\, dx = x^4/4, \ l^3 = a^3, \ \int l^3 = \int a^3\, dx = a^3x \quad [dx = a = 1].$

‡ There is clearly an error in the third term (3). I would like to extend the initial integral sign (as shown by the broken line in the initial equation, to cover $\overline{\int l^3}^{\,3}$. The third term thus becomes $\int \overline{\int l}^{\,3} = \int x^3\, dx = x^4/4$, as given by Leibniz. (There is, however, no sign of such a line in the MS.) This seems to be the most likely slip. Dimensional considerations have dictated the form of the original equation and the term $\overline{\int l}^{\,3}$ does not make sense dimensionally. With the line extended the question and the solution become logically coherent.

At this stage Leibniz was satisfied that these results were new and notable and would lead to a new calculus.[†]

The next step is to explore the contrary calculus: let $\int l \sqcap ya$, and suppose $l \sqcap ya/d$; then, just as $\int$ will increase the dimensions, so d will diminish the dimensions.[‡] $\int$ means a *sum*, d a *difference*. From the given y, ya/d, i.e. l, can be found (the difference of the y's).

For example, since
$$\int c\,\overline{\int l}^{2} = \frac{c\,\overline{\int l}^{3}}{3a^3},$$
then
$$c\,\overline{\int l}^{2} = \frac{c\,\overline{\int l}^{3}}{3a^3 d}.$$
Also,
$$\int x^3/b + \int x^2a/e = \int \overline{x^3/b + x^2a/e},$$
$$\frac{x^3}{db} + \frac{x^2a}{de} = \frac{x^3/b + x^2a/e}{d}.$$

It is worth noting that at this point Leibniz turns aside to interpret his general transmutation in terms of the new calculus.[§]

Some days later (11th November 1675) he returns to the inverse tangent problem, or the solution of differential equations. All these are not correctly resolved but the work shows increasing fluency in the development of operational rules relating to the new calculus.

For example, to determine the curve in which the subnormal BP (see Fig. 7.19) is inversely proportional to the ordinate BC, let $BC = y$, $AB = x$, $BP = w$, $B(B) = z\ (\delta x)$.

† "Satis haec nova et notabilia, cum novum genus calculi inducant", *Analysis tetragonistica* (Part II): *Briefwechsel*, p. 155.

‡ It is once again dimensional considerations which supply the form $\int x = x^2/2$, and so on: $\int$ *increases* the dimensions. To take a difference is to *lower* the dimensions and (probably by analogy with the division process) the symbol d is placed below.

§ *Analysis tetragonistica* (Part II): *Briefwechsel*, pp. 155–6.

We have, from previous results, $\int wz \sqcap y^2/2,$†

$$wz \sqcap y^2/2d.$$

But, from the quadrature of the triangle, $y^2/2d \sqcap y$, hence, $wz \sqcap y$. Since, however, $w \sqcap b/y$, $bz/y \sqcap y$, and $z \sqcap y^2/b$. But, $\int z \sqcap x \sqcap \int y^2/b \sqcap y^3/3ba$, and consequently, $x \sqcap y^3/3ba$.‡

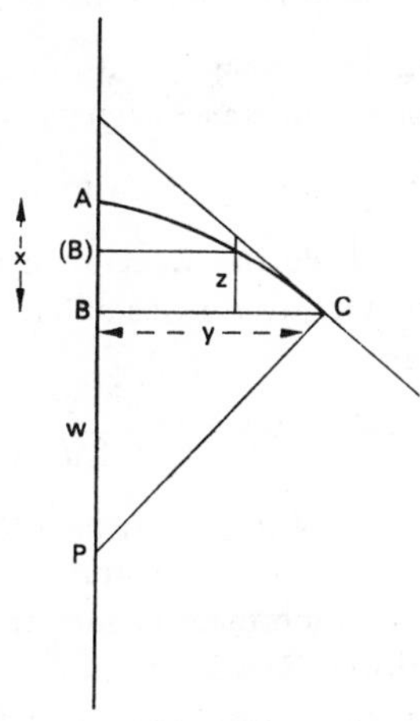

FIG. 7.19

Another more difficult example leads to interesting and fundamental investigations: let AP (see Fig. 7.19) be inversely proportional to the ordinate BC, then we have, $x+w \sqcap a^2/y$. As before, $wz \sqcap y^2/2d$, $\int z \sqcap x$, $z \sqcap x/d$,

$$\text{and} \qquad wx/d \sqcap y^2/2d, \quad w \sqcap \frac{y^2/2d}{x/d}, \quad w+x \sqcap a^2/y \sqcap x+\frac{y^2/2d}{x/d}.$$

The expression on the right is full of pitfalls: since we now have to deal with the *differences* of the x's and also with the *differences* of the y's it is necessary to investigate the relation between them. If the x's are in arithmetical pro-

† $\int y(dy/dx)\,dx = \int y\,dy = y^2/2.$

‡ $y\,dy/dx = b/y$, $dy/dx = b/y^2$, $\int y^2\,dy = \int b\,dx$, $y^3/3 = bx$, $x = y^3/3b$ $(a = 1)$.

gression ($\delta x = a = 1$), then the differences are constant and $x/d = 1$; if, however, the x's are in arithmetical progression then the y's are not and the interpretation of $y^2/2d$ is left open; if, on the other hand, the y's are taken to be in arithmetical progression, then $y^2/2d = y$; x/d in this case remains undetermined.†

The door has now been opened for a deeper penetration into the nature of the d operator. It is curious to note that under the stress of the inquiry and without further comment Leibniz is now writing, $d\overline{x^2+y^2}$ and $dy \int y, dx \int x$, and so forth. Investigating the same questions in terms of moments we have the sum of the moments of the differences equals the moment of the sum of the differences leading to $\int y\,dy \sqcap y^2/2$, with the clear implication that the 'dy' should be placed inside the integral sign. When, on the other hand, the integral arises as the inverse of a difference the written form is not so obvious. Thus, if we have

$$\sqrt{(x^2+y^2)} = w = \frac{dy^2}{2\,dx},$$

then, $\sqrt{(x^2+y^2)} = y$, and there is nothing to indicate that integration is with respect to x.‡ In the progress of this discussion Leibniz comments that if the x's are in arithmetical progression then the motion (in describing the curve) parallel to the x-axis is uniform. He finds, however, the concept of continuous motion inacceptable and prefers to think of finite differences.§

Leibniz now investigates the interpretation to be given to $dx\,dy$ and dx/dy. Is $dx\,dy$ equal to $d\,\overline{xy}$ and dx/dy equal to $d(x/y)$? Eventually, by the use of

† Leibniz explores many different alternatives here. The final solution given runs roughly as follows: $x+y\,dy/dx = a^2/y$, $\int x\,dx+\int y\,dy = \int (a^2/y)\,dx$, so that $x^2+y^2 = 2\int (a^2/y)\,dx$. This latter integral Leibniz gives as a logarithm having failed effectively to distinguish between $\int dy/y$ and $\int dx/y$.

‡ $d(y^2/2) = y$, $dx = 1$, hence, $\dfrac{d(y^2/2)}{dx} = y$.

§ We know that Leibniz had heard of the determination of tangents to curves by considerations of motion (he noted this during his visit to England). Roberval's tangent methods were discussed in Schooten. Leibniz had also purchased Barrow's book and it would be surprising if he were unaware that the basis of Barrow's early chapters was, in fact, the description of curves by motion. There are many passages here and there in Leibniz' scribbled notes which suggest that he had been dipping into Barrow's *Lectiones*. The above remarks make it clear that he was unable to accept the concept of motion as having any immediate relevance for his own work.

small differences, he correctly concludes that $d\,\overline{xy}$ is not, in general, equal to $dx\,dy$ nor is dx/dy equal to $d(x/y)$.

These, then, are the famous papers in which Leibniz, in a search for a simple and effective representation of the basic elements and operations in infinitesimal processes, invented the notation of the infinitesimal calculus. He was not, at this stage, essentially concerned with the creation of new mathematics and only incidentally with problem solving but rather with a purposeful construction of a set of symbolic representations through which special problems could be simply expressed in a standardised notation. The first decisive steps had been taken and a line of approach established which he was to develop in the years which followed.

It would be profitless to discuss the limitations of this primitive calculus devised to represent existing processes rather than to develop new ones. Leibniz' approach to geometry was based essentially on the concept of finite sums and differences,† which, in continuous curves, could be allowed to become infinitely small. He was concerned, not to develop new theorems but to establish a method through which, without diagrams, these things which depend on a figure can be derived by a calculus.

The work of Cavalieri, Saint-Vincent, Wallis, Gregory and Barrow all represented material which he hoped to translate into formulae and incorporate in his new calculus. Although the concept of sums and differences provided him with direct knowledge of some simple formulae, he was prepared to take on trust from Wallis, or anyone else who provided it, such general results as

$$\int x^n\,dx = \frac{x^{n+1}}{n+1}.$$

As for differentiation, although he understood how to calculate dy and dx by the use of small increments, the only means he had of differentiating anything other than the simplest algebraic functions was Sluse's Rules, the mathematical foundation for which he only dimly grasped. The machinery required for operating with the new calculus was at this stage almost entirely lacking.

Leibniz made outstanding contributions in metaphysics and in logic as well as in mathematics. All his activities were systematically interrelated and his contributions in one field cannot be fully appreciated without taking into consideration all his other activities. To separate his diverse productions is to mutilate his thoughts. For Leibniz mathematics was an important and fasci-

† See particularly Hofmann, J. E., Wieleitner, H. and Mahnke, D., Die Differenzenrechnung bei Leibniz, *op. cit.*

nating activity of the human mind and, from his student days, he cherished the idea of a universal characteristic closely associated with a language and symbolism through which all knowledge could be assembled and related through analysis to certain primitive logical elements.

Underlying all Leibniz' varied mathematical activities during his Paris years was this project and it was out of it that came the desire to create a new higher analysis based on symbolising the basic elements and operations. This he called a calculus. It is in relation to his own avowed aim and purpose that his achievement should be assessed rather than in terms of his immediate contribution to seventeenth-century mathematics.

Bibliography

A. PRIMARY SOURCES

I. Manuscripts

Textual references to manuscript sources have been given from the catalogues listed below.

Newton MSS.

Cambridge University Library, Portsmouth Collection. *Catalogue of Portsmouth Collection*, Cambridge, 1888.

Leibniz MSS.

Catalogue critique des manuscrits de Leibniz, II (ed. Rivaud, A.), Poitiers, 1914–24.
Die Leibniz-Handschriften der Königlichen Öffentlichen Bibliothek zu Hannover (ed. Bodemann, E.), Hanover and Leipzig, 1895.

A collection of thirty-five reels of microfilm of the letters and manuscripts of G. W. Leibniz, at present in the Lower Saxony State Library, Hanover: selected by Paul Schrecker on the basis of the Bodemann catalogue, and Concordance of Leibniz Microfilm with this catalogue prepared by Sterling, C. G., on behalf of the University of Pennsylvania Library, Philadelphia, 1955. (Supplementary microfilm was supplied by the Librarian of the Lower Saxony State Library, Hanover.)

II. Collected Works

Aristotle, the Works of (ed. Ross, W. D. and Smith, J. A.), 11 vols., Oxford, 1908–31.
Birch, T.: *The History of the Royal Society of London*, 4 vols., London, 1756–7.
Descartes, R.: *Oeuvres* (ed. Adam, C. and Tannery, P.), 12 vols. and supplement, Paris, 1897–1913.
Fermat, P.: *Oeuvres* (ed. Tannery, P. and Henry, C.), 4 vols., Paris, 1891–1912; *Supplément* (ed. Waard, C. de), Paris, 1922.
Galilei, Galileo: *Opere* (edizione nazionale), 20 vols., Florence, 1890–1909.
Huygens, C.: *Oeuvres complètes*, published by the Société Hollandaise des Sciences, 22 vols., The Hague, 1888–1950.
Kepler, J.: *Opera omnia* (ed. Frisch, C.), 8 vols., Frankfort-on-Main, 1858–71.
Leibniz, G. W.: *Der Briefwechsel mit Mathematikern* (ed. Gerhardt, C. I.), Berlin, 1899.
Leibniz, G. W.: *Mathematische Schriften* (ed. Gerhardt, C. I.), 7 vols., Berlin, 1849–63.
Leibniz, G. W.: *Die philosophischen Schriften* (ed. Gerhardt, C. I.), 7 vols., Berlin, 1876–90.

LEIBNIZ, G. W.: *Sämtliche Schriften und Briefe* (ed. Preuss. Akad. d. Wiss. Berlin), Darmstadt and Leipzig, 1924– (6 vols. to date).

MERSENNE, M.: *Correspondance* (ed. Waard, C. de), 7 vols., Paris, 1932–62.

NEWTON, I.: *The Correspondence of Isaac Newton* (ed. Turnbull, H. W.), Cambridge, 1959– (3 vols. to date).

NEWTON, I.: *Unpublished Scientific Papers of Isaac Newton* (ed. Hall, A. R. and Hall, M. B.), Cambridge, 1962.

PASCAL, B.: *Oeuvres* (ed. Brunschvicg, L. and Boutroux, P.), 14 vols., Paris, 1908–25.

RIGAUD, S. P.: *Correspondence of Scientific Men of the 17th Century*, 2 vols., Oxford, 1841.

ROBERVAL, G. P. DE: *Divers Ouvrages de Mathématique et de Physique, par Messieurs de l'Académie Royals des Sciences*, Paris, 1693.

STEVIN, S.: *The Principal Works of Simon Stevin* (ed. Dijksterhuis, E. J.), 5 vols., Amsterdam, 1955.

STEVIN, S.: *Les Oeuvres mathématiques de Simon Stevin de Bruges* (ed. Girard, A.), Leyden, 1634.

STEVIN, S.: *Hypomnemata mathematica* (ed. Snell, W.), Leyden, 1608.

TACQUET, A.: *Opera mathematica*, Antwerp, 1669.

TORRICELLI, E.: *Opere di Evangelista Torricelli* (ed. Loria, G. and Vassura, G.), 3 vols., Faenza, 1919.

VIÈTE, F.: *Opera mathematica* (ed. Schooten, F. van), Leyden, 1646.

WALLIS, J.: *Opera mathematica*, Oxford, 1693–9 (3 vols.).

The Correspondence of Sir Isaac Newton and Professor Cotes (ed. Edleston, J.), London, 1850.

Commercium epistolicum D. Johannis Collins, et aliorum de analysi promota, London, 1712.

III. JOURNALS

Acta eruditorum, Leipzig, 1682–1731.

Histoire et mémoires de l'Académie Royale des Sciences, Paris, 1699–1790.

Journal Literaire, The Hague, 1713–.

Journal des Sçavans, Paris, 1665–1792.

Philosophical Transactions, London, 1666–.

IV. SINGLE WORKS

ANDERSON, A.: *Vindiciae Archimedis in stereometriam Kepleri*, Paris, 1616.

BARROW, I.: *Lectiones geometricae*, London, 1670.

BARROW, I.: *Lectiones mathematicae*, London, 1683.

BRADWARDINE, T.: *Tractatus de proportionibus* (ed. Crosby, H. L.), Madison, 1955.

CAVALIERI, B.: *Geometria indivisibilibus continuorum nova quadam ratione promota*, Bologna, 1635 (2nd ed. 1653).

CAVALIERI, B.: *Exercitationes geometricae sex*, Bologna, 1647.

COMMANDINO, F.: *Liber de centro gravitatis solidorum*, Bologna, 1565.

DESCARTES, R.: *Discours de la méthode*, Leyden, 1637.

DESCARTES, R.: *The Geometry of René Descartes* (ed. Smith, D. E. and Latham, M. L.), Chicago, 1925.

DESCARTES, R.: *Geometria a Renato des Cartes* (ed. Schooten, F. van), Leyden, 1649.

DESCARTES, R.: *Renati Descartes Geometria: editio secunda*, Amsterdam, 1659.

FAILLE, J. C. DE LA: *De centro gravitatis partium circuli et ellipsis*, Antwerp, 1663.

FAULHABER, J.: *Miracula arithmetica*, Augsburg, 1622.

FAULHABER, J.: *Academia algebrae*, Ulm, 1631.

GALILEI, G.: *Discorsi e dimostrazione matematiche*, Leyden, 1638.
GALILEI, G.: *Dialogue concerning two new sciences* (trans. Crew, H. and De Salvio, A.), New York, 1914.
GALILEI, G.: *Dialogue on the Great World Systems* (Salusbury trans. ed. Santillana, G. de), Chicago, 1953.
GREGORY, J.: *Vera circuli et hyperbolae quadratura*, Padua, 1667.
GREGORY, J.: *Geometriae pars universalis*, Padua, 1668.
GREGORY, J.: *Exercitationes geometricae*, London, 1668.
GREGORY, J.: *Tercentenary Memorial Volume* (ed. Turnbull, H. W.), London, 1939.
GREGORIUS A S. VINCENTIO (Saint-Vincent, Grégoire de): *Opus geometricum quadraturae circuli et sectionum coni*, Antwerp, 1647.
GULDIN, P.: *Centrobaryca*, Vienna, 1635–41.
HOBBES, T.: *Six Lessons*, London, 1656.
KEPLER, J.: *Nova stereometria doliorum vinariorum*, Linz, 1615.
LALOVERA, A. DE: *Quadratura circuli et hyperbolae segmentorum*, Toulouse, 1651.
MAUROLICO, F.: *Admirandi Archimedis Syracusani monumenta omnia mathematica, quae extant*, Palermo, 1685.
MENGOLI, P.: *Novae quadraturae arithmeticae*, Bologna, 1650.
MENGOLI, P.: *Geometria speciosa*, Bologna, 1659.
MERCATOR, N.: *Logarithmotechnia*, London, 1668.
MERSENNE, M.: *Universae geometricae mixtaeque mathematicae synopsis*, Paris, 1644.
MERSENNE, M.: *Cogitata physico-mathematica*, Paris, 1644.
ORESME, N.: *Der "Algorismus proportionum" des Nikolaus Oresme* (ed. Curtze, M.), Berlin, 1868.
OUGHTRED, W.: *Arithmeticae in numeris et speciebus institutio*, London, 1631.
PAPPUS OF ALEXANDRIA: *Pappus d'Alexandrie: La Collection Mathématique* (trans. Ver Eecke, P.) (2 vols.), Paris, 1933.
PASCAL, B.: *Traité du triangle arithmétique*, Paris, 1665.
PROCLUS: *The Philosophical and Mathematical Commentaries of Proclus* (trans. Taylor, T.), 2 vols., London, 1788.
PSEUDO-ARISTOTLE: *The Mechanica* (trans. Forster, E. S.), Oxford, 1913.
RICCI, M.: *Exercitatio geometrica*, Rome, 1666.
ROSEN, F.: *The Algebra of Mohammed Ben Musa*, London, 1831.
RUDOLFF, C.: *Die Coss*, Konigsberg, 1554.
SARASA, A. A. DE: *Solutio problematis a Mersenne propositi*, Antwerp, 1649.
SCHOOTEN, F. VAN: *Exercitationes mathematicae*, Leyden, 1657.
SLUSE, R. F. DE: *Mesolabum*, Liège, 1659 (2nd ed. 1668).
STEVIN, S.: *L'arithmétique*, Leyden, 1585.
STIFEL, M.: *Arithmetica integra*, Nuremberg, 1544.
SUISETH, R.: *Suiseth, Ricardus: Calculator* (ed. Trinchavellus, V.), Venice, 1520.
TACQUET, A.: *Cylindrica et annularia*, Antwerp, 1651.
TACQUET, A.: *Opera mathematica*, Antwerp, 1669.
TORRICELLI, E.: *Opera geometrica*, Florence, 1644.
TORRICELLI, E.: *De infinitis spiralibus* (ed. Carruccio, E.), Pisa, 1955.
VALERIO, L.: *De centro gravitatis solidorum*, Rome, 1604.
WALLIS, J.: *A Treatise of Algebra*, Oxford, 1685.

B. SECONDARY SOURCES

I. Collected Works and General Histories of Mathematics

Becker, O.: *Das mathematische Denken der Antike*, Göttingen, 1957.
Bell, E. T.: *The Development of Mathematics*, New York, 1945.
Bourbaki, N.: *Éléments d'histoire des mathématiques*, Paris, 1960.
Boyer, C. B.: *The Concepts of the Calculus*, New York, 1949.
Brunschvicg, L.: *Les Étapes de la philosophie mathématique*, Paris, 1912.
Burtt, E. A.: *The Metaphysical Foundations of Modern Physical Science*, London, 1925.
Cantor, M.: *Vorlesungen über Geschichte der Mathematik*, Leipzig, 1894–1908 (2nd ed.)
Clagett, M.: *The Science of Mechanics in the Middle Ages*, Madison, 1959.
Crombie, A. C.: *Augustine to Galileo*, London, 1952.
Datta, B. and Singh, A. N.: *History of Hindu Mathematics*, Lahore, 1935–8 (2nd ed. Bombay, 1962).
Diels, H.: *Die Fragmente der Vorsakratiker*, Berlin, 1951–2 (3 vols., 6th ed.).
Dijksterhuis, E. J.: *Archimedes*, Copenhagen, 1956.
Freeman, K.: *The Pre-Socratic Philosophers*, Oxford, 1946.
Heath, T. L.: *A History of Greek Mathematics*, 2 vols., Oxford, 1921.
Heath, T. L.: *Apollonius of Perga*, Cambridge, 1896.
Heath, T. L.: *The Works of Archimedes*, Cambridge, 1897 (Supplement, 1912): references given from the Dover ed., New York, n.d.
Heath, T. L.: *The Thirteen Books of Euclid's Elements*: references given from the Dover edition, New York, 1956 (3 vols.).
Hofmann, J. E.: *Geschichte der Mathematik*, 3 vols., Berlin, 1953–7.
Lasswitz, K.: *Geschichte der Atomistik vom Mittelalter bis Newton*, 2 vols., Hamburg and Leipzig, 1890.
Marie, M.: *Histoire des sciences mathématiques et physiques*, 12 vols., Paris, 1883–8.
Montucla, J. E.: *Histoire des mathématiques*, Paris, 1799–1802.
Neugebauer, O.: *The Exact Sciences in Antiquity*, Copenhagen, 1951.
Neugebauer, O.: *Mathematical Cuneiform Texts* (with Sachs, A.), New Haven, 1945.
Scott, J. F.: *The History of Mathematics*, London, 1958.
Van der Waarden, B. L.: *Science Awakening*, Groningen, 1954.
Ver Eecke, P.: *Les Coniques d'Apollonius de Perge*, Paris, 1924 (2nd ed. 1963).
Zeuthen, H. G.: *Histoire des mathématiques*, Paris, 1902.

II. Articles and Other Works

Aubry, A.: Sur l'histoire du calcul infinitésimal entre les années 1620 et 1660, *Annaes Scientificos da Academia Polytechnica do Porto* 6 (1911), pp. 82–9.
Auger, L.: *Un Savant méconnu: Gilles Personne de Roberval* (1602–75), Paris, 1962.
Becker, O.: Eudoxos-Studien, *Quellen und Studien zur Geschichte der Mathematik, Astronomie und Physik*, Studien (B) 2 (1933), pp. 311–33, 369–87; 3 (1936), pp. 236–44, 370–410.
Bochner, S.: *The Role of Mathematics in the Rise of Science*, Princeton, 1966.
Bopp, K.: Die Kegelschnitte des Gregorius a S. Vincentio, *Abhandlungen zur Geschichte der Mathematischen Wissenschaften* 20 (1907), pp. 83–314.
Bortolotti, E.: La memoria "De infinitis hyperbolis" di Torricelli, *Archeion* 6 (1925), pp. 49–58, 139–52.
Bortolotti, E.: La scoperta e le successive generalizzazioni di un teorema fondamentale di calcolo integrale, *Archeion* 5 (1924), pp. 204–27.

BOSMANS, H.: Le calcul infinitésimal chez Simon Stevin, *Mathésis* 37 (1923), pp. 12–18, 55–62, 105–9.

BOSMANS, H.: Les démonstrations par l'analyse infinitésimale chez Luc Valerio, *Annales de la Société Scientifique de Bruxelles* 37 (1913), pp. 211–28.

BOSMANS, H.: Le mathématicien Anversois Jean-Charles della Faille de la Compagnie de Jesus, *Mathésis* 41 (1927), pp. 5–11.

BOSMANS, H.: La notion des "indivisibles" chez Blaise Pascal, *Archivio di Storia della Scienza* 4 (1923), pp. 369–79.

BOSMANS, H.: Sur l'oeuvre mathématique de Blaise Pascal, *Mathésis* 38 (1924), *Supplément*, pp. 1–59.

BOSMANS, H.: Sur quelques exemples de la méthode des limites chez Simon Stevin, *Annales de la Société Scientifique de Bruxelles* 37 (1913), pp. 171–99.

BOSMANS, H.: Grégoire de Saint-Vincent, *Mathésis* 38 (1924), pp. 250–6.

BRUINS, E. M.: On the system of Babylonian geometry, *Sumer* 11 (1955), pp. 44–9.

CAJORI, F.: Who was the first inventor of the calculus?, *Amer. Math. Monthly* 26 (1919), pp. 15–20.

CAJORI, F.: History of Zeno's arguments on motion, *Amer. Math. Monthly* 22 (1915), pp. 1–6, 39–47, 77–82, 109–15, 143–9, 179–86, 215–20, 253–8, 292–7.

CHILD, J. M.: *The Geometrical Lectures of Isaac Barrow*, Chicago and London, 1916.

CHILD, J. M.: *The Early Mathematical Manuscripts of Leibniz*, Chicago and London, 1920.

CLAGETT, M.: Archimedes in the Middle Ages: the *De mensura circuli*, *Osiris* 10 (1952), pp. 587–618.

CLAGETT, M.: The *Liber de motu* of Gerard of Brussels and the Origins of Kinematics in the West, *Osiris* 12 (1956), pp. 73–175.

CLAGETT, M.: *Archimedes in the Middle Ages*, Madison, 1964.

COOLIDGE, J. L.: The unsatisfactory story of curvature, *Amer. Math. Monthly* 59 (1952), pp. 375–9.

DE MORGAN, A.: *Essays on the Life and Work of Newton* (ed. Jourdain, P.), Chicago and London, 1914.

DEHN, M. and HELLINGER, E.: On James Gregory's *Vera quadratura*, in *Gregory Tercentenary Memorial Volume*, pp. 468–78.

DÖRRIE, H.: *Triumph der Mathematik*, Breslau, 1933.

DRABKIN, I. E.: Aristotle's Wheel: notes on the history of a paradox, *Osiris* 9 (1950), pp. 162–98.

DUHAMEL, J. M. C.: Mémoire sur la méthode des maxima et minima de Fermat et sur les méthodes des tangentes de Fermat et Descartes, *Mémoires de l'Académie des Sciences de l'Institut Impérial de France* 32 (1864), pp. 269–330.

ENESTRÖM, G.: Sur un théorème de Kepler équivalent à l'intégration d'une fonction trigonometrique, *Bibl. Math.* (N.S.) 3 (1889), pp. 65–6.

ENESTRÖM, G.: Über die angebliche Integration einer trigonometrischen Funktion bei Kepler, *Bibl. Math.* (3) 13 (1913), pp. 229–41.

EVANS, G. W.: Cavalieri's Theorem in his own words, *Amer. Math. Monthly* 24 (1917), pp. 447–51.

FRITZ, K. VON: The discovery of incommensurability by Hippasus of Metapontum, *Annals of Maths.* 46 (1945), pp. 242–64.

GERHARDT, C. I.: *Die Entdeckung der Differentialrechnung durch Leibniz*, Salzwedel, 1848.

GOLDBECK, E.: Galileis Atomistik und ihre Quellen, *Bibl. Math.* (3) 3 (1902), pp. 84–112.

GÜNTHER, S.: Über eine merkwürdige Beziehung zwischen Pappus und Kepler, *Bibl. Math.* N.S. 2 (1888), pp. 81–7.

HAAS, K.: Die mathematischen Arbeiten von Johann Hudde (1628–1704), *Centaurus* 4 (1956), pp. 235–84.

HASSE, H. and SCHOLZ, H.: *Die Grundlagenkrisis der Griechischen Mathematik*, Berlin, 1928.
HOBSON, E. W.: *Squaring the Circle*, Cambridge, 1913 (2nd ed. 1953).
HOFMANN, J. E.: Erste Quadratur der Kissoide (with Hofmann, J.), *Deutsche Mathematik* 5 (1941), pp. 571–84.
HOFMANN, J. E.: Das Opus geometricum des Gregorius a S. Vincentio und seine Einwirkung auf Leibniz, *Abh. d. Preuss. Ak. d. Wiss.*, 1941, Math-naturw. Kl. 13, Berlin, 1941.
HOFMANN, J. E.: Studien zur Vorgeschichte des Prioritätstreites zwischen Leibniz und Newton um die Entdeckung der höheren Analysis, I: Materialien zur ersten Schaffensperiode Newtons (1665–75), *Abh. d. Preuss. Ak. d. Wiss.*, 1943, Math.-naturw. Kl. 2, Berlin, 1943.
HOFMANN, J. E.: *Die Entwicklungsgeschichte der Leibnizschen Mathematik während des Aufenthaltes in Paris* (1672–6), Munich, 1949.
HOFMANN, J. E.: Über Gregorys systematische Näherungen für den Sektor eines Mittelpunktkegelschnittes, *Centaurus* 1 (1950), pp. 24–37.
HOFMANN, J. E.: Der junge Newton als Mathematiker (1665–75), *Math.-phys. Sem.-Ber.* 2 (1951), pp. 45–70.
HOFMANN, J. E.: Descartes und die Mathematik, in *Descartes* (Scholz, H., Kratzer, A. and Hofmann, J. E.), Munster, 1951.
HOFMANN, J. E.: Die Differenzenrechnung bei Leibniz, *Sitz. d. Preuss. Ak. d. Wiss.* 1931, Phys.-Math. Kl. 26, Berlin, 1931, pp. 562–600 (with Wieleitner, H. and Mahnke, D.).
HOFMANN, J. E.: *Frans van Schooten der Jüngere*, Wiesbaden, 1962.
HOPPE, E.: Zur Geschichte der Infinitesimalrechnung bis Leibniz und Newton, *Jahr. Deutsche Math.-Verein.* 37 (1928), pp. 148–87.
JACOLI, F.: Evangelista Torricelli ed il metodo delle tangenti detto metodo del Roberval, *Bull. di Bibl. e di Stor. delle Sc. Mat. e Fis.* 8 (1875), pp. 265–304.
KARPINSKI, L. C.: *Robert of Chester's Latin Translation of the Algebra of Al-Khowarismi*, New York, 1915.
KOYRÉ, A.: A documentary history of the problem of fall from Kepler to Newton, *Trans. Amer. Phil Soc.* 45 (4) (1955), pp. 329–95.
LACOMBE, D.: L'Axiomatisation des mathématiques au IIIe siècle avant J.-C., *Thalès* 6 (1949–50), pp. 37–58.
LE PAIGE, C.: Correspondance de René Francois de Sluse, *Bull. di Bibl. e di Stor. delle Sc. Mat. e Fis.* 17 (1824), pp. 427–554, 603–726.
MAHNKE, D.: Neue Einblicke in die Entdeckungsgeschichte der höheren Analysis, *Abh. d. Preuss. Ak. d. Wiss.* 1925, Phys.-Math. Kl. 1, Berlin, 1926, pp. 1–64.
MAHNKE, D.: Zur Keimesgeschichte der Leibnizschen Differentialrechnung, *Sitz. d. Gesell. z. Beförd. d. Ges. Naturw. z. Marburg* 67 (1932), pp. 31–69.
MAIER, A.: *Die Vorläufer Galileis im 14 Jahrhundert*, Rome, 1949.
MARAR, K. MUKUNDA and RAJAGOPAL, C. T.: On the Hindu quadrature of the circle, *Journal of the Bombay Br. of the Royal Asiatic Society* 20 (1944), pp. 65–82.
MILHAUD, G.: *Descartes savant*, Paris, 1921.
MONTUCLA, J. E.: *Histoire des recherches sur la quadrature du cercle*, Paris, 1754 (2nd ed. 1831).
MORE, L. T.: *Isaac Newton*, London, 1934.
NAPOLI, F.: Intorno alla vita ed ai lavori di Francesco Maurolico, *Bull. Boncamp.* 9 (1876), pp. 1–156.
OSMOND, P. H.: *Isaac Barrow*, London, 1944.
PRAG, A.: John Wallis, *Quellen und Studien zur Gesch. d. Math.*, Studien (B) 1 (1931), pp. 381–402.
REY, A.: *La Science Orientale avant les Grecs*, Paris, 1942.

RIGAUD, S. P.: *Historical Essay on the First Publication of Sir Isaac Newton's Principia*, Oxford, 1838.

ROSENFELD, L.: René-Francois de Sluse et le problème des tangentes, *Isis* 10 (1928), pp. 416–34.

RUDIO, F.: *Der Bericht des Simplicius über die Quadraturen*, Leipzig, 1907.

RUDIO, F.: Das Problem von der Quadratur des Zirkels, *Vierteljahrsschrift der Naturforschenden Gesellschaft in Zürich* 35 (1890), pp. 1–51.

RUSSELL, B. A. W.: *The Principles of Mathematics*, 1, Cambridge, 1903.

SARTON, G.: Simon Stevin of Bruges, *Isis* 21 (1934), pp. 241–303.

SCHRAMM, M.: *Die Bedeutung der Bewegungslehre des Aristoteles für seine beiden Lösungen der zenonischen Paradoxie*, Frankfort-on-Main, 1962. Discussed by Mau, J. in *History of Science* 3 (1964), pp. 127–31.

SCOTT, J. F.: The Reverend John Wallis, F.R.S., *Notes and Records of the Royal Society of London* 15 (1960), pp. 57–67.

SCOTT, J. F.: *The Mathematical Work of John Wallis*, London, 1938.

SCRIBA, C. J.: *James Gregorys frühe Schriften zur Infinitesimalrechnung*, Giessen, 1957.

SCRIBA, C. J.: Zur Lösung des 2. Debeauneschen Problems durch Descartes, *Archive for History of Exact Sciences* 1 (1961), pp. 406–19.

SCRIBA, C. J.: The Inverse Method of Tangents: a dialogue between Leibniz and Newton (1675–77), *Archive for History of Exact Sciences* 2 (1964), pp. 113–37.

SENGUPTA, P. C.: History of the infinitesimal calculus in ancient and mediaeval India, *Jahresbericht der Deutschen Mathematiker Vereinigung* 40 (1931), pp. 223–7.

SINGH, A. N.: On the use of series in Hindu mathematics, *Osiris* 1 (1936), pp. 606–28.

STEIN, W.: Der Begriff des Schwerpunktes bei Archimedes, *Quellen und Studien*, Studien (B) 1 (1931), pp. 221–44.

STRUIK, D. J.: Kepler as a mathematician, in *Johann Kepler, 1571–1630*, Baltimore, 1931.

SUTER, H.: Die Abhandlung über die Ausmessung des Paraboloides von el-Hasan b. el-Hasan b. el-Haitham, *Bibl. Math.* (3) 12 (1911–12), pp. 289–332.

TANNERY, P.: *Pour l'histoire de la science hellène*, Paris, 1887.

THUREAU-DANGIN, F.: *Textes mathématiques Babyloniens*, Leyden, 1938.

TURNBULL, H. W.: *The Mathematical Discoveries of Newton*, London, 1945.

VACCA, G.: Maurolycus, the first discoverer of the principle of mathematical induction, *Bull. Amer. Math. Soc.* 16 (1909), pp. 70–3.

VAN DER WAARDEN, B. L.: Zenon und die Grundlagenkrise der Griechischen Mathematik, *Math. Ann.* 117 (1940), pp. 141–61.

VER EECKE, P.: Le théorème dit de Guldin considéré au point de vue historique, *Mathésis* 46 (1932), pp. 395–7.

WALLNER, C.: Über die Entstehung des Grenzbegriffes, *Bibl. Math.* (3) 4 (1903), pp. 246–59.

WALLNER, C.: Die Wandlungen des Indivisibilienbegriffs von Cavalieri bis Wallis, *Bibl. Math.* (3) 4 (1903), pp. 28–47.

WHISH, C. M.: On the Hindu quadrature of the circle, *Trans. Royal Asiatic Society* 3 (1835), pp. 509–23.

WHITESIDE, D. T.: The expanding world of Newtonian research, *History of Science* 1 (1962), pp. 16–29.

WHITESIDE, D. T.: Patterns of mathematical thought in the later seventeenth century, *Archive for History of Exact Sciences* 1 (1961), pp. 179–388.

WHITESIDE, D. T.: Newton's discovery of the General Binomial Theorem, *Math. Gaz.* 45 (1961), pp. 175–80.

WHITESIDE, D. T.: Henry Briggs: the Binomial Theorem anticipated, *Math. Gaz.* 45 (1961), pp. 9–12.

WHITMAN, E. A.: Some historical notes on the cycloid, *Amer. Math. Monthly* 50 (1943), pp. 309–15.

WIELEITNER, H.: Das Fortleben der Archimedischen Infinitesimalmethoden bis zum Beginn des 17 Jahrhundert insbesondere über Schwerpunktbestimmungen, *Quellen und Studien*, Studien (B), 1 (1931), pp. 201–20.

WIELEITNER, H.: Über den Funktionsbegriff und die graphische Darstellung bei Oresme, *Bibl. Math.* (3) 14 (1914), pp. 193–248.

WIELEITNER, H.: Zur Geschichte der unendlichen Reihen im christlichen Mittelalter, *Bibl. Math.* (3) 14 (1914), pp. 150–68.

WIELEITNER, H.: Bemerkungen zu Fermats Methode der Aufsuchung von Extremwerten und der Bestimmung von Kurventangenten, *Jahresbericht der Deutschen Mathematiker-Vereinigung* 38 (1929), pp. 24–35.

WIENER, P. P.: The Tradition behind Galileo's Methodology, *Osiris* 1 (1936), pp. 733–46.

WINTERBERG, C.: Der Traktat Franco's von Luettich *De quadratura circuli*, *Abh. z. Gesch. d. Math.* 4 (1882), pp. 135–90.

Index

In this index references only appear which are likely to prove of significant value to the reader. For example, many proper names occur in the text which are not given in the index. Bibliographical references are given in the bibliography and not in the index.

Academy of Plato 16
Algebraic notation 8, 123, 195
Algebraic tangent method 249–51
Analogy 2, 109
Analytic quantity 8
Anaxagoras of Clazomenae (*ca.* 420 B.C.) 31
Angle
  of contact 137–8
  Greek concept of 53–5
Antiphon (*ca.* 430 B.C.) 31
Apollonius of Perga (*ca.* 230 B.C.) 17
  treatment of normals 51–2
Application of areas 16–17
Arabs 61, 65
  mathematics of 66–9
  rotation of parabolic segments 66–9
  summation of series by 70
Arc
  length 232
  rectification of 223–8, 265
  relation to tangent 4–5, 231–2, 249–50
Archimedes of Syracuse (d. 212 B.C.) 17
  Axiom of 35
  centre of gravity of conoid 49–50
  compression method 37–40
  convexity lemmas 7, 37
  definition of spiral 55
  discovery method 46–50
  exhaustion method 34–6
  influence of 89
  measurement of circle 30, 33
  quadrature of parabola 36–7, 45, 47–8
  quadrature of spiral space 43–4
  surface area of sphere 38–41
  tangent to spiral 55–6
  volume of conoid 42–3, 48
Areas, early methods of calculation 11
Aristotle of Stagira (384–322 B.C.)
  influence on West 69, 71
  kinematics 76–7
  views on continuum 72
  Wheel of 117–18
Arithmetic progression, sums of powers of terms in 197–8
Atomists 20, 74–5

Babylonian mathematics 11, 13–14
Barrow, Isaac (1630–77) 239–52
  algebraic tangent method 249–51
  arc-tangent relation 249–50
  concept of tangent 244
  fundamental theorem of calculus 248
  inverse relation between tangents and quadratures 242–3
  and Newton 249
  use of means 246–8
  views on Time and Motion 240
Beaugrand, Jean de (*ca.* 1640) 172–3
Binomial coefficients 197–8
Binomial series
  Huygens and 222–3
  Newton and 256, 267

Calculators 3, 5, 14, 81–2
Cavalieri, Bonaventura (1598–1647) 122–35
  indivisible techniques 125–35
  solution of Kepler's problem 132
  views on continuum 125
Cavalieri's Theorem 21, 126
Centre of gravity
  of conoid 49–50, 91–6, 177–8

determinations in the sixteenth century 90–107
and Roberval 179
of triangle 97–99
Circle *see under* Quadrature
Circumference
of circle 30, 110
of ellipse 109
Circular motion
early views on 76–7, 78–80
generation of curves by 241
Cissoid of Diocles 57, 58
Classification
of angles 53
of curves 56–7
Commandino, Federigo (1509–75) 94–6
Compression method 37–40
Conchoid of Nicomedes 57
tangent to 175–77
Conoid 42
centre of gravity of 49–50, 91–6, 177–8
volume of 42–3, 48, 104–5
Continued proportions *see* Geometric series
Continuum 71–5
Aristotle's views on 72
Cavalieri's views on 125
paradoxes of 22
point structure of 73–4
Convergent double sequence 229–30
Convexity 37, 227
Curvature 54–5, 262–3, 264
Curves
application of moments to 282–3
classification of 56–7
evolutes and involutes 233–4
generated by motion 55–7, 240–4
known to Greeks 51–9
Newton's investigations of 257–66
Cycloid 156–61
contest on 199
and Descartes 158–60
and Fermat 157–8
and Huygens 220–1
and Leibniz 278–9
and Roberval 156–7
tangent to 163–4, 175, 176

Democritus of Abdera (*ca.* 430 B.C.) 18, 20–2
treatment of horn angle 53–4
Democritian theory 74–5
Descartes, René (1596–1650) 163
investigations on cycloid 158–60
and Newton 257, 261
normal to algebraic curves 165–6
tangent controversy 169–72
tangent to cycloid 163–4
Discovery method
centre of gravity of conoid 49
fundamental basis of 47
quadrature of parabola 47–8
volume of conoid 48

Euclid of Alexandria (*ca.* 330 B.C.) 16, 17
definition of angle 53–4
definition of tangent 51
Elements 25–6
geometric progressions 29
relation between means 29–30
theory of proportion 27–30
Eudoxus of Cnidus (*ca.* 370 B.C.) 16, 17
definition of rationals 27
method of Exhaustion 34
theory of proportion 27–30
Evolute
and Apollonius 52
and Gregory 233
and Newton 265
Exhaustion method 17, 34–6
generalisation of, by Valerio 101–4

Faille, Jean-Charles de la (1597–1652) 147
Fermat, Pierre de (1601–65) 150–1
centre of gravity of conoid 177–8
investigations on cycloid 157–8
method of extreme values 166–8
quadrature of curves 151–3, 162
rectification of arcs 226–8
tangent method 168–70
Figurate numbers 19
Fluxions 84, 267
Function 6–8
concept of 7, 13, 87–8
exponential 88–9
logarithmic 8

Galilei, Galileo (1564–1642) 116–22
mean speed theorem 120–1
mediaeval paradoxes 117–20
path of a projectile 5, 121–2
potential and actual infinite 118

Geometric algebra 13, 16–17
Geometric model 4
Geometric progression
and Euclid 29
and Wallis 206
Geometric series
and Barrow 246
and Saint-Vincent 136–7
and Torricelli 183–5
Graphical representation 5
development by Oresme 82–7
Greek mathematics
axiomatisation of 25–6
background to 11–4
chronological development of 15–18
classification of angles 53
classification of curves 56–7
concept of angle 53–5
discovery method in 46–50
exhaustion method in 34–6
kinematic methods in 55–9
known curves in 51–9
quadrature of lunes 31–2
recovery of, in West 14–15, 60–1, 90
relation to physical world of 18–25
synthetic character of 18
tangent concept in 51
theory of proportion in 27–30
Gregory, James (1638–75) 228–39
fundamental theorem of calculus 233
geometric approach to calculus 231–3
geometric integration methods 234–8
relation between tangent and arc 231–2
use of evolutes by 233–4
Guldin, Paul (1577–1643) 133, 147

Harmonic triangle 271–2
Heuraet, Hendrick van (1633–60), rectification of arc 223, 225–6
Hindu mathematics 61–5
concept of differentiation 65
inverse tangent series in 63–5
operations with zero in 62
place-value system in 62, 65
Hippias of Elis (*ca.* 430 B.C.) 32, 57, 59
Hippocrates of Chios (*ca.* 430 B.C.) 31–2
Hudde, Johann (1628–1704) 217
tangent method of 217–19
Hudde's Rules 217–19, 220, 225, 257, 260
Huygens, Christiaan (1629–95) 219–23
approximation to binomial series 222
quadrature of cycloidal segment 220–1
rectification of arcs 224–5
views on indivisibles 221–3
Hyperbola, rectangular
quadrature of, 138–9, 267
relation to logarithm 138, 147
tangent to 175, 176
Hyperbolas, higher 151
quadrature of 151, 162, 185–8
tangents to 189–90

Ibn-al-Haitham (Alhazen) (*ca.* A.D. 1000)
summation of series 70
volumes of revolution 67–9
Index notation 209–10
Indivisibles 108–48
and Barrow 243–4
and Cavalieri 122–35
and Galileo 116–22
and Huygens 221–3
and Kepler 108–16
in the Middle Ages 71–5
and Pascal 199–200
and Roberval 153–4
and Torricelli 183
and Wallis 206–7
Induction 197–8
Infinite 22
actual and potential 72, 118
Infinitesimals *see* Indivisibles
Infinity, use of symbol for 206–7
Inflexion 59, 220, 264
Instantaneous centre of rotation 164
Inverse tangent problem 115, 263–4, 283–6
Irrational numbers 16

Kepler, Johann (1571–1630) 108–16
astronomical problems 108–9
maxima and minima 115–16
solids of revolution 111–15
use of infinitesimals 110
Kinematics
in Aristotle 76–7
of Gerard of Brussels 78–81
graphical methods in 75–6
in Greek mathematics 55–9
growth of, in West 75–81

Lalovera, Antoine de (1600–64) 147
Latitude of Forms 81–7

Leibniz, Gottfried Wilhelm (1646–1716) 268–90
background of study 268–70
development of calculus 280, 281–90
difference series 270–1
exploration of symbolism 283–9
harmonic triangle 271–2
inverse tangent problem 281, 283–6
–Newton controversy 65, 279–80
transmutation 275–9
use of moments 282–3
visit to London 272–4
Limit 34, 167–9, 178, 183, 185
Limit sum of integral powers 208–9
Logarithmic function 8
Longitude 82
Low countries 214–23
development of notation 214
infinitesimal methods in 214–23
tangent methods in 214–19
Lune 31

Maurolico, Francesco (1494–1575) 91–4
Maxima and minima
and Apollonius of Perga 51–2
and Kepler 115–16
Fermat's method 167–8
Hudde's method 217
Mean speed theorem 82, 84–5
Galileo's proof 120–1
Means
definitions of 28–9
use of 246–8
Mersenne, Marin (1588–1648) 149–50
and cycloid 156
Method of Exhaustion 17, 34–46
Method of moments 91–4, 95–6, 282–3
Motion
Aristotle's statements on 76–7, 88
and Barrow 240–4
circular 76–7
curvilinear 87
Newton's use of 263–6
Torricelli's use of 190–3
uniform and difform 81–3
work of Oresme on 82–7
Motions, composition of 174–7

Neile, William (1637–70), rectification of arc 223–5
Newton, Isaac (1642–1727) 255–68
background of study 255–7
and Cartesian subnormal 257, 259, 261
and curvature 260, 262–3, 264
differential process 263
and fluxions 267–8
inflexion 264
integration 263–4
–Leibniz controversy 65, 279–80
and motion 263–6
notation 261, 262, 266–7
rectification of arc 265
Normals
and Apollonius 51–2
and parabola 165–6
Notation 8–9, 195, 261–4, 266–7, 283–9

Onglet 203
Oresme, Nicole (*ca.* 1323–82) 82–7

Pappus of Alexandria (*ca.* A.D. 320) 3, 18, 56–7
Pappus–Guldin Theorem 57, 80–1, 111, 147, 180–2, 187
Parabola
normal to 165–6
quadrature of 36–7, 45, 47–8, 183, 185, 190
tangent to 168–9, 174–5, 207–8
Parabolas, higher 151, 162
quadrature of 151–2, 162, 181–2, 190–1
tangents to 181–2, 191, 247
Paradoxes 22, 71
of Zeno 22–4, 137
mediaeval 74, 117–20
Pascal, Blaise (1623–62) 196–205
binomial coefficients 197–8
concept of sums 199–200
general integral forms 201–4
Place-value system 62, 65
Platonists 73
Point structure of continuum 73–4
Powers of line elements 127–9
Principle of Interpolation 210
Proclus (A.D. 410–485) 54
Proof by induction 197–8
Pythagorean School 16

Quadratrix of Hippias 32, 57, 59

Quadrature
of circle 11–12, 110–11, 135–6, 210–11, 256
in Greek mathematics 30–3
and Francon of Liège 60
and Huygens 219–20
and Leibniz 276–7
of cycloidal segment 157–60, 220–1, 278–9
of lunes 31–2
of hyperbola, rectangular 138–9, 267
of hyperbolas, higher 185–8
of parabola 36–7, 45, 47–8, 155, 183, 185, 190
of parabolas, higher 151–2, 162, 181–2, 190–1
of spiral space 43–4, 132–3, 152–3, 193

Radius of curvature 260–3
Rectification of arcs 4, 56, 223–8, 265
Regula 124–7, 129
Rings 111
volumes of 112–14
Roberval, Gilles Personne de (1602–75) 150
area under curve 154–6
centres of gravity 179
composition of motions 174–7
concept of surface 154
and indivisibles 153–4
investigations on the cycloid 156–7
volumes of revolution 160–1, 179–82

Saint-Vincent, Grégoire de (Gregorius a S. Vincentio) (1584–1667) 135–47
geometric integration method 139–47
quadrature of hyperbola 138–9
solution of Zeno's paradox 137
treatment of angle of contact 137–8
treatment of geometric series 136–7
Sarasa, Alphons Anton de (1618–67) 147
Schooten, Frans van (1615–60) 214
Simple geometrical transformations 11–12
Sluse, René François de (1622–85) 214
tangent method of 215–17
Solids of revolution 66, 111–15
and Roberval 160–1, 182
Spirals
kinematic generation of 55
quadrature of space of 43–4, 132–3, 162, 193
rectification of arc 56
tangent to 55–6, 191–3
Spiral of Galileo 152
quadrature of space 152–3
Squaring the circle *see* Quadrature of circle
Stevin, Simon (1548–1620) 96–101
centre of gravity of triangle 97–9
water pressure on wall 99–100
Summation
of arithmetic series 19, 29, 63
of difference series 271–2
of geometric series 29, 136–7, 162, 183–5, 206
of integral powers 41, 151, 197–8, 199–200
Surface
construction of 154
generated by motion 240–4
Surface area of sphere 38–41

Tacquet, André (1612–60) 147, 199
Tangent
to circle 51
concept of 51, 163, 171, 244
to conchoid of Nicomedes 175–7
to cycloid 163–4, 175–6
to ellipse 172–3, 175
to hyperbola, rectangular 175, 176
to hyperbolas, higher 189–90
to parabola 168–9, 174–5, 207–8
to parabolas, higher 181–2, 191, 247
to spiral 55–6, 191–3
Tangent controversy 166–73
Tangent methods
and Barrow 244–51
and Descartes 163–6
and Fermat 168–9
and Hudde 217–19
and Newton 257–62
and Roberval 174–7
and Sluse 215–17
and Torricelli 191–3
and Wallis 207–8
Theaetetus (d. 368 B.C.) 16
Theory of Proportion and Means 27–30
Torricelli, Evangelista (1608–47) 182–94
concepts of motion 190–1
and geometric series 183–5
quadrature of curves 185–8
and spirals 191–4
tangent methods 189, 191–3

Transmutations 234, 275–9
Triangular sums 19, 200–1
Trigonometric functions 38, 160–1, 250–1

Ungula 145–7, 211–12

Valerio, Luca (1552–1618) 101–6
  generalisation of exhaustion method 101–4
  volume of conoid 104–5
  volume of hemisphere 105–6
Variable, small change of 167–9
Velocity-time graph 190
Volume
  of conoid 42–3, 48, 104–5
  of hemisphere 105–6
  of solids of revolution 67–9, 111–15, 131–2, 160–1, 187–8

Wallis, John (1616–1703) 205–13
  and indivisibles 206–7
  limit sum of integral powers 208–10
  Principle of Interpolation 210
  rectification of arcs 224–5
  summation of geometric series 206
  tangent to parabola 207–8
  use of ungula 211–12
Western European mathematics
  advances in motion theory 81–7
  Aristotle's influence on 69, 71
  function concept in 87–9
  growth of kinematics in 75–81

Zeno of Elea 22–4
Zeno's paradox 22, 137
Zero 61–2

A CATALOG OF SELECTED
DOVER BOOKS
IN ALL FIELDS OF INTEREST

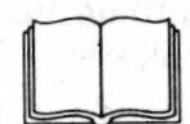

A CATALOG OF SELECTED DOVER BOOKS IN ALL FIELDS OF INTEREST

CATALOG OF DOVER BOOKS

REASON IN ART, George Santayana. Renowned philosopher's provocative, seminal treatment of basis of art in instinct and experience. Volume Four of *The Life of Reason*. 230pp. 5⅜ × 8. 24358-3 Pa. $4.50

LANGUAGE, TRUTH AND LOGIC, Alfred J. Ayer. Famous, clear introduction to Vienna, Cambridge schools of Logical Positivism. Role of philosophy, elimination of metaphysics, nature of analysis, etc. 160pp. 5⅜ × 8½. (USCO) 20010-8 Pa. $2.75

BASIC ELECTRONICS, U.S. Bureau of Naval Personnel. Electron tubes, circuits, antennas, AM, FM, and CW transmission and receiving, etc. 560 illustrations. 567pp. 6½ × 9¼. 21076-6 Pa. $8.95

THE ART DECO STYLE, edited by Theodore Menten. Furniture, jewelry, metalwork, ceramics, fabrics, lighting fixtures, interior decors, exteriors, graphics from pure French sources. Over 400 photographs. 183pp. 8⅜ × 11¼. 22824-X Pa. $6.95

THE FOUR BOOKS OF ARCHITECTURE, Andrea Palladio. 16th-century classic covers classical architectural remains, Renaissance revivals, classical orders, etc. 1738 Ware English edition. 216 plates. 110pp. of text. 9½ × 12¾. 21308-0 Pa. $11.50

THE WIT AND HUMOR OF OSCAR WILDE, edited by Alvin Redman. More than 1000 ripostes, paradoxes, wisecracks: Work is the curse of the drinking classes, I can resist everything except temptations, etc. 258pp. 5⅜ × 8½. (USCO) 20602-5 Pa. $3.95

THE DEVIL'S DICTIONARY, Ambrose Bierce. Barbed, bitter, brilliant witticisms in the form of a dictionary. Best, most ferocious satire America has produced. 145pp. 5⅜ × 8½. 20487-1 Pa. $2.50

ERTÉ'S FASHION DESIGNS, Erté. 210 black-and-white inventions from *Harper's Bazar*, 1918-32, plus 8pp. full-color covers. Captions. 88pp. 9 × 12. 24203-X Pa. $6.50

ERTÉ GRAPHICS, Erté. Collection of striking color graphics: *Seasons, Alphabet, Numerals, Aces* and *Precious Stones*. 50 plates, including 4 on covers. 48pp. 9⅜ × 12¼. 23580-7 Pa. $6.95

PAPER FOLDING FOR BEGINNERS, William D. Murray and Francis J. Rigney. Clearest book for making origami sail boats, roosters, frogs that move legs, etc. 40 projects. More than 275 illustrations. 94pp. 5⅜ × 8½. 20713-7 Pa. $2.25

ORIGAMI FOR THE ENTHUSIAST, John Montroll. Fish, ostrich, peacock, squirrel, rhinoceros, Pegasus, 19 other intricate subjects. Instructions. Diagrams. 128pp. 9 × 12. 23799-0 Pa. $4.95

CROCHETING NOVELTY POT HOLDERS, edited by Linda Macho. 64 useful, whimsical pot holders feature kitchen themes, animals, flowers, other novelties. Surprisingly easy to crochet. Complete instructions. 48pp. 8¼ × 11. 24296-X Pa. $1.95

CROCHETING DOILIES, edited by Rita Weiss. Irish Crochet, Jewel, Star Wheel, Vanity Fair and more. Also luncheon and console sets, runners and centerpieces. 51 illustrations. 48pp. 8¼ × 11. 23424-X Pa. $2.50